THERMODYNAMIC FOUNDATIONS OF THE EARTH SYSTEM

Thermodynamics sets fundamental laws for all physical processes and is central to driving and maintaining planetary dynamics. But how do Earth system processes perform work, where do they derive energy from, and what are the ultimate limits?

This accessible book describes how the laws of thermodynamics apply to Earth system processes, from solar radiation to motion, geochemical cycling and biotic activity. It presents a novel view of the thermodynamic Earth system that explains how it functions and evolves, how different forms of disequilibrium are being maintained, and how evolutionary trends can be interpreted as thermodynamic trends. It also places human activity into a new perspective in which it is treated as a thermodynamic Earth system process.

This book uses simple conceptual models and basic mathematical treatments to illustrate the application of thermodynamics to Earth system processes, making it ideal for researchers and graduate students across a range of Earth and environmental science disciplines.

AXEL KLEIDON leads a research group in Biospheric Theory and Modelling at the Max-Planck-Institute for Biogeochemistry, Jena, Germany. He uses thermodynamics to quantify natural energy conversions within the Earth system and their limits, and applies this approach to understand atmosphere-biosphere interactions, Earth system responses to global change, and the natural limits of renewable energy.

THERMODYNAMIC FOUNDATIONS OF THE EARTH SYSTEM

AXEL KLEIDON

Max-Planck-Institut für Biogeochemie

CAMBRIDGE
UNIVERSITY PRESS

CAMBRIDGE
UNIVERSITY PRESS

University Printing House, Cambridge CB2 8BS, United Kingdom

Cambridge University Press is part of the University of Cambridge.

It furthers the University's mission by disseminating knowledge in the pursuit of education, learning and research at the highest international levels of excellence.

www.cambridge.org
Information on this title: www.cambridge.org/9781107029941

First published 2016

A catalog record for this publication is available from the British Library

Library of Congress Cataloguing in Publication Data
Names: Kleidon, A. (Axel), author.
Title: Thermodynamic foundations of the Earth system / Axel Kleidon (Max-Planck-Institut für Biogeochemie, Jena, Germany).
Description: New York, NY : Cambridge University Press, 2016. | ©2016 | Includes bibliographical references and index.
Identifiers: LCCN 2015034136| ISBN 9781107029941 (Hardback : alk. paper) | ISBN 1107029945 (Hardback : alk. paper)
Subjects: LCSH: Thermodynamics. | Earth sciences. | Entropy.
Classification: LCC QC311 .K58 2016 | DDC 550.1/5367–dc23
LC record available at http://lccn.loc.gov/2015034136

ISBN 978-1-107-02994-1 Hardback

Contents

Preface

This book is about how thermodynamics applies to the Earth system. It is less about thermodynamics itself, but rather about how it applies to Earth system processes, their interactions, and the operation of the Earth system as a whole.

The motivation for writing this book stems from my interest in gaining a better, and more profound understanding of the Earth system, of the role that life plays within the system, and of how human activity changes the Earth system at a time when humans increasingly alter the operation of the planet. One way to deal with this challenge is to build increasingly comprehensive, yet also increasingly incomprehensible models of the Earth system. The other way is to search for a fundamental missing constraint that describes in comparably simple terms how systems operate and evolve. Since my doctoral work I have increasingly concentrated on this search. I looked into optimality approaches in vegetation, the Gaia hypothesis, and worked on the proposed principle of maximum entropy production (MEP). Over the years, I had many discussions with colleagues and took part in several workshops on these topics. I am tremendously thankful for these stimulating discussions, as these ultimately helped to shape my understanding that is now described in this book.

Today I think the answer to this missing constraint lies in the second law of thermodynamics. This law formulates a fundamental direction in physics that requires entropy to increase, at the small scale of an engine as well as at the scale of the whole Universe. Yet, its application to Earth system processes is almost absent, particularly when dealing with the whole Earth system. The second law, jointly with a thermodynamic formulation of the different processes yields a foundation to Earth system science that expresses processes in the same units of energy; it allows us to describe evolutionary dynamics as a thermodynamic direction imposed by the second law, and it sets fundamental limits and constraints on the emergent dynamics and interactions within the system. These limits can be quantified and yield estimates for Earth system processes that are largely consistent with observations, but require hardly any empirical parameters, substantiating that the second law provides missing constraints. It thus yields a grand picture of the Earth system in which its dynamics and evolution are a manifestation of the second law, a picture that is largely consistent with current descriptions yet yields a few critical insights that are not apparent from common formulations of the Earth system.

I think that these profound insights from thermodynamics should be accessible to a broad audience in the geosciences. Unfortunately, most books on non-equilibrium thermodynamics are only accessible to a highly specialized readership. Over the years I encountered several colleagues who studied thermodynamics yet still found it difficult to grasp, and this includes myself. Yet, I find this really unfortunate because thermodynamics can be fun and provides an elegant and simple way to look at the Earth system. For this view, it does not require much thermodynamic details to recognize its relevance and to use it for first-order estimates. In this book I aim to make thermodynamics accessible and thus describe only the bare essentials that are needed to formulate Earth system processes in thermodynamic terms.

To accomplish such an interdisciplinary, thermodynamic description of the whole Earth system, from radiation to human activity, poses a challenge as it requires a broad range of processes to be described. I therefore decided to focus on the mere minimum of thermodynamics and of Earth system processes to understand how thermodynamics applies to them and how these processes relate to each other. The book is thus not a comprehensive review of thermodynamics and its applications to Earth system processes. The text then includes references to related literature, and I apologize to those that I may have missed or that I may not have represented adequately. This led to a structure in which after the introduction, Chapters 2–5 provide the background in thermodynamics while the major processes of the Earth system are covered in Chapters 6–11. Chapter 12 closes with a synthesis to yield the perspective of the thermodynamic Earth system and how it can yield insights for Earth system science. Each chapter aims to be relatively self-contained and follows a similar format. It starts with a general introduction and closes by placing the material of the chapter back into the context of the Earth system and describes the linkages to the other chapters. By describing a broad range of processes across disciplines, one practical challenge was the mathematical formulation, as the letters of the alphabet are used for different variables in different disciplines. The letter G, for instance, is used for the gravitational constant, but also for Gibbs free energy. I tried to compromise and used mostly the convention of the different disciplines, so that some symbols refer to different aspects in different chapters. To help avoid confusion, the symbols are summarized in a table at the beginning of the book. Furthermore, a glossary includes brief explanations of the most central terms.

Even though the book was not written as a textbook, it is written at a level accessible to an audience in Earth and environmental sciences and is suitable for a course at the graduate level. As it involves the physical formulation of the different Earth system processes, the reader does require a certain level of familiarity with basic physics. The book illustrates the basic concepts associated with thermodynamics at a qualitative level supported by illustrations, and then uses comparatively simple models to demonstrate the application of thermodynamics and to estimate limits that predict magnitudes of different Earth system processes. These simple models are certainly not meant to be complete, but rather provided as an illustration of how thermodynamics is applied and how it can be used to establish magnitudes of Earth system processes.

This book would have been impossible to write without the substantial support and many stimulating discussions on various aspects of the Earth system as well as thermodynamics with colleagues and within my research group over the years. The number of colleagues are too many to list here, but I am very thankful for the stimulating discussions we had, for the disciplinary knowledge they provided, and for answering the seemingly strange questions that I sometimes asked. From my research group, I particularly thank James Dyke, Fabian Gans, Lee Miller, Philipp Porada, Maik Renner, Stan Schymanski, Eugenio Simoncini, and Nathaniel Virgo for the many discussions we had on entropy, life, Earth, and the universe. I thank Uwe Ehret and Christian Reick for thoroughly reading through the draft of the book, providing constructive feedback, identifying unclear passages, and finding errors. I thank Cambridge University Press, particularly Susan Francis and Zoë Pruce, for their support and insistence to bringing this book to completion. Last, but not least, I thank my partner, Anke Hildebrandt, for her support at critical points and times in this and other projects. She and our kids were very patient, tolerated entropy discussions at the dinner table over the years, and accepted the time I spent in the last year to complete this book.

I hope you will find this book useful in providing a starting point to more applications of thermodynamics to Earth system science. I would be curious to hear back from the reader about any comments, suggestions, or activities to which this book may have helped to contribute.

Axel Kleidon

Symbols

Overview of the most frequently used symbols in the book, which may be supplemented by additional indices. For those symbols that are used to describe more than one property, the section or chapter where the respective symbol is being used is also given. Note that some variables, such as fluxes, are also used in reference to unit area.

Symbol	Description	Units	Value	Primary use
α	albedo	frac.	-	sec. 6.3.3
A	chemical affinity of a reaction	J mol^{-1}	-	sec. 9.2
A	area (typically surface area)	m^2		
B	geometric factor	-	-	
c	speed of light	m s^{-1}	$3 \cdot 10^8$	chap. 6
c	heat capacity	J K^{-1}		
c_p	specific heat capacity at constant pressure	J kg^{-1} K^{-1}	-	
c_v	specific heat capacity at constant volume	J kg^{-1} K^{-1}	-	
C	energy conversion rate (within Lorenz cycle)	W	-	sec. 7.3
C	condensation rate	kg m^{-2} s^{-1}	-	sec. 8.2
C_d	drag coefficient	-	-	sec. 7.3
d_e	mean distance of Earth to Sun	m	$150 \cdot 10^9$	
D	dissipation rate	W	-	
e	partial pressure of water vapor	Pa	-	
e_{sat}	partial pressure of water vapor at saturation	Pa	-	
E	evaporation rate	kg m^{-2} s^{-1} or mm d^{-1}	-	
ϵ	dilution factor	-	-	chap. 6
ϵ_{lue}	light use efficiency	μmol CO_2 (μmol PAR)$^{-1}$	-	sec. 10.6

Symbol	Description	Units	Value	Primary use
ϵ_{wue}	water use efficiency	g CO_2 (kg $H_2O)^{-1}$	-	sec. 10.6
η	efficiency (= power/flux)	frac.	-	
f	fraction	frac.	-	
f	feedback factor	-	-	chap. 5
f	Coriolis parameter	s^{-1}	-	sec. 7.6
f_w	water limitation factor	frac.	-	sec. 10.4
F	Helmholtz free energy	J	-	sec. 3.5
F	force	kg m s^{-2}	-	sec. 4.6
ϕ	geopotential	m^2 s^{-2}		sec. 2.3.4
ϕ	latitude	$^\circ$		chap. 7
g	gravitational acceleration	m s^{-2}	9.81	
G	gravitational constant	m^3 kg^{-1} s^{-2}	$6.67 \cdot 10^{-11}$	sec. 2.3.4
G	generation rate (power)	W	-	
G	Gibbs free energy	J	-	sec. 3.5, chap. 9
ΔG_r	Gibbs free energy of a reaction	J	-	chap. 9
γ	psychrometric constant	Pa K^{-1}	≈ 65	
Γ_d	dry adiabatic lapse rate	K m^{-1}	$9.81 \cdot 10^{-3}$	
Γ_{dew}	lapse rate of the dew point	K m^{-1}	$1.8 \cdot 10^{-3}$	
H	enthalpy	J	-	sec. 3.5
H	sensible heat flux	W m^{-2}	-	chap. 10
h	Planck's constant	J s	$6.63 \cdot 10^{-34}$	
I	current	A	-	
i	van't Hoff factor	-	-	
J	heat flux	W	-	
J_{in}	influx of energy	W	-	
J_{out}	outflux of energy	W	-	
J_s	entropy flux	W K^{-1}	-	chap. 2
J_m	mass flux	kg s^{-1}	-	
J_{mom}	momentum flux	kg m s^{-3}	-	
k	conductivity	W K^{-1}	-	
k	friction coefficient	(depends)	-	sec. 4.6
k_b	Boltzmann's constant	J K^{-1}	$1.38 \cdot 10^{-23}$	
k_f	forward constant for chemical reactions	mol l^{-1} s^{-1}	-	sec. 9.2
k_r	reverse rate constant for chemical reactions	mol l^{-1} s^{-1}	-	sec. 9.2
k_r	radiative linearization constant	W m^{-2} K^{-1}	-	
K_ν	spectral energy density	J sr^{-1} m^{-2}	-	sec. 6.2
K_{eq}	equilibrium constant	-	-	sec. 9.2
L	length	m	-	
L_ν	spectral entropy density	J sr^{-1} m^{-2} K^{-1}	-	sec. 6.2
λ	wavelength	m	-	chap. 6
λ	latent heat of vaporization	J kg^{-1}	$2.5 \cdot 10^6$	chap. 8, chap. 10
λE	latent heat flux	W m^{-2}	-	

Symbol	Description	Units	Value	Primary use
μ	chemical potential	J mol^{-1}		
m	mass	kg	-	
n	molar mass	kg mol^{-1}	-	
N	number of particles	- or mol	-	sec. 2.4.1
N_ν	distribution function	-	-	sec. 6.2
N	Nusselt number	-	-	sec. 7.7
ν	frequency	s^{-1}	-	
Ω	solid angle	sr	-	chap. 6
Ω_{sun}	solid angle of the Sun in the Earth's sky	sr	$6.8 \cdot 10^{-5}$	chap. 6
Ω	Earth's angular velocity	s^{-1}	$7.27 \cdot 10^{-5}$	sec. 7.6
p	pressure	Pa	-	
p_s	surface pressure	Pa	$1.01325 \cdot 10^5$	
p	radiation pressure	Pa	-	sec. 6.2
p	probability	-	-	sec. 3.3
p	precipitation	kg m^{-2} s^{-1} or mm d^{-1}	-	chap. 8
π	osmotic pressure	Pa	-	
q	specific humidity	kg kg^{-1}	-	
Q	amount of heat added or removed	J	-	
Q	charge	C	-	sec. 4.6
Q	runoff	kg m^{-2} s^{-1}	-	sec. 10.4
ρ	density	kg m^{-3}	-	
r_{sun}	radius of the Sun	m	$695.8 \cdot 10^6$	
r_e	radius of the Earth	m	$6.372 \cdot 10^6$	
R	ideal gas constant	J kg^{-1} K^{-1}	287 (air)	
		J kg^{-1} K^{-1}	461 (water vapor)	
		J mol^{-1} K^{-1}	8.314 (general)	
R_{in}, R_{out}	radiative flux	W m^{-2}	-	
R_l	flux of terrestrial radiation	W m^{-2}	-	
$R_{l,\text{up}}$	flux of terrestrial radiation (upwards) at the surface	W m^{-2}	-	
$R_{l,\text{down}}$	flux of terrestrial radiation (downwards) at the surface	W m^{-2}	-	
$R_{l,\text{net}}$	net flux of terrestrial radiation at the surface	W m^{-2}	-	
$R_{\text{sun,tot}}$	solar luminosity	W	$7.6 \cdot 10^{26}$	
$R_{s,\text{in}}$	influx of solar radiation at the top of the atmosphere	W m^{-2}	1370	
R_s	flux of solar radiation	W m^{-2}	-	
$R_{s,a}$	absorbed solar radiation in the atmosphere	W m^{-2}	≈ 75	
$R_{s,s}$	absorbed solar radiation at the surface	W m^{-2}	≈ 165	
$R_{s,\text{toa}}$	total absorbed solar radiation	W m^{-2}	≈ 240	

Symbol	Description	Units	Value	Primary use
R	resistance	Ω	-	sec. 4.6
s	slope of the saturation vapor pressure curve, $s = de_{\text{sat}}/dT$	Pa K^{-1}	-	
S	thermal entropy	J K^{-1}		sec. 2.3.2
S	radiation entropy	J K^{-1}	-	sec. 6.2
σ	Stefan-Boltzmann constant	W m^{-2} K^{-4}	$5.67 \cdot 10^{-8}$	sec. 6.2
σ	entropy production	W K^{-1}	-	
t	time	s	-	
Δt	time interval	s	-	
T	temperature	K	-	
T_a	atmospheric temperature	K	-	
T_e	engine temperature	K	-	
T_r	radiative temperature	K	≈ 255	
T_s	surface temperature	K	-	
T_{sun}	emission temperature of the Sun	K	5760	
θ	potential temperature	K	-	
τ	residence time or time scale	s	-	chap. 2, chap. 5
τ	optical depth	-	-	chap. 6
u	energy density	J m^{-3}	-	chap. 6
u	velocity (zonal component)	m s^{-1}	-	chap. 7
U	internal energy	J	-	
U_{rad}	radiative energy	J	-	
U_{te}	thermal energy	J	-	
U_{pV}	uncompensated heat	J	-	
U_{pe}	potential energy	J	-	
U_{ke}	kinetic energy	J	-	
U_{be}	binding energy	J	-	
U_{other}	other, non-thermal form of energy	J	-	
U	voltage	V	-	sec. 4.6
U	heat storage	J m^{-2}	-	sec. 10.3
v	velocity (general, or meridional component)	m s^{-1}	-	chap. 7
v	reaction velocity	mol l^{-1} s^{-1}	-	sec. 9.2
V	volume	m^3	-	
W	number of possible arrangements	-	-	sec. 2.4.1
W	work	J	-	
W_s	soil water content	kg m^{-2}	-	sec. 10.4
x	horizontal dimension	m	-	
ξ	extent of reaction	mol	-	
X	dilution effect on entropy	-	-	sec. 6.3
y	horizontal dimension	m	-	
Δz	vertical thickness	m	-	
z	vertical coordinate	m	-	
z_0	scaling height in the barometric formula	m	≈ 8425	

1

Thermodynamics and the Earth system

1.1 A thermodynamic basis for Earth system science

The Earth is a vastly complex system. This complexity is reflected in the broad range of processes that it entails, from the solar radiative forcing to the highly dynamic circulatory patterns in the atmosphere, ocean and the interior, to high level and diversity of metabolic activity of life, and to human activities. The complexity is further enhanced by strong interactions by which processes alter their own drivers. Atmospheric motion, for instance, transports such vast amounts of heat that it alters the radiative exchange with space. The activity of the Earth's biosphere, the sum of all living organisms, has strongly altered the chemical composition of the atmosphere, as for instance reflected in its high abundance of molecular oxygen, resulting in altered physical and chemical conditions. And finally, human activity over the last century has released such large amounts of buried organic carbon by its industrial activities that it has substantially altered the global carbon cycle resulting in enhanced concentrations of carbon dioxide in the atmosphere and global climate change. With such complexities in mind, it would seem almost impossible to make robust predictions of magnitudes, the strength of interactions, and the overall evolutionary direction of the Earth system as a whole in order to get a robust, physical understanding of how the whole Earth system functions and responds to change.

Yet there is a range of fundamental, practical, and relevant questions that require such a robust understanding. What determines, for instance, the strength of the atmospheric circulation and its ability to transport and mix heat and mass? The answer to this question would help us to make better predictions of the magnitude of climate system processes and how these would respond to perturbations and change. Does the climate system, and the planet as a whole, regulate its climatic state to some particular reference level? Is climate even regulated to a point that is most suitable to life, because of the presence of life, as proposed by the Gaia hypothesis (Lovelock 1972b,a; Lovelock and Margulis 1974)? If this is so, how would human activity play into such a planetary regulation? A better understanding of these questions would provide information about the role of the biosphere at the planetary scale and the factors that shape planetary habitability. What are the limits to human activity, for instance, limits related to food production and the availability of renewable energy and how do these relate to the functioning of the planet? What are the associated human impacts on the system, and can these impacts tip the planet off some edge into an inhabitable state? The answer to this question would help us to develop better scenarios of a sustainable future. What these questions have in common is that they require a perspective on the whole Earth system, as the limits that these questions involve are of a physical nature, and ultimately relate to the planetary forcing represented by solar radiation and the cooling of the interior Earth.

The goal of this book is to provide a fundamental basis rooted in physics that allows to approach these questions. This book will show that a central component for this basis is described by thermodynamics, a fundamental theory in physics that deals with conversions of energy and their direction towards states of higher entropy. It is particularly the latter aspect, known as the second law of thermodynamics, that has intrigued many scientists, that provides a fundamental direction for processes, and that sets fundamental limits. Energy and entropy are very basic quantities that apply to practically all processes, from radiation to metabolic and human activity. Thermodynamics thus provides a way to formulate all Earth system processes in comparable quantities, thus providing a general accounting basis.

Yet, it is not just the thermodynamic formulation that is important. Equally important is to place the thermodynamic formulation into a systems perspective of the whole Earth system. This combination allows us to link processes to their ultimate driver of solar radiation and interior heat through sequences of conversions. These conversions are associated with converting energy of different forms, and the laws of thermodynamics constrain these conversions by thermodynamic limits. When thermodynamic limits are then applied to these sequences, we can quantify the limits on their rates and on the interactions that result from these processes. As we will see in the book, these limits provide basic and robust estimates for a range of Earth system processes that compare well with observations. This is not because these processes are organized in a simple, predictable way, but rather likely because they are so complex and have evolved so far that they operate near their thermodynamic limit. Even if not all processes may necessarily operate near their limit, it nevertheless provides us with an evolutionary "target" that can be used to interpret evolutionary dynamics, and it can be used to yield relatively simple and transparent estimates for the magnitude of processes that typically require only a mere minimum of empirical knowledge. This view may then already provide sufficient information to understand Earth system processes, how they respond to change, how life and human activity fit in and what a sustainable future may look like.

In the remaining part of this chapter, the basics of thermodynamics are described qualitatively, with the details being described in Chapter 3. It is illustrated how systems are maintained in an ordered state by fluxes of different entropy going through the system, and how this relates to living organisms and the Earth system as a whole. It is then explained how these basic components of thermodynamics result in limits to the ability to perform work, and how this work feeds sequences of energy conversions of different processes. At the end of this chapter, an overall view of the thermodynamic Earth system is given and it is described how this view is partitioned into the different chapters of the book.

1.2 Thermodynamics in a nutshell

Thermodynamics sets the rules for the conversions of energy from one form into another and sets the general direction into which these conversions take place. These two aspects are described by the first and second law of thermodynamics. The first law essentially states the conservation of energy. When energy is converted from one form into another, overall no energy is lost or gained. As energy is converted, its concentration, or reversely, its dispersal is altered. The extent of energy dispersal during a conversion is described by entropy, with more dispersed forms of energy corresponding to a higher entropy. The second law states that energy is, overall, increasingly being dispersed. This dispersal of energy is, for instance, reflected when a heated object such as a hot cup of coffee cools down and approaches the temperature of its surroundings. Here, the first law would tell us that the heat given off by the object is added to the heat content of the surroundings so that the total energy of the object and its surroundings is conserved. The dispersal of heat that is associated with the cooling of the object, however, is not captured by the first law. This tendency is rather the manifestation of a profound direction of nature to spread energy, mass, and other physical attributes into uniform concentrations. Such states of uniformity are described in thermodynamics as states of maximum entropy, or thermodynamic equilibrium. The natural direction of processes to spread energy and to increase entropy is described by the second law of thermodynamics.

These two laws of thermodynamics, the conservation of energy, and the increase in entropy, are so general, that Albert Einstein once said that (Klein 1967):

"[a] theory is more impressive the greater the simplicity of its premises, the more different are the kinds of things it relates, and the more extended its range of applicability. Therefore, the deep impression which classical thermodynamics made on me. It is the only physical theory of universal content, which I am convinced, that within the framework of applicability of its basic concepts will never be overthrown."

The second law in particular sets such a profound direction for physical processes that it has been labeled the "arrow of time" (Eddington 1928). The increase in entropy tells us something quite specific about how we would expect the dynamics within systems to take shape. No matter how complicated a system is, how large it is, or how many types of processes and constituents it involves, the overall dynamics that take place within the system must obey the laws of thermodynamics. Overall, energy needs to be conserved, and entropy needs to increase. How this increase in entropy is accomplished within a system is non-trivial as it also needs to consider how exchange fluxes across the system alter the entropy of the system.

One important characteristic of Earth system processes is that they typically operate far from thermodynamic equilibrium as gradients and fluxes are maintained within the system. This disequilibrium can be maintained in a steady state, in which the mean properties of the system do not change in time, without violating the laws of thermodynamics. The actual formulation of these laws is somewhat different and needs to account for the exchanges between the system and its surroundings across the system boundaries. These exchanges do not only exchange energy, mass, or other physical quantities, they also exchange entropy. When energy is added to warm places, and the same amount of energy is removed from cold places, the total energy within the system does not change, but a gradient in temperature is being maintained, reflecting disequilibrium. As this energy was added and removed at different temperatures, the exchange of entropy does not cancel out, but results in a net export of the entropy that is produced within the system so that the system can be maintained in a state of thermodynamic disequilibrium. In the application of the second law, this exchange of entropy needs to be taken into account. This exchange with the surroundings has important consequences for the state of a system: It allows a system to maintain a state away from thermodynamic equilibrium, and the entropy exchanges across the boundary reflects important information on the extent to which a system is maintained in a state of thermodynamic disequilibrium.

To illustrate this critical point in more detail, let us consider the two systems shown in Fig. 1.1. The top row of this figure shows a "system A" in which no exchange with its surroundings takes place. In such a setting, an initial internal difference in temperature would fade in time. Expressed differently, the processes within the system are directed to deplete this temperature difference. The total energy of the system during this redistribution of heat remains unchanged, but its distribution has changed. This latter aspect reflects the increase of entropy within the system. In the final state of a uniform temperature distribution, energy is distributed most uniformly within the system, the entropy is at a maximum, and this state corresponds to a state of thermodynamic equilibrium. This situation is akin

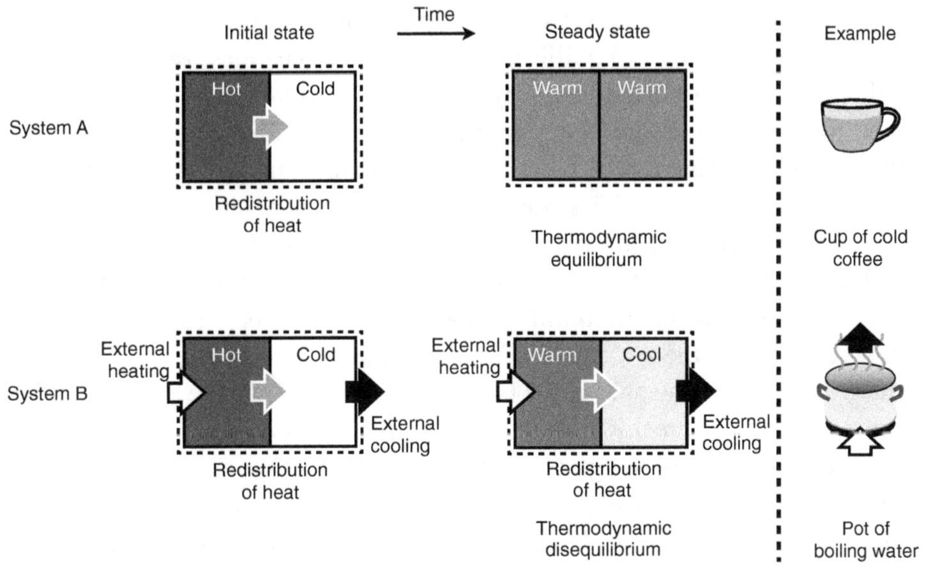

Figure 1.1 Two different types of systems (A: an "isolated" system; B: a "non-isolated" system) develop from an initial state to a final, steady state. An everyday example for such systems are given on the right.

to a hot cup of coffee in a room that would cool down to the temperature of the surroundings.

The bottom of Fig. 1.1 shows a "system B" in which heat is added to one side of the system, while it is cooled at the other side of the system. This setting is comparable to a pot of water on a stovetop that is heated from below, and cooled from above. Just as in the situation described earlier, the processes within the system are directed to deplete the temperature difference within the system, attempting to spread the heat uniformly within the system. However, since energy is continuously added and removed at different parts of the system boundary, the system is maintained in a state of disequilibrium. This disequilibrium manifests itself in the temperature difference that is being maintained within the system and that we can observe, but is also reflected in the dynamics that take place in the system. In the case of the pot of water, these dynamics are simply the convecting motion of boiling water within the pot.

To sum up, the two systems shown in Fig. 1.1 may look the same in terms of the amounts of heat that they contain, but they differ significantly in terms of the internal dynamics and their thermodynamic state. System A describes a system in which the final, steady state is a static state of thermodynamic equilibrium. The properties of the system do not change in time, there is no exchange in energy or entropy, and the system does not show any dynamics. System B also reaches a steady state in which the properties of the system do not change in time. However, its steady state is characterized by disequilibrium and reflects dynamics associated with the heat flux within the system. These dynamics are maintained by exchange fluxes, and the trend to deplete the temperature gradient within the system is mirrored by the entropy exchange of the system with its surroundings. This latter aspect, that the dynamics within the system are directed to deplete gradients, is not a consequence of energy conservation, but of the second law of thermodynamics. This steady state in which fluxes and gradients are being maintained in system B reflect thermodynamic disequilibrium and this state is being maintained by the entropy exchange of the system with its surroundings.

These considerations apply to the Earth system as well. As the Earth is a thermodynamic system that is maintained in a state with fluxes and gradients like system B, this thermodynamic view suggests that the dynamics that we can observe within the Earth system result as a consequence of the second law as well. A necessary foundation to implement these considerations is the formulation of the dynamics entirely in terms of energy and entropy exchange. This is not just captured by the Earth's energy balance, which is the common starting point in climatology. It also requires that all other processes are described in energetic terms, ranging from atmospheric motion to geochemical cycling, biotic and human activity. Furthermore, it requires a description of the entropy fluxes that are associated with these dynamics, which are rarely considered in Earth system science, yet central if we want to interpret the dynamics of the Earth system in terms of the second law.

1.3 Disequilibrium, life, and Earth

Entropy considerations are central when we want to understand how disequilibrium is being maintained and how highly complex phenomena such as life or the Earth system as a whole do not violate the second law, at each and every process and at the scale of the whole planet. In his seminal book on "What is life," Erwin Schrödinger (1944) described that a living organism satisfies the second law by consuming low-entropy food and producing high-entropy waste (Fig. 1.2a). Averaged over some time period, the food uptake by the organism roughly balances its waste, so that there is no net gain or loss of mass. The mass flux in itself does not contain the relevant information that would tell us that the organism is alive. That the influx of mass balances the outflux of mass is simply a consequence of the overall conservation of mass. The relevant information comes from the fact that the influx of mass is of a different constitution than the outflux. It is this difference in constitution that is captured by entropy, and it is this difference in entropy in the exchange of the living organism with its environment that allows the living organism to extract energy to run its metabolism. This metabolic energy is then dissipated and released as heat.

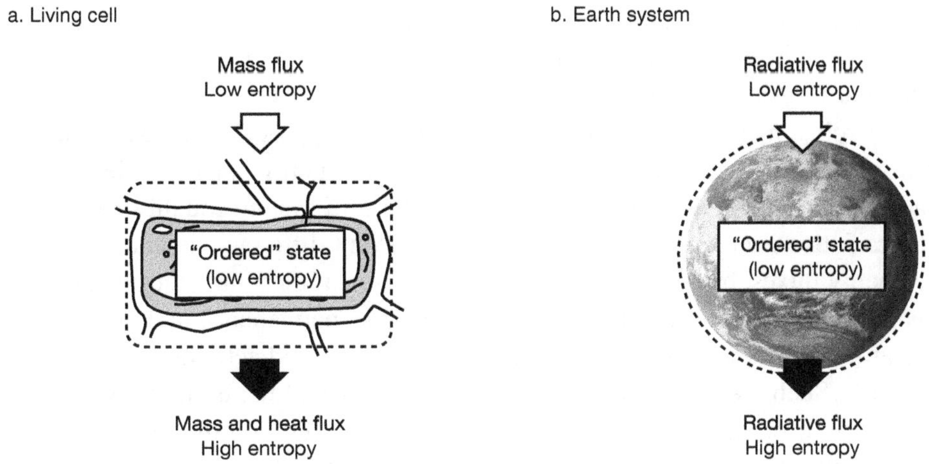

Figure 1.2 (a) A living cell and (b) the whole Earth system as examples of dissipative systems that are maintained far from thermodynamic equilibrium by the exchange of entropy. Source of Earth image: NASA.

A living organism is just one example of a so-called dissipative structure, a term introduced by Ilja Prigogine, a chemist who extensively worked on non-equilibrium thermodynamics and structure formation. His many accomplishments include textbooks on non-equilibrium thermodynamics (Prigogine 1962; Kondepudi and Prigogine 1998), and his major contributions are summarized in an article published on the occasion of the Nobel prize in Prigogine (1978). There are many more examples for such dissipative structures, ranging from the patterns that form in chemical reactions, convection cells in fluids, living organisms, and, ultimately, the whole Earth system. These structures have in common that they maintain states far away from thermodynamic equilibrium, and that these states are being maintained by the entropy exchange with the surroundings, just like the simple system B that was depicted in Fig. 1.1.

In the Earth system (Fig. 1.2b), the entropy exchange of all dissipative structures is ultimately integrated at the planetary scale to the entropy exchange with space. Just like a living organism is maintained by the entropy exchange associated with the fluxes of "food" and "waste," so is the activity of the whole Earth system being maintained by entropy exchange. The similarity between a living cell and the whole Earth as being dissipative structures was noted by James Lovelock and Lynn Margulis. It was this thermodynamic consideration that led Lovelock and Margulis to view the Earth system as a "superorganism" and to formulate the controversial Gaia hypothesis that compared the functioning of the Earth system to a living organism (Lovelock and Margulis 1974).

For the Earth system, the vast majority of entropy exchange is accomplished by the radiative exchange with space (Fig. 1.2b). The "food" of the Earth system is solar radiation, which has a low entropy because the radiative energy is composed of relatively short wavelengths, an aspect we get back to in Chapter 6. The "waste" of the Earth system is exported by the radiation emitted to space, which is radiation of relatively long wavelength and thus has a high entropy. As in the case of the living cell, we deal with a system in which the planetary energy balance is balanced, so that the total radiative energy absorbed by the Earth system roughly balances the total radiative energy emitted to space. The relevant difference is not contained in the energy fluxes, but rather in the associated fluxes of planetary entropy exchange. It is this planetary entropy exchange that allows the maintenance of thermodynamic disequilibrium as well as all the dissipative structures and activities that take place on Earth. Such a system with dissipative structures and net entropy exchange we refer to as a dissipative system, and to the processes that take place within the system and cause an increase in entropy as dissipative activity.

While living organisms and the Earth system both constitute dissipative systems, they do not act independently. In a system's view, the dissipative structures of living organisms feed on the boundary conditions set by the Earth system, yet the products of living organisms will affect their surroundings, and thereby alter the planetary system. This connection between life and the Earth system was formulated very nicely by Ludwig Boltzmann (1844–1906), a physicist who set much of the statistical foundation of thermodynamics. He expressed this connection as (Boltzmann 1886):

"The general struggle for existence of living organisms is therefore not a struggle for the basic materials – these materials are abundantly available for organisms in air, water and soil – nor for energy, which is abundant in form of heat in any body, albeit unfortunately unavailable, but a struggle for entropy, which through the transformations of energy from the hot sun to the cold Earth becomes available. To fully explore this transformation, plants spread their leaves in unimaginable extent and force the solar energy to perform chemical synthesis in yet unexplored ways before this energy sinks to the temperatures of the Earth's surface. The products of this chemical kitchen forms the object of struggle for the animate world."

Note that Boltzmann formulated this relationship before much of the details involved in photosynthesis were discovered. This description of living organisms places biotic activity into a planetary context. Yet, before we describe these interdependencies and their implications in greater detail, we first explore what else thermodynamics can tell us about dynamics, interactions, and evolution.

1.4 Thermodynamic limits

So far, we described the first and second law of thermodynamics and how these provide a constraint and a direction for Earth system processes. When these two

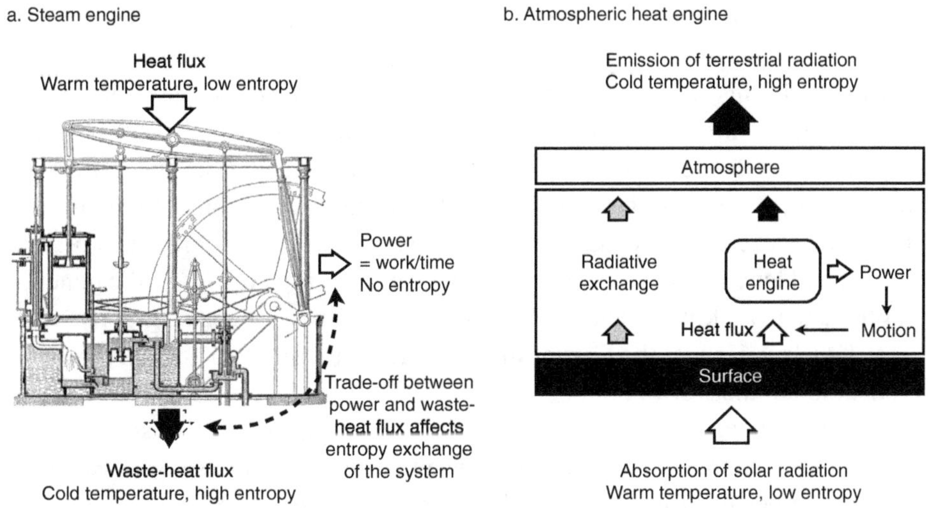

Figure 1.3 (a) Illustration of a heat engine using Watt's original steam engine as a template. Only a fraction of the heat flux into the engine can be converted to mechanical power because of the condition imposed by the second law. (b) An illustration of the surface-atmosphere system in which a heat engine generates motion from radiative heating of the surface and the cooling of the atmosphere. Steam engine image from Meyer (1886), available on www.wikipedia.org.

laws are being combined, they yield a fundamental limit on how much work can be performed, or, more generally, how much energy can be converted from one form into another. This limit constrains, for instance, how much of the radiative heating of the Earth's surface can be converted into the kinetic energy associated with atmospheric motion. The two laws thus set firm constraints on the magnitude of the dynamics that can take place within the Earth system as a result of the planetary forcing.

The limit for converting heat into mechanical work is known as the Carnot limit, named after the engineer Sadi Carnot (1796–1832), who was one of the pioneers in developing thermodynamics at the times at which steam engines were invented. The limit is illustrated in the following using the steam engine shown in Fig. 1.3a. This steam engine operates by the addition of heat at a high temperature, typically by the combustion of a fuel. This is shown in the figure by the white arrow at the top. The first law states that this addition of heat is balanced by the removal of "waste" heat through its exhaust at a colder temperature, shown by the black arrow at the bottom, and by the mechanical work (the white arrow on the right). The second law requires that the engine cannot decrease the overall entropy, but that it at best remains unchanged. This leads to a constraint on the entropy exchange of the engine, which is accomplished by the heat exchange at different temperatures.

The waste-heat flux plays a critical role, as it exports heat at a colder temperature and exports higher entropy to the surroundings of the engine. When the engine performs work, this must come at the expense of a reduced waste-heat flux. This follows from the first law. With a reduced waste-heat flux, less entropy is being exported, so that the net entropy exchange of the engine decreases the more work is being performed by the engine. The Carnot limit marks the point at which there is no net entropy exchange by the system, which is the absolute limit that is permitted by the second law. The work performed by the system then results in the generation of motion, that is, kinetic energy, or the lifting of mass against gravity, that is, potential energy. Hence, the Carnot limit describes the limit to energy conversions from heat to another form.

As already imagined by Carnot, this limit does not just apply to steam engines, but to "natural" engines that drive Earth system processes. In his book "Reflections on the motive power of fire" (Carnot 1824), he writes that:

> "The vast movements which take place on the Earth are ... due to heat. It causes the agitations of the atmosphere, the ascension of clouds, the fall of rain and of meteors, the currents of water which channel the surface of the globe, and of which man has thus far employed but a small portion. Even earthquakes and volcanic eruptions are the result of heat."

Carnot's writing suggests already back in the early nineteenth century that one should view these Earth system processes as if these operate like steam engines, or, more generally, heat engines, hypothetical devices that convert thermal energy into physical work. These engines would be subjected to the same Carnot limit as a heat engine in how much work these could perform and thus result in the dynamics that we can observe.

When formulating thermodynamic limits for Earth system processes, there are a few critical differences to the setup of a Carnot heat engine that need to be considered. An example of this application to atmospheric convection is shown in Fig. 1.3b. Firstly, the generation of atmospheric motion can be viewed as the result of an atmospheric heat engine that is constrained by the heating and cooling associated with radiative exchange. The setup of the Carnot limit does not, however, account for the fluxes of energy that bypass the engine. This aspect becomes important for the atmosphere, where radiation transports energy from the heated surface to the cold atmosphere and which cannot be utilized by the convective heat engine of the atmosphere. This bypass results in unavoidable entropy production within the system which can only be reduced, but not avoided. Secondly, the mechanical work that is derived from the engine actually feeds back upon itself. Atmospheric motion, for instance, results from this work being performed, yet atmospheric motion is associated with the heat flux that drives the engine. In other words, there are internal dynamics that take place within the system that we need to consider in the steady state that describe the strength of that engine. Thirdly, when the engine performs work and transports large amounts of heat, its driving temperature gradient does not remain unchanged, but is reduced. This effect is well known. The Earth's surface, for instance, is cooled rather substantially by the presence of convective heat fluxes, which implies that the vertical temperature difference is depleted by the convective flux. These effects have implications for the Carnot limit as they alter the entropy exchanges of the engine. As will be shown in Chapter 4, when these effects are being considered, the resulting limit to work is much lower than what would be expected from a naive application of the Carnot limit to Earth system processes.

1.5 Thermodynamics, evolutionary dynamics, and structures

Thermodynamic limits set a rigid, physical bound on Earth system processes, yet they say nothing about whether, why, and how a system would operate at this limit. This, then, becomes a question about what thermodynamics can tell us about a general, evolutionary direction of systems and how systems would need to be organized such that they reach their limits. This is quite a profound question. If the evolution of systems is such that the processes within evolve towards their thermodynamic limit, the final, steady state of the system becomes predictable simply by considering the thermodynamic limit. Then, the emergent dynamics may be extremely complex in detail, but the overall behavior of the system becomes simple as it is dominated by the overall constraint imposed by the thermodynamics of the system. This constraint specifically concerns the conditions at the system boundary as well as the entropy exchange of the system. The complexity reflected in the "details" within the system should still show some form of regularity, for instance in form of characteristic structures such as networks (Fig. 1.4), turbulent, or wave-like patterns. We can thus formulate a hypothesis that thermodynamic systems evolve towards their limit by the formation of structures, a hypothesis that we refer to here as "maximization by structures." This hypothesis intimately links the presence of such structures to differences in the dissipative behavior of the system, and links these to a general evolutionary direction of thermodynamic systems.

This question about the link among thermodynamics, evolution, and structures has long been explored. Several hypotheses or principles have been proposed, and some of these may sound contradictory to each other, but this contradiction can be resolved upon a closer look. Perhaps the first to formulate such a principle was Alfred Lotka (1880–1949), an engineer mostly known for his contribution to population biology. In two papers published in 1922 (Lotka 1922a,b), Lotka extended Boltzmann's notion of the "struggle for existence of living organisms" and argued that for biological evolution, "... natural selection tends to make the

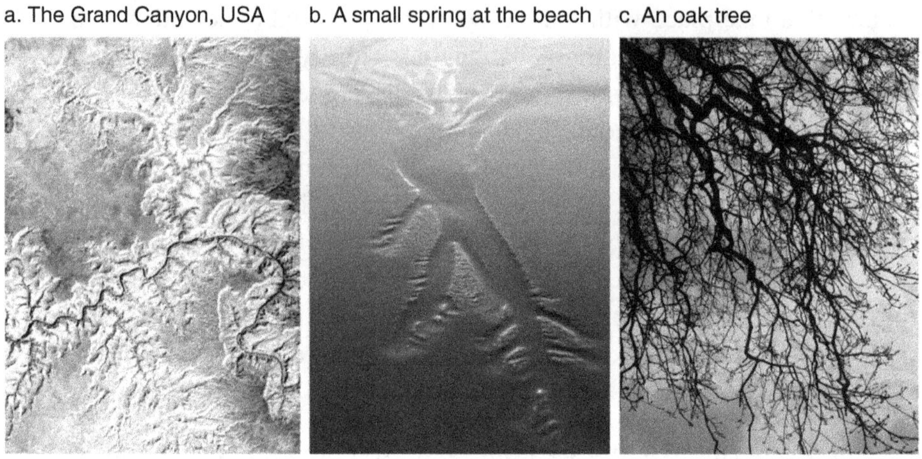

Figure 1.4 Examples of structures that illustrate non-uniform distribution of gradients and flows: (a) a section of the Grand Canyon; (b) a spring at a Danish beach; (c) branches of an oak tree. Sources: (a) terraprints.com/NASA; (b) and (c) the author.

energy flux through the system a maximum so far as compatible with the constraints to which the system is subject." Hence, evolution should tend to make this energy flux a maximum. He noted that this energy flux relates to the power of the process, the physical work performed through time, and that this formulation of evolution and natural selection should make this power a maximum. Lotka focused in his description on the biosphere and its evolution, yet a similar, more general argument could be made for other, physical processes as well (Odum and Pinkerton 1955). This formulation of the maximum power principle combines a physical limit of maximum power with a general evolutionary trend towards this maximum.

There are a series of related thermodynamic extremum principles that have been proposed. There is, for instance, the statement that systems close to thermodynamic equilibrium minimize their entropy production ("MinEP") that was formulated by Prigogine, and which has also been used in engineering (e.g., Bejan 1996). Seemingly contrary to this, a principle of maximum entropy production ("MEP") has been formulated, originally in atmospheric sciences by the works of Paltridge (1975, 1978). This proposed principle of MEP has been shown to be rather successful in predicting the poleward heat transport and turbulent phenomena (e.g., Ozawa et al. 2003; Kleidon et al. 2010). On the other hand, the minimization of frictional dissipation or energy expenditure has been rather successful in explaining the fractal nature of vascular networks in biological systems (West et al. 1997) and river networks (Rinaldo et al. 1992; Rodriguez-Iturbe and Rinaldo 1997), such as those shown in Fig. 1.4. Minimization or maximization sounds rather contradictory, but when evaluating these propositions in more detail, these simply represent alternative views on the same system. If there are two different kinds of processes in a system, for instance radiative and convective heat transfer, but the boundary conditions of the system are fixed, then the maximization of the entropy production associated with one process inevitably results in the minimization of the other process. Likewise, the MEP hypothesis is closely related to Lotka's maximum power principle. As the performance of work results, in the end, in some form of dissipative heating as any motion is associated with frictional losses, it also results in entropy production. When such a system is viewed in a steady state in which its power is balanced by dissipation, the maximization of power is equivalent to the maximization of dissipation, which in turn is almost the same as the maximization of entropy production. This link among different thermodynamic optimality principles is described in Chapter 4 in greater detail.

While formulating a target, these extremum principles still say nothing about why and how systems should evolve to these states. A partial answer to this question was already formulated by Prigogine, who coined the expression of "order through fluctuations" (e.g., Prigogine et al. 1972). Even though Prigogine stated that there is no general thermodynamic extremum principle that describes systems far from equilibrium (e.g., Kondepudi and Prigogine 1998), he described a thermodynamically based, evolutionary feedback by which a thermodynamic system would increase its entropy production. Once a system surpasses a threshold, this would generate instabilities by fluctuations in the system, which in turn would cause greater entropy production by the system. A greater entropy production is maintained by greater entropy export and this would allow for the maintenance of a more ordered state, hence expressing these dynamics by the phrase "order through fluctuations." Such thresholds are found in many Earth system processes, from the threshold that characterizes the transition to turbulent flow, the threshold associated with the detachment of sediment in river systems, or the threshold of a minimum population size in ecology. We could thus envision that this basic thermodynamic mechanism of Prigogine applies to a whole range of Earth system processes up to the planetary scale and that would form a positive, dynamical feedback that enhances the growth of dissipative structures within the Earth system.

Yet, what is missing is a dynamic feedback that limits this growth so that a steady state is reached within the system. This is where thermodynamic limits come into play. They set a physical boundary to the dissipative behavior of a system through the entropy exchange with the surroundings. As dissipative structures grow and make up an increasing share of the entropy production within the system, they ultimately deplete the driving gradient that feeds their growth. This forms a relatively slow-acting negative feedback that should set limits to the growth of dissipative structures in their natural environment.

We can thus envision a general template of a sequence of dynamics that take place towards the thermodynamic limit, similar to those described in Kleidon et al.

(2013) for the thermodynamic evolution of river networks. This sequence starts with the system being in a state close to or at thermodynamic equilibrium. In case of water flow over a land surface, this initial state would be represented by a surface with a relatively uniform slope with no network that channels runoff to an outlet. When this system is perturbed, a small perturbation or fluctuation can cause the initiation of dynamics that further feeds its growth. For instance, a rainfall event may supply sufficient water flow that starts to detach sediment, starting the process of forming a channel. The removal of sediment at one site then locally steepens the gradient, which allows for more detachment at this site in relation to other parts of the slope. By developing such heterogeneous structures as in the case of river networks, the system can then channel its flows and reduce frictional losses within the structure, thereby further feeding the growth of structures and its dissipation. At one point, the structures would enhance fluxes to an extent such that the driving gradients are being depleted. Finally this last aspect is advanced to a point that further growth is counteracted by the depletion of gradients, resulting in a state of maximum dissipative behavior of the dissipative structures. One requirement for such dynamics to take place is, of course, that there are sufficient fluxes across the boundary so that the threshold of sediment detachment is achieved, and there needs to be sufficient degrees of freedom so that the structure can arrange in space in an optimum way. This sequence then combines the feedbacks that describe the dynamics with structure formation and thermodynamic limits. It can be formulated in sufficiently general terms that it should apply to the different types of dissipate structures that are found in the Earth system. In this book, this description of structure formation and its relation to dynamic feedbacks and thermodynamic limits remains qualitative and is thus still speculative.

The combination of dissipative structures and their interactions with the boundary conditions of the system can thus provide an extremely powerful theory to understand evolutionary dynamics in general terms. Evolutionary dynamics would reflect the overarching acceleration of the direction imposed by the second law, making dissipative structures more dissipative through the interaction with their boundary conditions and making processes evolve towards their thermodynamic limit. It would allow us to understand the evolutionary target of thermodynamic systems, from atmospheric convection to river networks, terrestrial vegetation, and the Earth system as a whole. The highly complex spatiotemporal organization of processes, as for instance reflected in the structures shown in Fig. 1.4, could explain how a process can reach its thermodynamic limit by forming structures that affect the dissipative behavior of the process. Dissipative structures would then form a central element to understand the emergent, thermodynamic state of the Earth system. In this case, the emergent state of the system can then be predicted by the thermodynamic limit, not because the system is organized in a simple way, but rather the contrary – it is so complex that its emergent state is characterized by the thermodynamic limit. To evaluate the potential of this theory, it requires that processes are linked to each other, particularly regarding their energy-, mass-, and entropy exchange, ultimately up to the planetary radiative forcing. This would then formulate a thermodynamic theory of the whole Earth system that can be evaluated by deriving quantitative estimates that can be tested with observations.

1.6 Connecting Earth system processes

A critical component in formulating thermodynamic limits of Earth system processes and to evaluate the theory just described is to systematically link processes to each other and to include the effects and interactions in the evaluation. These linkages are schematically shown in Fig. 1.5. This picture closely mirrors a view of the Earth system that was already described by Lotka in his book "Elements of physical biology" (Lotka 1925), in which he wrote that:

"the picture we must keep before us ... is that of a great world engine ... composed of a multitude of subsidiary units, each separately, and all together as a whole, working in a cycle. It seems, in a way, a singularly futile engine, which ... carefully and thoroughly churns up all the energy gathered from the source. It spends all its work feeding itself and keeping itself in repair ... ."

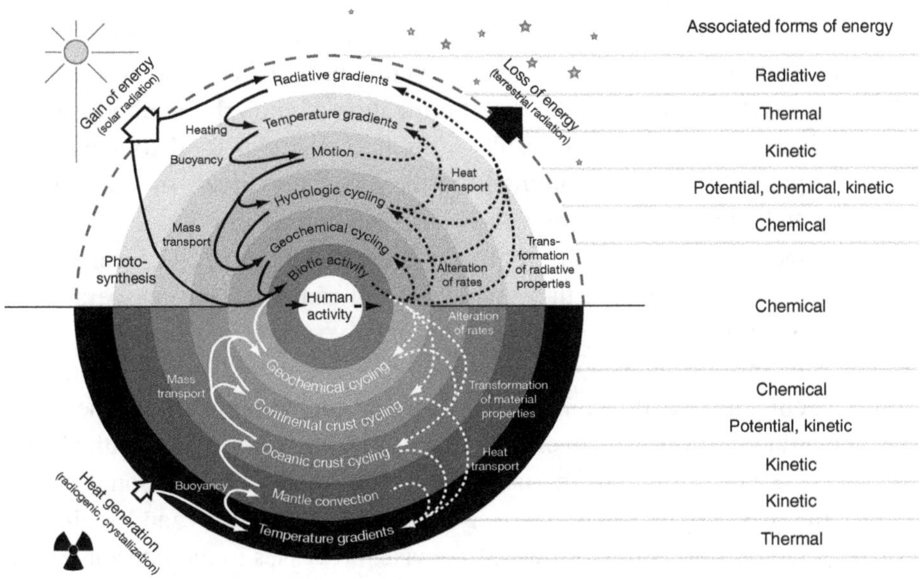

Figure 1.5 A planetary view of the thermodynamic Earth system, with its cascades of energy conversions (left, solid lines), its effects (right, dashed lines), and associated forms of energy shown on the right. The description of this planetary view and the quantification of thermodynamic limits of the conversion rates are the main objectives of this book. Modified from Kleidon (2010, 2012).

The picture shown in Fig. 1.5 embraces Lotka's view. It shows the processes of the surface-atmosphere system (top half) and those of the Earth's interior (bottom half) as well as the two major drivers, solar radiation and interior cooling, of the "great world engine." The first driver results from the spatial and temporal variation in solar radiation at the system boundary. The black arrows in Fig. 1.5 show how this forcing then acts to cause further gradients and dynamics in the system. The gradients in radiative forcing generates uneven heating of the surface, resulting in temperature gradients. These temperature gradients create density differences in the near surface air and buoyancy forces that result in atmospheric motion. Atmospheric motion provides the means to transport water vapor and other constituents. By doing so, it lifts moist air to greater heights, brings the water vapor to condensation and precipitation, so that the descending air is dehumidified. The dehumidification of the atmosphere is associated with gradients that drive evaporation and desalination of the evaporated water at the surface. Thus, atmospheric motion is mostly responsible for maintaining the global hydrologic cycle. The transport of water to land drives river runoff and sediment transport, but also provides the means to chemically dissolve the continental crust, thus driving geochemical cycling of dissolved substances. A similar sequence is shown for interior processes, where heat generation in the interior as well as secular cooling provide the means to generate temperature gradients in the interior. These temperature gradients cause density differences that generate mantle convection, plate tectonics, continental uplift, and thereby the means for the geochemical cycling of material mostly in solid form. Hydrological and geochemical cycling provide the basic resources that feed biotic activity, jointly with the direct utilization of sunlight by photosynthesis. Human activity, in turn, uses the results of biotic activity as food sources. Hence, we get a sequence of transformations from the planetary drivers of solar radiative energy flux and interior cooling down to biotic and human activity.

The linkages between the different processes shown in Fig. 1.5 relate to the different forms of energy that they involve. With radiation and temperature differences, these forms are straightforward and involve radiative energy and thermal energy. Forms of motion are associated with kinetic energy. Hydrologic and geochemical cycling involve different forms of energy, including potential energy, e.g., associated with water droplets in clouds; chemical energy associated with chemical composition, e.g., carbon in different forms such as methane or carbon dioxide; in form of binding energies, such as water bound to a soil matrix; and with kinetic energy, e.g., associated with river flow. Likewise, processes in the interior are associated with different forms of energy. These involve the thermal energy stored in the interior, the kinetic energy associated with the motion of mantle convection and plate tectonics, the potential energy associated with the continental crust, and chemical energy of the geochemical composition. The sequences described here that link the planetary drivers to the different processes within the Earth system are thus intimately linked with energy conversions among the different types of energy. This is where the linkage to thermodynamics comes in as it constrains these rates of energy conversions.

Also shown in Fig. 1.5 are the consequences of the different processes back to the drivers by the dashed arrows. These effects can broadly be understood in terms of altered rates of material transformations, transport of heat, and in terms of altering the radiative properties of the system. To start at the center, biotic activity alters the rates of geochemical and hydrologic cycling and it alters the radiative properties of the surface. A vegetated surface, for instance, is generally darker, thus absorbs more solar radiation, and it can maintain evapotranspiration for longer periods than a bare surface in the same climatic environment. The rates of geochemical cycling in the associated soil are also typically enhanced compared to a bare soil substrate. These altered rates affect the chemical composition of the atmosphere which results in modified radiative properties. Furthermore, hydrologic cycling as well as motion in general involves substantial rates of heat transport, which affects temperature gradients. As temperature is directly linked to the radiative exchange, these effects on temperature gradients affect emission of radiation and radiative gradients. This, ultimately, alters the radiative entropy export to space, the planetary boundary conditions, and the thermodynamic state of the planet.

Thermodynamics applies to these sequences and their effects, and sets firm limits to the magnitude of the associated rates of energy conversions and thus to the overall operation of the "great world engine." By following the planetary forcing down along these sequences, thermodynamic limits allow us to estimate the rates of energy conversions, the associated magnitudes of the fluxes and rates and the constraints on effects, interactions, and feedbacks in the Earth system. We can then use these limits to explore feedbacks and evolutionary directions, evaluate the conditions and consequences of life, and place human activity into the context of the operation of the whole Earth system. The description and quantification of this approach is the main scope of this book.

1.7 Structure of this book

This book describes the thermodynamic foundation to understand the functioning of the Earth system. Thermodynamics in combination with simple formulations are used to describe energy conversions, to estimate the magnitude of these conversions as well as the associated processes, and to illustrate the thermodynamic implications of the resulting dynamics on the thermodynamic state of the Earth system. What this book does not provide is a complete and detailed description of thermodynamics. Here, the reader is referred to introductory physics textbooks (e.g., Feynman et al.

1966), general books on non-equilibrium thermodynamics (e.g., Kondepudi and Prigogine 1998), or specific books on thermodynamics of Earth system processes (e.g., Ambaum 2010; Bohren and Albrecht 1998; Verhoogen 1980). This book focuses on the extent to which thermodynamics is needed to gain a better understanding of how processes operate and interact within the Earth system context and thereby shape the functioning of the whole, thermodynamic Earth system.

The first part of the book provides the basics of thermodynamics that are needed to make this planetary view quantitative: Chapter 2 deals with the different forms of energy and entropy that are associated with Earth system processes. The laws of thermodynamics in the context of the Earth system are described in Chapter 3. The limits that follow from these laws are derived and demonstrated with a few examples in Chapter 4. Chapter 5 contains a general description of the basic feedbacks that are associated with the dynamics that would result in the tendency of natural systems to evolve to their thermodynamic limit.

The second part of the book deals with the formulation of Earth system processes in terms of energy conversions. Radiative exchange is described in Chapter 6 and it is estimated how much energy can maximally be converted from radiation. Chapter 7 starts with the description of buoyancy as the main process that generates motion from a perspective of energy conversions, relates these to the various forms of motion, and estimates the magnitudes of energy conversions involved. In Chapter 8, the hydrologic cycle is related to these energy conversions. Geochemical transformations and resulting cycles as well as life as a biogeochemical process are described in Chapter 9. Chapter 10 sets a special focus on processes on land, where processes are strongly shaped and interact with terrestrial vegetation. This part describes the setting in which most of human activity takes place and where it affects the Earth system most directly. Human activity is then treated to the extent to which it relates to energy conversions in Chapter 11. Each of these chapters closes by placing the particular set of processes back into the overall thermodynamic setting of the Earth system.

This book closes with the integration of the various processes back to the planetary scale and discusses its implications. Chapter 12 summarizes the various energy conversions associated with Earth system processes described in the specific chapters. The synthesis of the thermodynamic foundation of the Earth system is provided by discussing the implications of this view with respect to the topics raised here, specifically regarding the habitability of planetary environments, the evolutionary dynamics of the biosphere and the Earth system, and a sustainable future of the Earth system. It closes with a perspective of how this thermodynamic theory of the Earth can be further developed and what the implications are for Earth system science.

2

Energy and entropy

2.1 The central roles of energy and entropy

Different forms of energy and conversions among these are central to the dynamics of the Earth system and to the application of thermodynamics. Energy is defined as a property of matter and radiation that is linked to the capacity to perform work. Performing work relates to the conversion of one form of energy to another. The actual capacity to perform work is described by free energy, which is linked to the dispersal of energy across the microscopic scale of atoms and molecules and is described by its entropy. This chapter focuses on the description of different forms of energy and entropy relevant to the Earth system; but we should keep in mind that the dynamics of the Earth system are not shaped by the magnitudes of energy or entropy in the system, but rather by the conversion rates that are related to differences in energy and entropy. These conversion rates are subject to the laws of thermodynamics and are dealt with in the following chapters.

Earth system processes involve different forms of energy. Solar and terrestrial radiation involve radiative energy. Atmospheric motion is associated with kinetic energy, while cloud droplets are associated with gravitational, or simply potential, energy. Soil moisture on land is linked to the energy associated with the binding energy of water to the soil matrix and with potential energy. The concentration of constituents in air, water, and solids as well as biomass is linked to forms of chemical energy. Likewise, any other process within the Earth system is associated with some form of energy. In this chapter, the major forms of energy are described and broad estimates of their magnitude are given to illustrate how these are quantified.

A more hidden aspect in these forms of energy is its spread at the scale of atoms and molecules that is described by entropy. At the microscopic scale, energy is stored in discrete, countable units. These units involve, for instance, photons, quanta of radiative energy, discrete energy levels of electrons in molecular bonds, and the distribution of kinetic energy over discrete number of molecules in a gas. When we describe Earth system processes at the planetary scale, such microscopic details are avoided to the extent possible, and aggregated, macroscopic variables, such as temperature, pressure, and density, are used instead that relate to macroscopic forms of energy. This is where entropy comes in as a means to scale up the dispersal of energy from the microscopic to the macroscopic scale at which variables such as temperature and density can be defined. Entropy allows us to distinguish different distributions of energy within a system even if the total amount of energy remains the same. For instance, the entropy of a system is different if a difference in temperatures is present within the system compared to the state with a uniform temperature. Likewise, differences in partial pressures of a substance within a system, or differences in elevation of water are captured by different values of entropy of the system. As we will see in the following chapters, specifically with respect to the second law of thermodynamics, changes in entropy govern the dynamics in systems. Hence, entropy is as important to describe Earth system processes as is the related form of energy. Before we can look at these changes in entropy, the forms of entropy associated with different forms of energy first need to be introduced and described.

A necessary starting point for a thermodynamic description of Earth system processes in terms of their energy and entropy is the definition of the thermodynamic system in which these processes take place. This definition is associated with the delineation of a boundary, and the description of its state in terms of variables. While this delineation can, in principle, be done in any way one wishes, there are delineations that are more practical than others. These delineations are important as they set the basis for where exchange fluxes of the system with its surroundings take place. These exchange fluxes jointly with the internal dynamics of the state variables of the system then formulate the accounting balances for those physical properties that are conserved, such as energy, mass, and momentum, and for the budgeting of entropy that is the necessary basis to apply the second law of thermodynamics.

This chapter provides a general description of the different forms of energy and entropy that are relevant to Earth system science. We first define different types of thermodynamic systems and the variables that describe their state to set the basis for the accounting of the different forms of energy and entropy. The different forms of energy that are most relevant to Earth system processes and how these are quantified are described in the following section. The concept of entropy is illustrated and linked qualitatively to the different forms of entropy relevant to Earth system processes. These forms are associated with radiative, chemical, and thermal processes. The forms of entropy that are associated with these processes are illustrated at the microscopic scale and it is described how they relate to each other. The chapter closes with a section on the formulation of energy and entropy budgets that form the basis for a thermodynamic description of the Earth system.

2.2 Defining thermodynamic systems

To account for the different forms and exchanges of energy and entropy that are associated with a particular process, we first need to define the necessary accounting basis. This is done in thermodynamics by defining the process in the context of a thermodynamic system. A thermodynamic system is defined by its boundary, by its state, and by its exchange fluxes. A schematic diagram of a thermodynamic system is shown in Fig. 2.1.

The boundary of the system delineates a certain volume of physical space and separates the system from its environment. Note that the term "environment" strictly refers to the surroundings of the system that is not being considered, and does not refer necessarily to the environmental conditions of, for instance, the Earth's surface. The boundary defines the volume over which the attributes of the system, such as the density or flow velocities, need to be integrated to yield the total mass or momentum of the system. It separates the parts of the dynamics that we consider to be internal dynamics of the system from the processes that take place in the environment of the system and that are not being described. We may also choose to consider only a subset of processes that deal with a specific form of energy when we define a system. We may, for instance, only describe the processes that deal with kinetic energy, and place other processes that take place in the same physical space, but deal with other forms of energy, such as thermal energy, in the system's environment. Hence, the definition of the boundary may also separate specific forms of energy that we consider from those that we are not interested in.

In principle, the delineation of the system boundary is arbitrary. A different delineation of a system results in a system of different size or shape. It may include, or exclude, different forms of energy, and this alters the types and forms of exchange fluxes that take place across the boundary. A different definition of a system boundary may lead to a different accounting of the state of the system. There are also

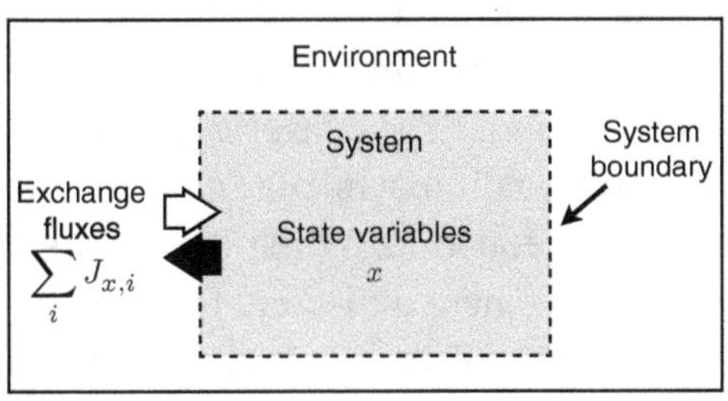

Figure 2.1 A thermodynamic system is defined by its boundary to the environment, by its state, and by its exchange fluxes $J_{x,i}$ with the environment. The state of the system is expressed by state variables (symbolized by x), which are subject to conservation laws.

some places for which it is quite convenient to draw a system boundary. When we think of the atmosphere, oceans, and the interior as thermodynamic systems, then we draw the boundaries between these along lines of different phases. The boundary separating the atmosphere from the ocean is the interface between the gaseous air and the liquid ocean. Likewise, the boundary between the atmosphere and the land surface represents the interface between the gaseous air and the solid Earth. Such a choice seems natural and makes it easier to integrate the properties of the system and to describe the exchange fluxes across the system boundary.

The state of a thermodynamic system is described by corresponding state variables. These variables describe physically observable quantities and characterize the contents of the system. Examples for such variables are the temperature T, pressure p, mass m, the height z at which the mass is located, the flow velocity v within the system, or the partial pressure of a substance. State variables fully characterize the state of the system independent of how the system evolved into a particular state. In thermodynamics, it is typically assumed that state variables are uniform within the system. To represent differences and gradients in variables within a system, we need to then split up the system into smaller subsystems in which the variables are then again uniform within the subsystem. This will be done in the following chapters, in which two box representations of a system are used to represent a difference in state variables. Note also that state variables are not necessarily independent of each other. For example, in a gas, pressure and temperature are closely related, with the relationship described by the ideal gas law. Exchange fluxes add or remove certain quantities to or from the system, such as energy, mass, and momentum. For instance, the absorption of radiation adds energy to a system, while the emission of radiation removes energy from a system. Changes in the state reflect the combination of the dynamics within the system and the exchange fluxes between the system and its environment.

These variables are typically constrained by physical conservation laws, most relevantly those of energy, mass, and momentum. The conservation laws balance the internal changes of a conserved quantity with the fluxes of this quantity through the system boundary. They yield the well-known equations that form the basis for describing the dynamics of systems. The conservation of mass applied to fluid flow is described by the continuity equation, while the conservation of momentum is represented by the Navier–Stokes equation. While these laws are typically expressed in their differential form, we use here a discrete form as we will mostly formulate these conservation laws in terms of budget equations for box models. Then, the conservation of a certain quantity x takes the form

$$\frac{dx}{dt} = \sum_i J_{x,i} \tag{2.1}$$

where x is the state variable, and the sum on the right hand side adds all exchange fluxes across the system boundary $J_{x,i}$ that affect the state variable. The sign convention for the exchange fluxes here is that a flux is positive when it adds a certain quantity to the system, that is, $J_{x,i} > 0$, while it is negative when it removes a certain quantity from the system. The definition of the system boundary then defines the volume over which these conservation laws apply. The associated budgets play a fundamental role as they link the state of the system to the exchange fluxes across the system boundary.

The exchange fluxes play a critical role in shaping the state of a system as these typically maintain gradients in systems that act as the drivers for the dynamics. Based on the different types of exchanges, we distinguish three different types of thermodynamic systems.

An isolated system is a thermodynamic system that exchanges neither energy, mass, nor any other quantity subjected to a conservation law with its environment. In such a system, all conservation laws of the form described by eqn. 2.1 are represented by the case in which all forms of exchange fluxes, $J_{x,i}$, are zero. Isolated systems are mostly of interest for theoretical considerations, because practically all systems of the Earth exchange some physical variables and are thus not isolated from their environment. The only exception is when the Universe as a whole is being considered (Lineweaver and Egan 2008). The consideration of isolated systems nevertheless plays a very useful role in defining a reference state of the system in the absence of exchange fluxes. Internal dynamics are possible in an isolated system when it is initially in a disequilibrium state. With such considerations, we can distinguish between the roles of the internal dynamics and the exchange fluxes in determining the thermodynamic state of a system.

A closed system is a thermodynamic system that only exchanges energy with its environment. In a closed system, the conservation law for energy has non-zero exchange fluxes of energy, so that for these fluxes, $J_{x,i} \neq 0$. The other conservation laws are then still described by exchange fluxes at the system boundary that are zero. A prime example for a nearly closed system is the Earth system as a whole. We draw the boundary of the Earth system at the interface of the upper atmosphere to space. The exchanges through this boundary are dominated by energy exchange in the form of radiative fluxes. In comparison, mass exchange with space and momentum exchanges with tides are much smaller, so that the Earth is a nearly closed system.

An open system is a thermodynamic system that exchanges energy and other physical properties such as mass and momentum with its environment. The conservation laws for energy, mass, and momentum then generally have non-zero exchange fluxes. The atmosphere, for instance, is such an open system. At the boundary to the surface of the Earth, there are exchange fluxes of energy, but also exchange fluxes of mass of, for instance, water vapor and carbon dioxide, and fluxes of momentum in the form of drag forces at the surface.

Most systems that we deal with in the context of Earth system processes are open systems and are tightly linked by the exchange of energy, mass, and momentum with their surroundings. One aspect that we need to keep in mind when defining an open system of a particular Earth system process is that the exchange fluxes of the system couple it to the Earth system and its forcing. This coupling can result in interactions and feedbacks that alter the conditions and fluxes at the system boundary. This aspect is described later in greater detail in Section 3.7, and we will encounter several examples of such strong interactions in the later chapters of the book.

2.3 Forms of energy

There are several forms of energy that are relevant to the Earth system and that are, in total, subject to energy conservation. These forms are summarized in Table 2.1, along with the variables that characterize these forms of energy, and the conservation laws that apply to these variables. There are also other forms, e.g., that associated with the magnetic field or with nuclear bonds in atoms, but these are not considered here.

The forms of energy listed in Table 2.1 are characterized by a pair of conjugate variables. Each form is composed of an intensive variable, which is independent of the size of the system, and an extensive variable which depends on the size of the system. To determine the amount of a particular form of energy in the system, the intensive variable is integrated with respect to the extensive variable to accomplish the integration over the size of the system. Since the extensive variable typically depends on the volume of the system, the integration typically is performed over the volume of the system. These main forms of energy and how they are calculated are briefly described in the following sections. In the variables that are being used, the index "*a*" refers to the atmosphere, "*o*" to the ocean, and "*i*" to the interior solid Earth.

2.3.1 Radiative energy

The first form of energy listed in Table 2.1 is radiative energy, which is the energy associated with fluxes of electromagnetic radiation of different wavelengths. Radiative energy is associated with the incident solar radiation as well as the thermal radiation that is emitted from the Earth to space. It is by far the most important exchange of energy of the Earth system with space and constitutes the primary forcing of the planet. Radiation can be described as a gas, consisting of quantized

Table 2.1 *Forms of energy relevant to Earth system processes, the conjugate pair of variables that describe them, the most relevant conservation law that applies to these variables and examples of related Earth system state variables or fluxes. A conjugate pair is formed by the combination of an extensive variable that depends on the size of the system and an intensive variable that is independent of the size of the system.*

Form of energy	Intensive variable	Extensive variable	Conservation law	Examples
Radiative energy	Radiative temperature, T Radiative pressure, p	Radiative entropy, S Volume, V	Energy	Solar radiation, terrestrial radiation
Thermal energy	Temperature, T	Thermal entropy, S	Energy	Surface temperature, air temperature
Uncompensated heat	Pressure, p	Volume, V	Energy	Air pressure
Potential energy (or gravitational energy)	Potential, ϕ	Mass, m	Mass	Cloud droplet, soil water
Kinetic energy	Velocity, v	Momentum, p	Momentum	Atmospheric flow, ocean currents, river flow
Binding energy	Chemical potential, μ	Number of particles, N or mass, m	Mass	Water vapor, soil water
Chemical energy	Affinity, A	Extent of reaction, ξ	Mass	Geochemical reactions
Electric energy	Voltage, U	Charge, Q	Charge	Lightning, also used for analogies

carriers of electromagnetic radiation called photons. This "photon gas" is described akin to an ideal gas in terms of radiative entropy S, radiative temperature T, as well as radiative pressure p and volume V. Radiative entropy describes the distribution of radiative energy across different wavelengths, an aspect that is described in more detail in the next section on entropy and in Section 6.2.

When we estimate the actual amount of radiative energy that resides within the Earth system, it is extremely small in relation to the magnitude of the fluxes, or, expressed differently, its residence time is very short. Using the magnitude of the incoming solar radiative flux of $R_{s,\text{in}} = 1370$ W m^{-2}, the cross sectional area $A = \pi r_{\text{earth}}$ and volume $V = 4/3\pi r_{\text{earth}}^3$ of the Earth (with $r_{\text{earth}} = 6372 \cdot 10^3$ m) to get a mean length scale $L = V/A$, as well as the speed of light ($c = 3 \cdot 10^8$ m s^{-1}), we can derive the residence time $\tau = L/c$ to be about 0.03 s, with the actual amount of radiative energy given by $U_{\text{rad}} = R_{s,\text{in}} A\tau$. Because of the very short residence time of radiative energy within the Earth system, we do not consider the amount of radiative energy in the Earth system, but only the radiative fluxes.

2.3.2 Thermal energy

Thermal energy is associated with the energy stored in mass that relates to its temperature. At the scale of molecules, this form of energy describes the kinetic energy stored in the random motion of these molecules, either when they move freely, as in a gas, or in terms of their rotation and vibrations, as in a liquid, a solid, or a gas. The associated energy at the macroscopic scale is described by temperature T, and the thermal entropy S. The thermal entropy S is defined by its increase in a system due to the addition of a certain amount of heat dQ, at a certain temperature, T (Clausius 1850):

$$dS = \frac{dQ}{T} \tag{2.2}$$

Note that technically speaking, the added heat is not a state variable of the system and thus cannot be represented as a differential. It is thus often written as δQ instead of dQ. Here, for simplicity, it is simply referred to as dQ. The entropy thus depends on the temperature of the system. Its microscopic interpretation is described in more detail in the next section.

A more common way to calculate the thermal energy of a system is through its heat capacity c, defined by the change in temperature dT, when a certain amount of heat dQ, is added to the system:

$$c = \frac{dQ}{dT} = T\frac{\partial S}{\partial T} \tag{2.3}$$

where the latter expression used the definition of entropy from eqn. 2.2. Note that for the derivative in eqn. 2.3, the variables that are kept fixed need to be specified. This specification is important because for gases, the specific heat capacity, that is, c normalized by mass, for a fixed volume, c_v, or a fixed pressure, c_p, have different values. This difference relates to how a gas reacts to the addition of heat. When the gas is allowed to expand (i.e., added while keeping the pressure fixed), some of the added energy appears as a change in $(p\,dV)$, a form of energy that is discussed later. If the gas is not allowed to expand, that is, the volume is kept fixed, the added heat increases only the thermal energy in the system. The specific heat capacities for fixed volume and pressure are thus different, but intimately linked. For an ideal gas, this relationship is given by $c_v + R = c_p$, where R is the ideal gas constant. For liquids and solids, where the thermal expansion is much smaller upon heating, the $(p\,dV)$ term is much smaller, so that the two specific heat capacities are almost the same.

The thermal energy of a system $U_{\rm te}$, is thus obtained from the specific heat capacity c, the mass m, and the temperature T, of the system:

$$U_{\rm te} = \int TdS = \int_V dQ \tag{2.4}$$

using eqn. 2.2. If the specific heat capacity and temperature are uniformly distributed within the system, we can use the specific heat capacity and eqn. 2.3 to perform this integration and obtain

$$U_{\rm te} = \int_V (c_v\rho)TdV = c_v\,T\int_V \rho dV = c_v \cdot m \cdot T \tag{2.5}$$

Estimates of the thermal energy stored in the atmosphere, ocean, and the interior are given in Table 2.2. The thermal energy storage in the atmosphere of about $1.8 \cdot 10^9$ J m^{-2} per unit surface area (Oort et al. 1989) reflects the specific heat capacity of air at constant volume, $c_a = 717$ J kg^{-1} K^{-1}, the atmospheric mass per unit area of $m_a = 10^4$ kg m^{-2}, and a mean temperature of $T_a = 255$ K. The thermal energy of about $4.3 \cdot 10^{12}$ J m^{-2} stored in the oceans (Oort et al. 1994) reflects the specific heat capacity of water of $c_o = 4180$ J kg^{-1} K^{-1}, the mass of the water column of about $\Delta z_o = 3700$ m depth of about $m_o = 3.7 \cdot 10^6$ kg m^{-2}, with a mean temperature of $T_o = 277$ K. The atmospheric storage of thermal energy per unit area thus corresponds to only about 1.6 m of ocean column. The thermal energy storage in the Earth's interior is estimated roughly using a specific heat capacity of $c_i = 1000$ J kg^{-1} K^{-1}, a total mass of the Earth of $m_i = 6 \cdot 10^{24}$ kg (which corresponds to about $1.2 \cdot 10^{10}$ kg m^{-2} when expressed in terms of surface area), and an approximate mean temperature of $T = 3000$ K (Elder 1976). This yields an estimate of about $2.8 \cdot 10^{16}$ J m^{-2} of thermal energy per unit surface area of the

Earth. This comparison shows that, not surprisingly, the largest storage of thermal energy by far is represented by the storage in the Earth's interior.

The major causes for changes in thermal energy relate to the absorption and emission of radiative energy, and to a much lesser extent, to radiogenic heating and secular cooling of the interior. Because these heating rates are relatively small compared to the heat capacities of the three examples shown in Table 2.2, thermal energy resides on relatively long time scales in the Earth system. This can be seen in the mean residence times of thermal energy, derived from the ratio of the thermal energy stored to the mean heating rate. For the atmosphere, which is heated by a mean absorption of solar radiation of about 240 W m^{-2}, this residence time corresponds to about 90 days. The ocean is heated by about 165 W m^{-2} of solar radiation absorbed at the surface, which yields a residence time of more than 800 years. The Earth's interior loses heat at a rate of less than 0.1 W m^{-2}, yielding a residence time of almost 9 billion years. What this demonstrates is that the magnitude of the thermal energy being stored in a system does not reflect much of the associated dynamics taking place within the system. In fact, the estimates illustrate that the atmosphere can react the fastest to external heating or cooling as it has the shortest residence time of the three examples.

2.3.3 Uncompensated heat

A form of energy that is associated with the pressure of a gas was referred to by Clausius (1850) as uncompensated heat, which at the microscopic scale reflects the aspect that molecules in a certain volume of space collide with each other and with the boundary of the volume. Changes in this form of energy are commonly referred to as mechanical work, but it is commonly not described as a separate form of energy. We will see in the following that this form of energy is described by other forms of energy, particularly the potential energy, so that it is not explicitly listed in Table 2.2.

This form of energy is measured by the pressure p of the gas and the volume V in which it is contained:

$$U_{pV} = \int dU_{pV} = \int_V p dV \tag{2.6}$$

For a gas, the pressure and volume are intimately linked to temperature, which can directly be seen in the ideal gas law,

$$pV = nRT \tag{2.7}$$

where n is the molar mass, R the ideal gas constant, and T is the temperature. This intimate connection to temperature also results in a pressure effect on the thermal

Table 2.2 *Estimates for major forms of energy in the Earth system. Estimates are stated with respect to the surface area of the Earth and for the whole planet. Based on estimates by Elder (1976) and Oort et al. (1989).*

Form of energy	Estimate $\mathrm{J\,m^{-2}}$	Estimate 10^{18} J
Radiative energy	$1.0 \cdot 10^{1}$	$1.6 \cdot 10^{-2}$
Thermal energy		
Atmosphere	$1.8 \cdot 10^{9}$	$9.3 \cdot 10^{5}$
Ocean	$4.3 \cdot 10^{12}$*	$1.6 \cdot 10^{9}$
Solid Earth	$2.8 \cdot 10^{16}$	$1.4 \cdot 10^{13}$
Potential energy		
Atmosphere	$7.3 \cdot 10^{8}$	$3.7 \cdot 10^{5}$
Ocean	$6.9 \cdot 10^{10}$*	$2.5 \cdot 10^{7}$
Solid Earth	$4.4 \cdot 10^{17}$	$2.2 \cdot 10^{14}$
Kinetic energy		
Atmosphere	$1.2 \cdot 10^{6}$	$6.1 \cdot 10^{2}$
Ocean	$7.8 \cdot 10^{3}$*	$2.9 \cdot 10^{0}$
Solid Earth	$5.4 \cdot 10^{-9}$	$2.8 \cdot 10^{-12}$
Binding energy		
Atmosphere (water vapor)	$7.0 \cdot 10^{7}$	$3.6 \cdot 10^{4}$
Ocean (dissolved salt)	$8.5 \cdot 10^{9}$*	$3.1 \cdot 10^{6}$
Solid Earth (core freezing)	$2.0 \cdot 10^{16}$	$1.0 \cdot 10^{13}$
Chemical energy		
Stratospheric ozone	$3.0 \cdot 10^{4}$	$1.5 \cdot 10^{1}$
Land biomass	$2.8 \cdot 10^{7}$**	$4.0 \cdot 10^{3}$
Ocean biomass	$5.4 \cdot 10^{4}$*	$2.0 \cdot 10^{1}$

* per unit area of ocean surface (72% of total surface area)
** per unit area of land surface (28% of total surface area)

entropy S. Using the definition of entropy given by eqn. 2.2 in combination with the first law, it can be shown that the entropy of an ideal gas is described by (e.g., Kondepudi and Prigogine 1998)

$$S = S_0 + c_v \ln T + R \ln V \tag{2.8}$$

where S_0 is a reference value. We will see later in the description of a thermodynamic cycle in Section 4.3 that this volume effect on the thermal entropy is important. It can result in the so-called adiabatic condition in which a gas simultaneously cools and expands, keeping its entropy fixed.

To illustrate this form of energy, we estimate this expression for the Earth's atmosphere and then compare it to other forms. We evaluate this form of energy per unit surface area of the Earth, so that $dV/A = dz$, with A being the surface area and z being the vertical coordinate. To express the variation of pressure with height, we use a well-known approximation known as the barometric formula for an isothermal atmosphere, that is, we neglect temperature variations with height. This formula describes the pressure dependence on height by $p = p_s \exp(-z/z_0)$, where p_s is the surface pressure and $z_0 = g/(RT_a)$ is a scaling height. When this expression is used in eqn. 2.6 and integrated over the vertical column, we obtain

$$U_{pV} = \int_0^\infty p_s e^{-z/z_0} dz = p_s \frac{RT_a}{g} = m_a R T_a \tag{2.9}$$

where m_a is the mass of the atmosphere per unit area, noting that the surface pressure represents the weight of the atmospheric column, $p_s = m_a g$. Note that when this form of energy is combined with the thermal energy, the combination can be described as if it were all thermal energy when using the specific heat capacity at constant pressure, since $c_v + R = c_p$. The uncompensated heat is thus proportional to the thermal energy of the atmosphere. When energy is added to or removed from a system, this form of energy changes together with the thermal energy of the system. These combined changes are described by changes in the so-called enthalpy, which is introduced later in Section 3.5, and which is more traditionally referred to as changes in the heat content of a system. We will see later that the uncompensated heat also constitutes the potential energy of the atmosphere.

This form of energy is altered when a gas expands or contracts which is the case when the mass of the atmospheric column is changed, mostly by horizontal advection, or, more importantly, by heating and cooling. In liquids and solids, this form of energy relates to the effects of compression or thermal expansion, which alter the volume of a system by the addition or removal of heat or mass, as in the case of the gas.

2.3.4 Potential energy

Potential, or gravitational, energy relates to the relative position of mass in a gravitational field. It is, for instance, associated with atmospheric mass at different heights above the surface, with water on land that is at an elevation greater than sea level, with water drops in clouds at some height above the surface of the Earth, or with solid mass at some distance to the core of the Earth. It is described by the mass as well as the geopotential $\phi = gz$ at which this mass is located at, with g being the gravitational acceleration and z the physical height above a reference level.

Potential energy is calculated by the integration of the geopotential over the mass. Since the mass is equal to its density times volume, $m = \rho V$, the potential energy is directly expressed as a volume integral over the size of the system

$$U_{\mathrm{pe}} = \int dU_{\mathrm{pe}} = \int_m \phi dm = \int_V (gz)\, \rho dV \tag{2.10}$$

When evaluating this energy per unit surface area, the volume integral is converted into an integral over the height of the system by $dV = A\, dz$, where A is the surface area and z is the height.

When the potential energy per unit surface area of the Earth is computed for the atmosphere, we note that the density ρ is a function of height and follows approximately the ideal gas law, $\rho = p/RT$. We obtain

$$U_{\mathrm{pe}} = \int_0^\infty (gz)\, \rho dz = -\int_{p_s}^0 z\, dp = \int_0^\infty p dz = U_{pV} \tag{2.11}$$

where the integration takes place from the surface to the top of the atmosphere. In this derivation, the hydrostatic balance, $dp = -\rho g dz$, that relates the vertical pressure difference, dp, with a difference in weight, $\rho g dz$, was used as well as $\int z dp = [zp] - \int p dz = -\int p dz$. Since the potential energy of the atmosphere directly relates to its uncompensated heat (cf. eqn. 2.6) and thus to its thermal energy, the sum of both is referred to as the total potential energy of the atmosphere (Holton 1992). The potential energy of the atmosphere is thus simply given by $R/c_p \cdot U_{\mathrm{te}}$ or 40% of the thermal energy.

For solids and liquids, the density ρ does not vary much with height, which simplifies the integration. For the ocean, we can estimate its potential energy with respect to the sea floor as a reference height (at which $z = 0$) by integrating over the mean depth of the ocean column:

$$U_{\mathrm{pe}} = \int_0^{\Delta z_o} \rho g z\, dz = \frac{1}{2} \rho g \Delta z_o^2 \tag{2.12}$$

Using a density of water, $\rho_o = 10^3$ kg m^{-3}, and a value of $\Delta z_o = 3700$ m yields an estimate for the potential energy. The resulting potential energy is about 100 times larger than for the atmosphere, which is less than the ratio of their masses. This is due to the lower density of the atmosphere, so that the mass is more spread out over the height leading to a higher value of its potential energy.

For the potential energy of the solid Earth, we need to consider the fact that the value of the gravitational acceleration g, is a function of depth and that we need to account for spherical aspects. This aspect was neglected for the atmosphere and ocean because their vertical extents are comparably small in relation to the distance to the center of the Earth. If we assume a uniform density in the interior, we obtain the potential energy by

$$U_{\text{pe}} = \int_V (gr)\rho dV = \int_0^{r_e} \left(\frac{4}{3}\rho G r^2\right)(4\pi r^2 \rho) dr = \frac{16}{15}\pi^2 G \rho^2 r_e^5 \tag{2.13}$$

where the general form of the gravitational force was used to express g as a function of depth ($g = 4/3\ \rho G r$, where G is the gravitational constant, $G = 6.67 \cdot 10^{-11}$ m^{-3} kg^{-1} s^{-2}). Using a mean density of $\rho_i = 5.5 \cdot 10^3$ kg m^{-3} and a radius of the Earth of $r_e = 6372 \cdot 10^3$ m yields the estimate shown in Table 2.2. Just as for thermal energy, the potential energy of the interior is, again, substantially higher than for the other systems. This is, of course, not surprising as it is simply reflecting the substantially different sizes of the considered systems. Note, however, that the values of potential energy are not directly comparable as they include different reference heights (surface, ocean floor, and center of the Earth), an aspect that would need to be considered as well.

Potential energy is altered when mass is added or removed, or when it is heated or cooled as this expands or contracts mass, thereby altering its vertical position. Such differences in potential energy can then be converted into kinetic energy, so that differences in this form of energy set an important intermediate step for the generation of kinetic energy associated with motion. We will revisit potential energy in Chapter 7 when we deal with buoyancy forces and the generation of motion.

2.3.5 Kinetic energy

Motion of mass with a certain velocity is associated with kinetic energy at the macroscopic scale. Kinetic energy is, for instance, associated with motion in the atmosphere, with river flow on land, the oceanic circulation, and with motion of the solid Earth in the form of mantle convection and plate tectonics. Motion plays a critical role for the material transport within the Earth system. Kinetic energy U_{ke}, is represented by the velocity v at which a certain mass m moves, as well as its momentum, $p = mv$.

$$U_{\text{ke}} = \int dU_{\text{ke}} = \int_p v dp \tag{2.14}$$

Note that p refers here to the momentum, and not to pressure. If we consider a certain, fixed mass associated with the system, its kinetic energy is obtained by integration from rest up to its velocity:

$$U_{\text{ke}} = \int_0^v (mv)\, dv = \frac{1}{2} m v^2 \tag{2.15}$$

Note also that we refer here to kinetic energy as a form of energy at the macroscopic level and not to the kinetic energy associated with the random motion of molecules at the microscopic level that is represented by the thermal energy.

Using this equation, we can use published estimates of kinetic energy that are listed in Table 2.2 in combination with the mass of the respective system to infer the typical velocity that is represented by the kinetic energy within the system. A mean kinetic energy of about $1.2 \cdot 10^6$ J m^{-2} and a mass of about $m_a = 10^4$ kg m^{-2} yields a typical velocity within the atmosphere of about $v_a = 16$ m s^{-1}. A kinetic energy of about $7.8 \cdot 10^3$ J m^{-2} with a mass of $m_o = 3.7 \cdot 10^6$ kg m^{-2} yields a mean velocity within the ocean of about $v_o = 0.07$ m s^{-1}. The kinetic energy of the interior of $5.4 \cdot 10^{-9}$ J m^{-2} reflects a mass of $m_i = 1.2 \cdot 10^{10}$ kg m^{-2} moving at a typical velocity of $v_l - 1.0 \cdot 10^{-9}$ m s^{-1}. These are merely rough estimates, yet they nevertheless provide an impression on the relative magnitudes of motion associated with the three different systems. As we will see later on, the comparatively small amounts of energy involved in motion nevertheless play a critical role in shaping patterns of heat and mass transport in the Earth system.

Kinetic energy is typically generated out of differences in potential energy that in turn relate to differential heating and cooling, as described earlier. When kinetic energy is dissipated by friction, it is converted back into thermal energy or can perform further work.

2.3.6 Binding energy

Binding energy summarizes several forms of energy that are involved in phenomena dealing with intermolecular binding forces that are involved in different phases of a substance, like liquid water or ice, the binding of liquid water to the soil matrix by adhesive and capillary forces, or surface tension of water droplets (in which case the energy can be referred to as a surface energy). In contrast to the vapor state, water molecules in their liquid state are bound to each other by hydrogen bonds. The energy associated with this intermolecular binding is described by a chemical potential μ and the number of particles, N, where the chemical potential is defined by the change in binding energy when a small amount of particles is added or removed. As the number of particles is related to mass, with the conversion given by the molar mass of the substance, this form of energy can also be formulated in terms of the chemical potential and mass, m. The potential energy described in Section 2.3.4. can also be seen as a form of binding energy, for which the chemical potential corresponds to the geopotential, $\mu = \phi$ (Kondepudi and Prigogine 1998). Also, dissolved substances can be described by this form of energy, with the chemical potential corresponding to the osmotic potential and the form of energy referred to as an osmotic energy. For instance, the energy associated with dissolved minerals in water is described by this form. The osmotic potential then depends on the concentration of the dissolved substance. We refer to these forms of energy as binding energy here.

The associated amount of energy is obtained by integration of the chemical potential over the mass of the system:

$$U_{be} = \int dU_{be} = \int_m \mu \, dm \tag{2.16}$$

For the integration, a functional relationship of the chemical potential, $\mu = f(m)$, for each addition of mass to the system is needed. This functional form is quite different for different forms of molecular binding energies.

Table 2.2 provides three examples of binding energies, two related to phase changes and one related to dissolved minerals. The first example shows the latent heat that would be released if the water vapor of the atmosphere would condense into liquid water. This energy corresponds to a mean mass of about $m_{\text{vapor}} = 28$ kg m^{-2} of water vapor in the atmosphere with a latent heat of vaporization of $\lambda \approx 2.5 \cdot 10^6$ J kg^{-1}. In this case, the binding energy is simply estimated by $U_{be} = \lambda\, m_{\text{vapor}}$. It should be noted, however, that this is a hypothetical estimate because water vapor cannot simply be released since the state of thermodynamic equilibrium of water vapor is its saturation state, as described in greater detail in Chapter 8. The second example involving energy associated with phase changes is associated with the freezing of the core in the Earth's interior, using the estimate of Elder (1976). The third example shown in Table 2.2 refers to the energy that is associated with the dissolved salts in sea water. The concentration of about 35 g kg^{-1} of dissolved salts is converted into an osmotic potential by using the Morse equation, $\mu = iMRT$, which represents an approximation for low concentrations. In this equation, i is the van't Hoff factor (which is $i = 2$ for single charged ions associated with sodium chloride), M is the molar concentration, $R = 8.314$ J K^{-1} mol^{-1} is the molar gas constant, and T is the temperature. For mean ocean water, $M = 500$ mol m^{-3} (computed with a molar weight of 70 g mol^{-1}), and $T = 277$ K, which yields a total energy of about $8.5 \cdot 10^9$ J m^{-2} that reflects the binding energy of dissolved salts in oceans.

Binding energy is altered when mixing and redistribution takes place within the system or when mass is added or removed, which requires mass exchange across the system boundary. This form of energy is considered in greater detail in Chapter 8 when we deal with hydrologic cycling.

2.3.7 Chemical energy

Chemical energy relates to the energy stored in chemical compounds and represents the binding energy of electrons in molecular bonds. This form of energy is described by two variables, the affinity A as well as the extent of the reaction ξ. affinity describes the difference in chemical potentials between reactants and products and thus measures the difference in binding energy between reactants and products. The extent of reaction describes the proportion of reactants and products and thus measures the mass that can still be transformed in the reaction before chemical equilibrium is reached. This form of energy is involved in photochemical and geochemical reactions as well as biotic and human activity. The application of thermodynamics to chemical reactions is described in greater detail in Chapter 9.

Two examples for chemical energy are given in Table 2.2. The first example relates to the energy associated with the presence of ozone in the Earth's stratosphere. With a concentration of about 5 ppm of ozone in the stratosphere (Cicerone 1987), a stratospheric mass of about 1700 kg m^{-2} and the energy that would be released when ozone is transformed to molecular oxygen of about 163.2 kJ mol^{-1}, this yields an estimate for the chemical energy in stratospheric ozone of about $1.5 \cdot 10^{19}$ J. The other example relates to the chemical energy stored in biomass on land and in the ocean. For the estimates, values for the terrestrial and marine biomass of $600 \cdot 10^{15}$ and $3 \cdot 10^{15}$ gC are used (Ciais et al. 2013). When assuming that biomass contains roughly the energy content of sugar of about 479 kJ mol^{-1}, this yields estimates of $4.0 \cdot 10^{21}$ and $2.0 \cdot 10^{19}$ J of energy stored in the living biomass of the terrestrial and marine biospheres.

Chemical energy is mostly generated from radiative energy by photochemistry, by hydrologic processes, from thermal energy by transport of mass at different temperatures in the Earth's interior, and by photosynthesis. During a chemical reaction, chemical energy is typically reduced and other forms are generated, particularly, thermal energy and uncompensated heat.

2.4 Forms of entropy

So far we described several forms of energy at a macroscopic level, but did not consider how these forms of energy relate to the microscopic representation of these forms of energy at the molecular scale. At this scale, energy is composed of discrete amounts, or quanta, that are distributed over a range of microscopic states by three different means that we consider here: photons, discrete amounts of energy that carry electromagnetic radiation; electronic states within and among molecules; and discrete states associated with vibrational, rotational, and translational modes of motion of molecules. The forms of energy described earlier relate to these three means to store energy, except for gravity: radiative energy relates to photons; thermal and kinetic energy relates to the modes by which molecules move; and binding and chemical energy relate to electronic states and related molecular binding mechanisms.

Within each of the three means, there are different ways by which energy can be stored across many microscopic states that are macroscopically indistinguishable. These microscopic configurations of how energy is stored in a system is captured by entropy. The term "entropy" is also used in other contexts, e.g., for measuring the information content in systems. In this book, entropy is only used as a measure for how uniformly or non-uniformly energy is stored across microscopic states. Essentially, entropy counts the different ways by which energy can be stored in the available microscopic states and maps these to a macroscopic characterization. The more ways that energy can be distributed across microscopic states to yield the same macroscopic value of this characterization, the higher its entropy is going to be. This is why entropy can be seen as a measure of how much energy is dispersed at the microscopic scale, an interpretation that is recently favored over the more traditional interpretation of disorder (Leff 1996; Lambert 2002; Leff 2007; Atkins and de Paula 2010).

Associated with the three microscopic means to store energy that we consider here, there are three associated forms of entropy that capture these distributions, which are illustrated in Fig. 2.2. The first means is associated with the way by which radiation is composed of different wavelengths. A certain amount of radiative energy can be represented by different distributions of this energy across photons of different wavelengths. The same amount of radiative energy can, for instance, be represented by radiation in the visible range, as in the case of the radiation

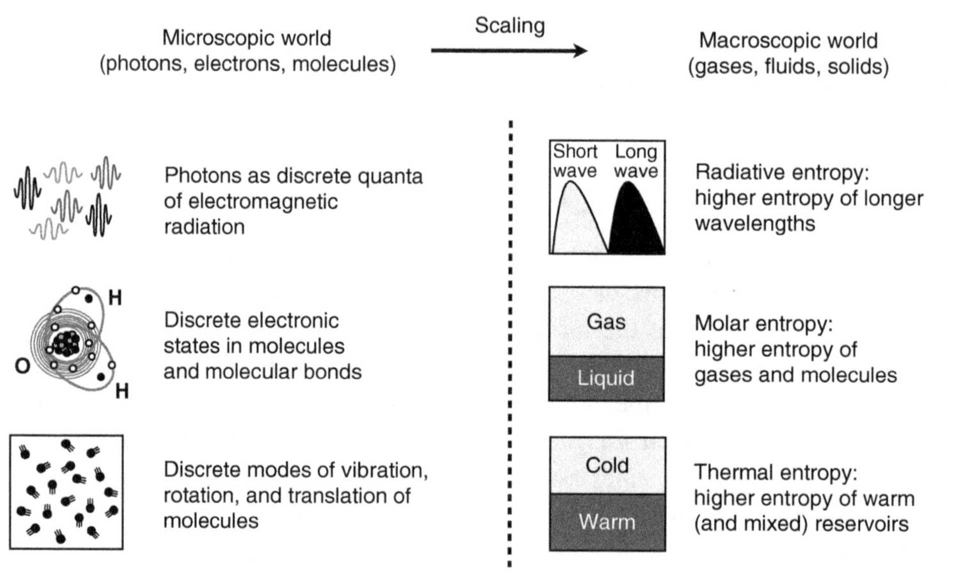

Figure 2.2 Illustration of forms of energy distribution at the microscopic scale and how it relates to macroscopic properties and different forms of entropy.

emitted by the Sun, or in the infrared, as in the case of the radiation emitted by the Earth. The entropy that captures this distribution of radiative energy is referred to as radiative entropy. The second means of distributing energy relates to energy associated with molecular bonds. Energy can, for instance, be stored in hydrogen bonds that hold water molecules in their liquid state, or this energy can be stored elsewhere. The ways in which energy can be distributed across molecular bonds is referred to as molar entropy here. It refers to the difference in the number of microscopic states that relate only to the way of how electronic states and molecules are being arranged, and is independent of how much heat is contained within the system. This property closely relates to the standard chemical characteristic of a substance that is referred to as its molar entropy. The third means to store energy relates to the means by which molecules move in terms of vibrations, rotations, or in space. A certain amount of energy can be associated with many molecules moving at similar speeds, or by few molecules moving much faster than the majority of molecules. This form of entropy is referred to as thermal entropy. This is the form of entropy that is most commonly considered in thermodynamics.

In the following, the microscopic interpretation of entropy is illustrated with a simple model of distributing a fixed number of energy quanta among many states. It allows for a qualitative interpretation of entropy as a measure of the probability of a certain distribution of energy at the microscopic scale, and it allows for the association of the most probable state with a state of maximum entropy. This simple model is then related to the three forms of entropy and how these can be calculated.

2.4.1 Entropy as probability

Entropy is simply a measure for the number of possible arrangements by which a given amount of energy can be distributed across microscopic states that result in the same macroscopic outcome. While the following considerations can be applied to any form of discrete aspects, such as particles, what we consider here is specifically the distribution of energy, as this relates to thermodynamic entropy. This statement is expressed by Boltzmann's famous formula, which relates the thermodynamic entropy S to the number of possible arrangements W of energy across microscopic states by

$$S = k_b \ln W \tag{2.17}$$

where k_b is the Boltzmann constant. What this relation tells us is that a greater number of ways to distribute energy, that is, a greater value of W, corresponds to a greater value of thermodynamic entropy S. Another way to look at this relationship is to relate a greater number of arrangements to a greater probability of such an arrangement taking place. A greater probability of distributing energy across microscopic states would then correspond to a greater entropy. The most probable arrangement then corresponds to a state of maximum entropy.

To illustrate entropy and its relation to microscopic distribution of energy, let us consider an isolated system that is made up of two reservoirs, A and B, as shown in Fig. 2.3, following an example given in Ambaum (2010). A certain amount of energy U, is distributed among the two reservoirs represented by N distinguishable quanta, so that each quantum represents an amount of energy $U_0 = U/N$. When we start with empty reservoirs and successively fill them randomly, we assume that each quantum has an equal probability to be placed in reservoir A or B, that is, we assume an equal probability $p_A = p_B$. After distributing the N quanta of energy, we consider an outcome in which k quanta are found in reservoir A and $N-k$ quanta in reservoir B. With N quanta being distributed among the two reservoirs with equal probability, there are 2^N different outcomes, each of these outcomes having the same probability. Among these 2^N outcomes are the extreme cases in which all quanta are located in A (i.e., $k = N$), equally distributed in A and B ($k = N/2$), or all located in B ($k = 0$). These outcomes are illustrated for a few cases in Fig. 2.3.

The 2^N outcomes represent the range of possible states in the microscopic view of this system in which we look at the specific details of how the individual quanta are partitioned among the two reservoirs. The macroscopic view represents the characterization of the system merely in terms of the total energy in each reservoir, U_A and U_B, or the respective fractions, $f_A = U_A/U = k/N$ and $f_B = U_B/U = (N-k)/N$. Since the total energy is given, that is, $f_A + f_B = 1$, the different outcomes merely differ in terms of the difference, $\Delta f = f_A - f_B = (2k - N)/N$. We can now ask how likely it is to find a certain macroscopic outcome, which is characterized by a certain value of Δf. To do so, we need to count the number of microscopic states that map into one particular value of Δf. The extreme case in which all energy is located in reservoir A and which is associated with the maximum value of $\Delta f = 1$ can only be found in one microscopic arrangement out of the 2^N possibilities. Hence, the probability of finding this extreme case is 2^{-N}. More generally, the number of microscopic arrangements W that correspond to a particular value of Δf is represented by the binomial distribution,

$$W(\Delta f, N) = \binom{N}{k} = \frac{N!}{(N-k)!\,k!} \tag{2.18}$$

noting that Δf is a function of k and N. This expression can be approximated by a simpler functional form for large values of N. By using Stirling's approximation to approximate the factorials by $n! \approx (2\pi n)^{1/2}\,(n/e)^n$, and taking the leading terms of a Taylor expansion (see Ambaum (2010) for details), W is approximated by:

$$W(\Delta f, N) \approx W(0, N)\, e^{-\frac{1}{2}\left(\frac{\Delta f}{2\sigma N}\right)^2} \tag{2.19}$$

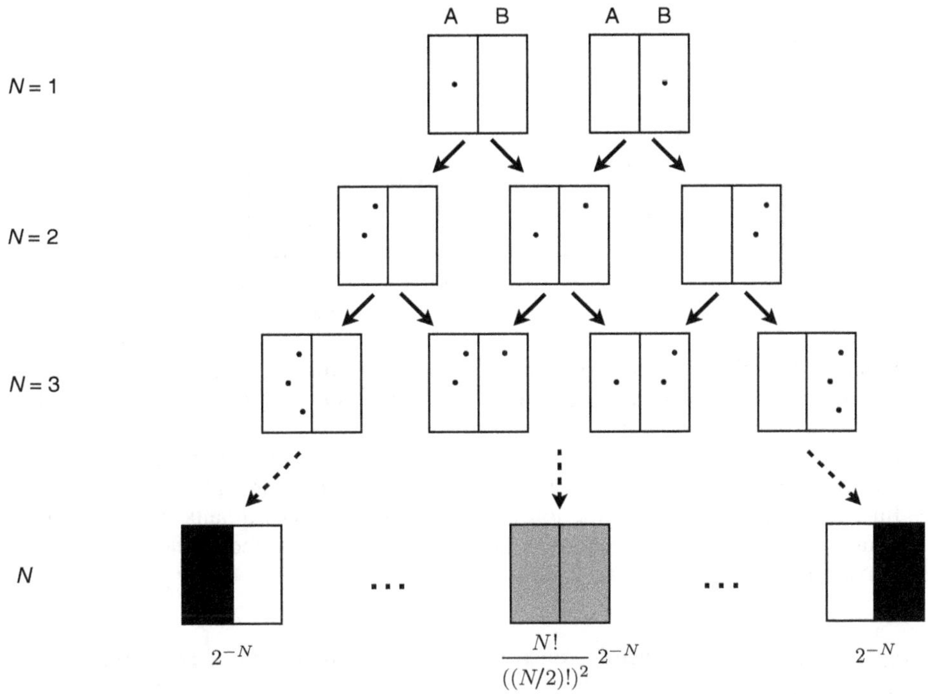

Figure 2.3 A thought experiment of filling two reservoirs ("A" on left, "B" on right of the shown square) with a given amount of energy with a certain number of quanta, N. Shown are the possible, "microscopic" outcomes of finding a certain number of quanta in the left and right reservoirs for $N = 1$, 2, and 3. The bottom row shows the general expression for the probability of finding all quanta in the left reservoir, an equal number in both reservoirs (middle), and to find all quanta in the right reservoir.

with a value of $W(0,N) = 2^{N+1}/(2\pi\ N)^{1/2}$ and a width of the distribution of $\sigma = (N/4)^{1/2}$. This expression of the number of possible arrangements can then be used in Boltzmann's expression for entropy given by eqn. 2.17:

$$S(\Delta f, N) = S(0,N) - \frac{k_b}{2}\left(\frac{\Delta f}{2\sigma N}\right)^2 \tag{2.20}$$

The distribution of states is shown in Fig. 2.4 for three different values of N. As would be expected, the most probable state is an equal distribution of energy among the two reservoirs, with a value of $W(0)$ and $k = N/2$. That is to say that when quanta of energy are randomly distributed among the two reservoirs, it is most likely that the energy is distributed uniformly. This most probable state becomes increasingly more probable with a greater number of quanta N to be distributed, which can be seen in Fig. 2.4 and in the above expression for σ in the narrowing of the distribution with greater values of N. In other words, the more quanta of energy are to be distributed, the more likely they are distributed in the most probable arrangement.

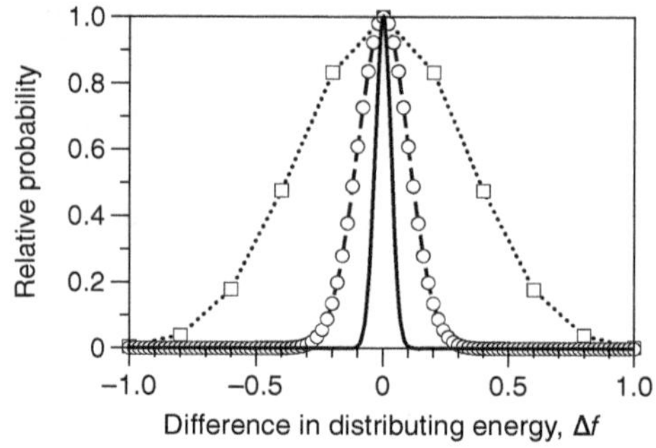

Figure 2.4 Probability of finding an uneven partitioning of the distributed energy among the two boxes, as measured by Δf. A value of $\Delta f = -1$ ($\Delta f = 1$) represents the case in which all energy is located in box B (box A) respectively. Energy is distributed equally in the case of $\Delta f = 0$. Three cases are shown in which $N = 10$ (dotted line), $N = 100$ (dashed line), and $N = 1000$ (solid line, for which the approximation of eqn. 2.19 was used) quanta of energy are being distributed. For comparison, the distributions are scaled to their respective maximum value.

To summarize this example, entropy is a measure for how probable a macroscopic outcome is that is associated with different ways to distribute quanta of energy across a range of microscopic states. In the example, this microscopic range of states refers to the arrangements of the individual quanta of energy in the two boxes, while the macroscopic state is characterized by the difference in energy distribution among the two boxes, which is described by Δf. In all of the cases, the same total amount of energy was distributed, emphasizing that entropy measures the distribution, or dispersal, of energy. The example also showed that a greater number of quanta distributed among the two boxes results in an increased probability for an equal distribution of energy. This most probable state is a state of maximum entropy and is associated with a state of thermodynamic equilibrium. The example will be used in the next chapter to describe the microscopic interpretation of the second law and states of thermodynamic equilibrium in greater detail.

When such a statistical formulation is applied to the three forms of entropy shown in Fig. 2.2, the distribution of microscopic states is not described by a binomial distribution, but by distributions that deal with specific aspects of quantum physics in terms of whether the particles are distinguishable and whether they are subjected to exclusion principles. The distribution of radiative energy by photons is described by Bose–Einstein statistics, energy stored in electronic states by the Fermi–Dirac statistics as electrons are subject to the Pauli exclusion principle, while energy stored in the motion of molecules are described by Maxwell–Boltzmann statistics. As a result, the formulation of entropy is somewhat different for these three forms. For photons, we get back to these statistics in Section 6.2, but otherwise, more details on these statistical treatment can be found in general physics textbooks that deal with quantum physics.

2.4.2 Illustrations of the three forms of entropy

As a next step, we use this illustration of entropy and describe how it applies to the three forms of entropy shown in Fig. 2.2 that are relevant to Earth system processes and use examples to illustrate their quantification.

We first consider radiative entropy. The microscopic level deals with the distribution of a given amount of radiative energy across a number of photons, the quanta that carry radiative energy, that differ mostly in terms of their frequencies. A certain amount of radiative energy can be obtained by assembling relatively few photons of high energy (e.g., ultraviolet radiation) or by many photons of low energy (e.g., infrared radiation). Radiative entropy refers to the probability of a particular distribution of radiative energy over different photons. The most probable distribution for a perfectly emitting body, a so-called blackbody, with a temperature T is the Planck's distribution of radiative energy, which when integrated yields the well-known radiation law of Stefan–Boltzmann. These radiative laws were originally derived from Planck (1906) using the assumption of thermodynamic equilibrium and maximum entropy, so that blackbody radiation represents radiation that is in thermodynamic equilibrium with the temperature of the emitting body.

We now consider the radiative flux of solar radiation of about 240 W m^{-2} that is on average absorbed by the Earth system, and compare it with the flux of emitted radiation to space. Solar radiation, the radiation emitted by the Sun, is representative of the radiation emitted by a blackbody with a temperature of about $T_{sun} = 5760$ K, which is approximately the surface temperature of the Sun. The radiation emitted by Earth into space, that is, terrestrial radiation, can also be approximated by the emission of a blackbody, but at a much lower temperature of about $T_r = 255$ K, which is the mean radiative emission temperature of the Earth. Both distributions represent the same radiative flux of 240 W m^{-2} in steady state, but the distributions differ strongly by their composition of photon energies, and in particular, in the number of photons. We have seen in the simple example described earlier that the probability of the most probable state increased with the number of quanta. Likewise, radiation can be composed of fewer, more energetic photons, or of more, but less energetic photons. This can be illustrated by the photon energy at which the peak emission takes place. For solar radiation, the peak emission is at a visible wavelength of about $\lambda_{max} = 500 \cdot 10^{-9}$ m, which is obtained from Wien's displacement law, $\lambda_{max} = b/T$, with $b = 2.8978 \cdot 10^{-3}$ m K and the Sun's emission temperature of $T_{sun} = 5760$ K. For the radiation emitted from Earth, the peak emission takes place at about $\lambda_{max} = 11.4 \cdot 10^{-6}$ m in the infrared range of wavelengths. The energy associated with each photon is described by $U_{photon} = hc/\lambda$, where $h = 6.63 \cdot 10^{-34}$ J s is Planck's constant and $c = 3 \cdot 10^8$ m s^{-1} is the speed of light. If we now simply consider the number of photons at these emission peaks of solar and terrestrial radiation, we notice that the Earth emits about twenty-three times as many photons than it absorbs. This is because the photons emitted from Earth are emitted at much longer wavelengths and thus the photons are on average associated with much less radiative energy. As we have seen in the example of the previous subsection, when we split a certain amount of energy into more quanta, the entropy of the most probable macroscopic state increases as there are more microscopic states over which energy can be distributed. Hence, the terrestrial radiation emitted from Earth has a much higher radiative entropy than the solar radiation that it absorbs. This difference in entropy associated with solar and terrestrial radiation plays an absolutely critical role in maintaining the dynamics of the Earth system, as it allows for net entropy export from the Earth system. We deal with radiative entropy in more detail in Section 6.2.

We next consider molar entropy which is associated with the different ways in which energy is stored in the distribution of electrons in atoms and molecules. The molar entropy of substances tend to be greater for heavier atoms and molecules, for molecules compared to the individual atoms, and greater for the vapor state compared to the liquid or solid state. These trends reflect a combined effect of the binding energies involved in how tightly electrons are bound to the nuclei, how tight molecules are bound to each other by intermolecular binding mechanisms such as hydrogen bonds, and the number of states over which the binding energies can be distributed. Heavier atoms and molecules have, for instance, more electrons and thus more states over which energy can be distributed, similar to the effect of a larger number N in the example of the previous subsection. The increase in entropy is then related to a greater availability of states, for instance, in a molecule compared to the compounds in atomic form, and to a typically stronger binding of electrons to the nuclei in molecules. Both effects contribute to the energy that is being released to or taken up from the environment when transformations occur such as phase transitions or chemical reactions. This energy we notice in the form of the heat being released by a chemical reaction or the latent heat taken up during vaporization. This aspect is described in greater detail in Chapters 8 and 9 when we deal with hydrologic and geochemical cycling.

A few examples for these trends can be seen in the values of molar entropies given in Table 2.3. The numbers show that the heavier oxygen (O) in atomic form has a greater molar entropy than the lighter atomic hydrogen (H) (161.01 vs. 114.71 J K^{-1} mol^{-1}), and that molecular hydrogen (H_2) has a greater molar entropy than atomic hydrogen (H) (130.68 vs. 114.71 J K^{-1} mol^{-1}). In this table, we also see

Table 2.3 *The molar entropies of hydrogen and oxygen in gaseous form as well as water in the three phases at a room temperature of 298 K. Data taken from Atkins and de Paula (2010).*

Substance	Phase	Molar entropy J K^{-1} mol^{-1}
H	Gas	114.71
H_2	Gas	130.68
O	Gas	161.06
O_2	Gas	205.14
H_2O	Solid	37.99
H_2O	Liquid	69.91
H_2O	Vapor	188.83

how water vapor has a much greater entropy than liquid water, which in turn has a greater entropy than ice.

Finally, we consider an example for thermal entropy. Thermal entropy describes the entropy change of a system when heat is added or removed in the absence of material transformations. It measures the number of states over which energy is distributed in the form of vibrations, rotations, and movement of molecules in solids, liquids, and gases. Thermal entropy is greater the more energy is stored within the system, which is again similar to the effect of a greater N on entropy illustrated in the example given earlier, as more energy yields more quanta to be distributed. We can immediately see this in the thermodynamic definition of entropy (eqn. 2.2). When $dS = dQ/T$ is combined with the definition of heat capacity, $c = dQ/dT$ (eqn. 2.3), we obtain $dS = cdT/T$ (note that a more general derivation includes a term Rdp/p). If we assume that c is approximately constant with temperature, this yields the solution $S(T) = S_{\text{ref}} + c \ln T$, where S_{ref} is an integration constant. Hence, a system that consists of a given substance with a greater thermal energy content generally has a greater thermal entropy.

To illustrate a calculation of thermal entropy, let us consider a system that consists of two reservoirs of air with equal mass and fixed volume that are at temperatures of $T_a = 300$ K and $T_b = 250$ K. With an assumed heat capacity of $c = 1004$ J K^{-1}, we find that the entropy of the warmer reservoir is greater than the entropy of the colder reservoir by $c[\ln(T_a) - \ln(T_b)] = 183$ J K^{-1}, or about 18% of c. When the thermal energy of both reservoirs is completely mixed, and the temperatures are both equal ($T'_a = T'_b = 275$ K), then the entropy of the whole system is greater by $c[\ln(T'_a) - \ln(T_a)] + c[\ln(T'_b) - \ln(T_b)] = 8.3$ J K^{-1}, or about 0.8% of c. This simple example illustrates how the warmer reservoir has a higher thermal entropy than the colder reservoir of equal size, but also that a well mixed system has a greater entropy than a system with a non-uniform distribution of thermal energy.

2.4.3 Linkages between forms of entropy

At the end, it is important to point out that the three forms of entropy described here are not independent, but rather closely related to each other, particularly for Earth system processes. The central process that links these forms of entropy is called thermalization. It describes the process by which the radiative temperature associated with radiative energy and the material temperature that describes the thermal energy of a system equilibrate. At the microscopic scale, this involves the absorption and emission of radiative energy associated with photons of different wavelengths by atoms and molecules that convert the radiative energy into other forms and back to radiation. In this book, we assume that thermalization is sufficiently rapid so that we do not need to distinguish between the two temperatures, which is generally the case on Earth. One notable exception where this assumption does not hold is at high elevations in the atmosphere near the transition to space where the air density is extremely low. Because of the high radiative forcing and the low air density, the process of thermalization is also slow and does not establish thermodynamic equilibrium between these forms.

As Earth system processes are driven by and deal with radiative energy, material transformations, and with heat, we will need to account for the associated forms of entropy throughout the book.

2.5 Energy and entropy budgets of the Earth system

When we aim to formulate Earth system processes in thermodynamic terms, energy and entropy budgets play a central role. Whatever kind of Earth system process we deal with, it relates to a certain form of energy that is expressed by a conjugate pair of variables, as shown in Table 2.1. To account for how the values of these variables develop, the boundary of the system needs to be defined, for which the energy and entropy budget are then formulated. It is important to note that the magnitude of a certain form of energy, as shown in Table 2.2, does not necessarily relate to its importance in shaping the dynamics of the Earth system. This is most clearly seen for radiative energy: while the solar radiative forcing clearly is the most important forcing of the planet, the stock of radiative energy within the Earth system is negligible. Likewise, the stocks of kinetic energy are very small as well, yet these are associated with the winds, currents, and flows in the Earth system that transport heat and matter around the planet. We should thus not disregard a form of energy because its stock is small.

What is critical for understanding the dynamics of the Earth system are not the stocks of energy, but rather the rates by which certain forms of energy are being generated out of some other forms, and what the effects of generating these forms are on other parts of the system. These generation rates, as well as the effects, are caused by the exchange fluxes of the system. When a system exchanges radiation, mass, or momentum, it not only alters the value of the state variable, it also affects the associated form of energy, and exchanges entropy with the surroundings. While at present, entropy budgets are rarely formulated for Earth system processes, we will see in the following chapters that entropy budgets set the critical basis for imposing the thermodynamic constraint on the system that originates from the second law of thermodynamics. In addition to conservation laws, it is particularly the second law that shapes the dynamics of the Earth system. Before we get to these dynamics, we first deal with the laws of thermodynamics and the limits that these result in in the following two chapters.

3

The first and second law of thermodynamics

3.1 The rules for energy conversions

The last chapter described how different Earth system processes are related to different forms of energy and entropy. This formulation in terms of energy sets a basis for making them comparable. The rules for converting one form of energy into another are described by the laws of thermodynamics. They ensure the conservation of energy during the conversion process, and set the direction into which these conversions occur.

Historically, these laws have grown out of the need to understand and improve the work output of steam engines in the mid-nineteenth century. Since then, their basis has been extended much beyond steam engines to all forms of energy transfer. The purpose of this chapter is to show those aspects of the laws of thermodynamics that have the most direct relevance to understand energy conversions by Earth system processes. The foundations set by the laws then allow us to make quantitative predictions of the direction in which the dynamics take place in Earth systems and set upper limits on energy conversion rates, as described in the following chapter.

In total, there are four laws of thermodynamics that are numbered from zero to three. They are summarized in Table 3.1.

The *zeroth law* sets the basis for comparing thermodynamic systems. It establishes the state of thermodynamic equilibrium as a reference state, which is the state of a system in which there is no net transformation or exchange of any physical quantity. The zeroth law formulates that if two systems are in thermodynamic equilibrium with a third system, then the two systems are also in thermodynamic equilibrium. As we will see in the following, the state of thermodynamic equilibrium serves as an important reference point, as it sets the "target" state for the dynamics that take place within a system and the exchanges with other systems. The zeroth law also comes into play when the equilibrium between different forms of energy is needed to describe energy conversion processes. This is, for instance,

Table 3.1 *An overview of the four laws of thermodynamics and how they are relevant to the Earth system.*

Law	Description	Relevance to the Earth system
0th	If two systems are in thermodynamic equilibrium with a third system, the two systems are also in thermodynamic equilibrium with each other.	Establishes thermodynamic equilibrium as a reference state for processes that convert energy (e.g., radiation, heat fluxes, or phase transitions).
1st	Energy is conserved when one form of energy is converted into another form.	Ensures energy conservation for any Earth system process that converts energy. This conservation may include exchange terms across the system boundary.
2nd	The entropy of an isolated system cannot decrease. In non-isolated systems, internal processes cannot decrease the entropy of the system.	For any Earth system process that converts energy, the entropy cannot decrease. This provides a direction for processes and limits to energy conversion.
3rd	The entropy of a system at 0 K is zero.	Sets absolute reference values for entropy.

the case for radiative processes, in which conversions between radiative and thermal energy are involved, or for phase transitions, in which conversions between liquid and gaseous phases are involved. However, as this law represents a more formal aspect, we will not deal with the zeroth law in greater detail here.

The *first law* formulates the conservation of energy within a system when energy is converted into different forms. It states that the internal changes of energy within the system are balanced by the exchanges with the surroundings and with the physical work that is done by (or on) the system. For instance, it connects the rate of heat exchange of the engine with its surroundings with the physical work done by a heat engine. As physical work generates another form of energy, e.g., kinetic or potential energy, the first law sets the accounting for energy forms in the conversion process. In the Earth system, there are many processes that convert energy from one form to another, and the first law in a broader sense ensures that, overall, no energy is created or lost during any conversion process.

The *second law* states that the entropy of an isolated system cannot decrease, so that energy can only be increasingly dispersed across microscopic states. This condition imposed by the second law represents a fundamental direction for energy transformations and results in a system evolving towards its state of thermodynamic equilibrium. This condition applies to every energy conversion process as well, and thereby sets limits on energy conversion, a topic that will be dealt with in the next chapter. The second law also applies to non-isolated systems that are driven by exchange fluxes with the surroundings. To ensure that the second law is obeyed in these systems, the exchange fluxes of the system need to be accounted for, not simply in terms of the associated fluxes of energy or mass, but also of their respective entropies. The second law applies to each energy conversion step associated with the various processes taking place within the Earth system, from the absorption of radiation to the dissipation of kinetic and chemical energy into heat.

Finally, the *third law* of thermodynamics formulates the behavior of systems near absolute zero temperature at which systems approach states of zero entropy. This law sets an absolute scale for entropy. However, as Earth system processes deal mostly with differences in entropy associated with energy conversions and not with absolute amounts, this law plays more of a formal role and has little practical significance for Earth system processes.

As can be seen from this qualitative introduction, it is the first and second law that are quantitative in their nature and that yield relevant constraints and predictions for energy conversions within the Earth system. This chapter describes these laws in greater detail, and their use is illustrated with some highly simplified examples. At the end of the chapter, the key elements are summarized and set in the context of energy conversions within the Earth system.

3.2 The first law of thermodynamics

In classical thermodynamics, the first law is formulated in terms of changes in internal energy that resides within the system dU, the heat added to or removed from the system dQ, and the work done by (or on) the system dW. The internal energy of a system U, refers, in thermodynamics, to the energy associated with motion and binding energies at the molecular level, which is represented by the thermal and binding energy described in the last chapter. Note that other forms of energy at the macroscopic level, particularly kinetic and potential energy, are not accounted for by the internal energy. In the following considerations, we refer to the internal energy as the thermal energy of the system, in contrast to the total energy of a system, U_{tot}, which refers to all types of energy that are being considered. The first law is schematically shown for a simple system in Fig. 3.1a. Mathematically, it is expressed as

$$dU = dQ - dW \tag{3.1}$$

In some texts, the symbols δQ and δW are used rather than dQ and dW, to indicate that these terms reflect processes that depend on the way in which energy is added and work is performed. This is typically referred to as path dependence, an aspect that we will not deal with here. We will rather focus on the first law in the following as a formulation of energy conservation.

The sign convention in this formulation is that when the system performs work, that is, it generates another form of energy that is not accounted for by U outside

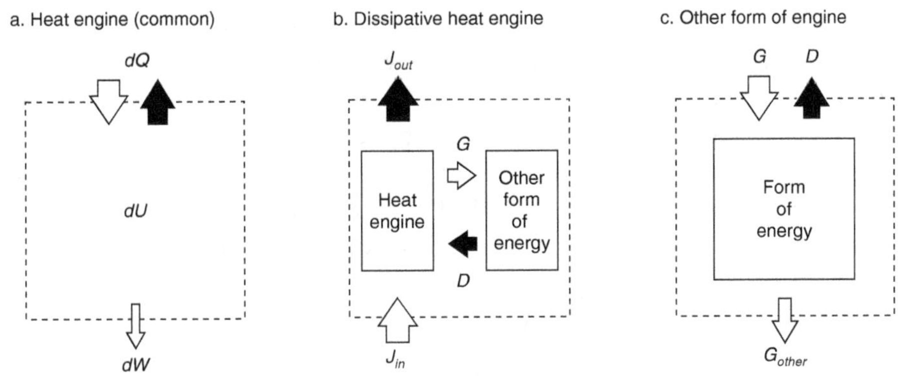

Figure 3.1 Schematic illustration of the first law for (a) a common heat engine, (b) a heat engine operating inside a closed system in which the work done relates to the generation of another form of energy that remains within the system, and (c) an engine that converts a non-thermal form of energy into another form. The system boundaries are shown by the dashed boxes. The diagrams illustrate eqns. 3.1, 3.2, and 3.4 using the respective symbols.

the system, then the term dW in eqn. 3.1 is positive. By this convention, the work done by the system removes dW from U, and creates another form of energy in the amount of dW. When we deal with a heat engine, for instance, heat is exchanged, which is described by dQ. The work done by the engine dW is removed as the engine performs external work that, for instance, results in the acceleration and the generation of kinetic energy. This form of energy is further converted, eventually turning back into heat. However, since this conversion process takes place outside the engine, it is not being accounted for in eqn. 3.1. This is fine when considering the heat engine as the system. For a description of processes that operate within the Earth system, however, it is essential to not just track the energy when it is converted by a process, but also to account for it in the subsequent energy balances so that it does not get lost. This requires a somewhat more complete and generalized form for the accounting of energy conversions that is not just focused on thermal energy, but also includes other forms like kinetic energy that make up the total energy of the system.

To account for energy conversions within the Earth system, we will use the first law in a somewhat different form. In most cases, we consider the heat engine in a state that is driven by heat fluxes, e.g., due to radiative heating, yet temperatures are in a steady state. In this case, the differentials in eqn. 3.1 become rates of thermal energy change dU/dt, heating and cooling rates dQ/dt, and work performed through time becomes power, $G = dW/dt$. The letter G is used to abbreviate power as it generates another form of energy rather than the letter P that is more traditionally used, but which is later used for precipitation as a major Earth system process.

Another addition that we need to consider in the context of Earth systems is that work is not performed externally, but inside the system, generating another form of energy, e.g., kinetic or potential energy. As this form of energy is eventually dissipated back into heat at a rate D, for instance, by friction, we need to account for this term as well when formulating how a heat engine operates within an Earth system. This is schematically shown in Fig. 3.1b. The first law applied to this heat engine is expressed in terms of rates and by the inclusion of a dissipative heating term:

$$\frac{dU}{dt} = \frac{dQ}{dt} - \frac{dW}{dt} = J_{in} - J_{out} - G + D \tag{3.2}$$

where $dQ/dt = J_{in} - J_{out}$ is the net heating rate with J_{in} being the heat flux into the system and J_{out}, the heat flux out of the system, G is the generation rate of some other form of energy U_{other}, that is not accounted for by the thermal energy U, and D the dissipation that converts U_{other} back into thermal energy U, thus resulting in a dissipative heating term. We then have an associated energy balance for this other form of energy U_{other}, in the form of

$$\frac{dU_{other}}{dt} = G - D \tag{3.3}$$

Note that we cannot tell from the energy balance (eqn. 3.2) if the dissipative heating is added to the heat flux that leaves the system, that is, $J_{out} + D$, or if it adds to the heat flux into the heat engine, that is, $J_{in} + D$, which is the case shown in Fig. 3.1b. Both cases satisfy the overall steady-state condition. For the following considerations, this does not matter as this does not make any difference to eqn. 3.2. We will see in the next chapter that it makes a notable difference when considering the limit on how much power can be generated by the engine (Section 4.4).

Let us consider the dynamics of this heat engine from an initial state in which $U_{other} = 0$, so that $D = 0$. At this state, the generation rate is simply given by the original form of the first law, which is $dU/dt = dQ/dt - G$, where $dQ/dt = J_{in} - J_{out}$. As the system gains the other form of energy U_{other}, $J_{in} > J_{out}$ with $G > 0$. Eventually, a steady state is achieved when the dissipative heating D, balances the generation rate of the engine G, so that $G = D$. Then, the system reaches a steady state and the net heating of the system is zero, $J_{net} = J_{in} - J_{out} = 0$. Note that $J_{net} = 0$ merely formulates the steady-state requirement of the system. It does not imply that nothing happens inside the system. Heating and cooling can take place at different places within the system, thereby causing work to be performed and dynamics to take place, so that $G = D > 0$. It is just that we cannot tell the difference from the steady-state energy balance as it merely acts as a constraint.

A further extension that we use here is that in a broader interpretation, the first law applies to other forms of energy conversions as well because they are also subject to energy conservation. When the heat engine performs work and, for instance, generates kinetic energy within the system, then the kinetic energy is subjected to budgeting as well. This was already expressed earlier in the form of eqn. 3.3. Yet, this form of energy can be converted into another form of energy as well, so that we would need to include an additional term that describes the conversion into another form. This form of conversion is shown in Fig. 3.1c and the first law can be expressed as

$$\frac{dU_{\text{other}}}{dt} = G - D - G_{\text{other}} \tag{3.4}$$

Note the equivalence to the common formulation of the first law: The left hand side describes changes of the form of energy being considered, just as the left hand side in eqn. 3.2 describes changes in thermal energy. The term $G - D$ on the right hand side describes the net generation and dissipation of the form of energy due to some exchange or conversion process, equivalent to the net heat flux, J_{net} in eqn. 3.2. The last term in eqn. 3.4, G_{other}, describes the energy conversion into another form. It corresponds to the work done by the heat engine, G, in eqn. 3.2. This term would then require a further energy budget equation for the form of energy being generated.

An example for such conversion is atmospheric motion. The heat fluxes J_{in} and J_{out} would represent radiative heating and cooling rates, and the generation rate G would represent the generation of kinetic energy, U_{other}, associated with motion. The dissipation rate D would represent frictional dissipation. The conversion in other forms G_{other} represents, for instance, the work done in lifting dust particles, the generation of turbulent motion or waves at the ocean surface, or the work done on wind turbines that convert the kinetic energy of wind into electric energy.

In summary, the first law states in a general way that during an energy conversion, energy is conserved. We use it here in a somewhat different form than how it is typically done in textbooks by explicitly accounting for the budgeting of the different forms of energy. This, of course, does not change the physics or the consequences of the first law, but it will make it easier to understand the energy conversions across different forms.

While the first law ensures the accounting for energy terms, it does not tell us anything about why and in which direction these energy conversions take place. This additional constraint is set by the considerations of the associated changes in entropy, which is where the second law of thermodynamics comes into play.

3.3 The second law of thermodynamics

The second law states that the entropy of an isolated system does not decrease. This implies that the energy of a system is dispersed by the processes that take place within the system, spreading it more across the states at the microscopic scale towards a state of maximum entropy. The law represents a condition, but it profoundly sets the direction in which processes proceed and, as we will see in the following chapter, is critical in setting the speed limits of energy conversion processes. Its manifestation is so fundamental that it has been named the "arrow of time" (Eddington 1928).

Mathematically, the second law is formulated for an isolated system simply by the condition that the entropy of the system S, can only increase:

$$dS \geq 0 \tag{3.5}$$

It is important to note that this formulation is done for an isolated system. Hence, what this formulation of the second law means is that whatever happens inside the system, for instance, the distribution of heat by conduction, these internal processes tend to increase the dispersal of energy within the system. The system will evolve towards states of higher entropy with time, until it finally reaches a state in which the entropy within the system is at the maximum. This state of maximum entropy corresponds to a state of thermodynamic equilibrium, something that we deal with in more detail later in this chapter.

The second law can be interpreted statistically by considering again the example of distributing quanta of energy across two reservoirs (see also Ambaum (2010) for a mathematical formulation of this example). In Section 2.4.1, it was shown that the most probable way to distribute a certain number of quanta was to equally partition them among the reservoirs. If the initial state of the system is not the most probable state, the dynamics taking place inside the system should nevertheless be directed towards the most probable state. This can be seen qualitatively in Fig. 3.2. The left part of the figure shows an initial state in which all quanta are situated in the left reservoir, a highly improbable case of low entropy. We consider a setup in which there is a small hole that connects the two reservoirs, and that a quantum has a certain probability, p_0, of passing through the hole. In the initial state, the probability of at least one of the N quanta to pass from the left to the right reservoir is then Np_0, but since no quantum is present in the right reservoir, there is zero probability for a quantum to pass through the hole in the opposite direction. Hence, in the initial state, the rate at which quanta pass from the left to the right reservoir should be proportional to Np_0. With time, quanta should thus spread into the right reservoir.

After some time, the system will reach a steady state in which there is no net exchange of quanta between the two reservoirs. This state is represented by the most probable state of thermodynamic equilibrium. In this state, the probability for quanta to pass from left to right is $(N/2)p_0$, because in the most probable state, there are $N/2$ quanta in the left reservoir. The probability for quanta to pass in

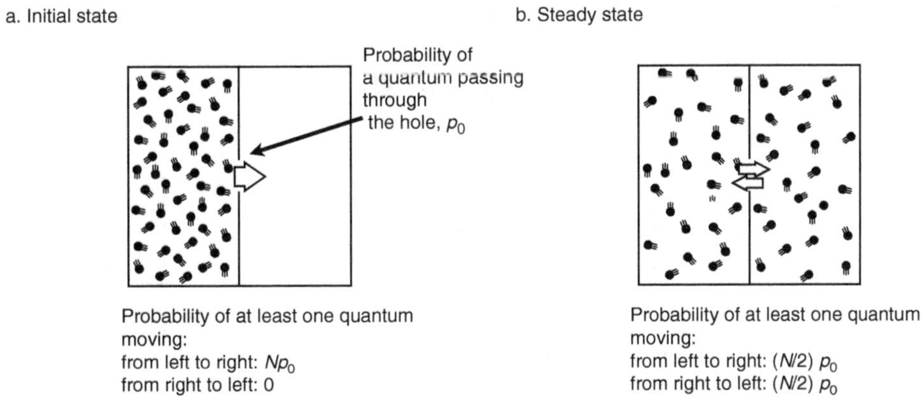

Figure 3.2 A simple example to illustrate the dynamics from an initial to a steady state at maximum entropy by using probabilities; here the example of Section 2.4.1 has been used.

the opposite direction is also $(N/2)p_0$, as there are also $N/2$ quanta in the right reservoir. Hence, there is no net exchange of quanta between the two reservoirs in the state of thermodynamic equilibrium, and the macroscopic state of the system would not change in time. Note that this does not mean that the system is static at the microscopic scale; it experiences some random fluctuations. There is still a non-zero probability for quanta to move from the left to the right reservoir, or vice versa. It is just that these two rates are approximately the same, so that there are no changes in the macroscopic state of the system. This state with an equal distribution of quanta in each reservoir is then the state of maximum entropy, as inferred in the example described in Section 2.4.1.

Most systems are, however, not isolated systems. They exchange energy, mass, or other quantities across the system boundary, and these exchange fluxes may exchange entropy of different values. The changes in entropy for such systems need to account for the changes that are due to internal changes, dS_i as well as exchange fluxes, dS_e. This is written as

$$dS = dS_i + dS_e \tag{3.6}$$

with the constraint of the second law represented by the condition that $dS_i \geq 0$. This formulation of the second law is illustrated in Fig. 3.3. In all cases shown in Fig. 3.3 internal processes produce entropy, $dS_i > 0$, and thereby tend to increase the entropy of the system, dS. In the case of a non-isolated system (Fig. 3.3b), exchange fluxes can, however, act to lower the entropy of the system, thereby maintaining it in a state of thermodynamic disequilibrium. We will see at the end of this section that this behavior of a non-isolated system can again be placed in the context of the second law applied to an isolated system when we consider a

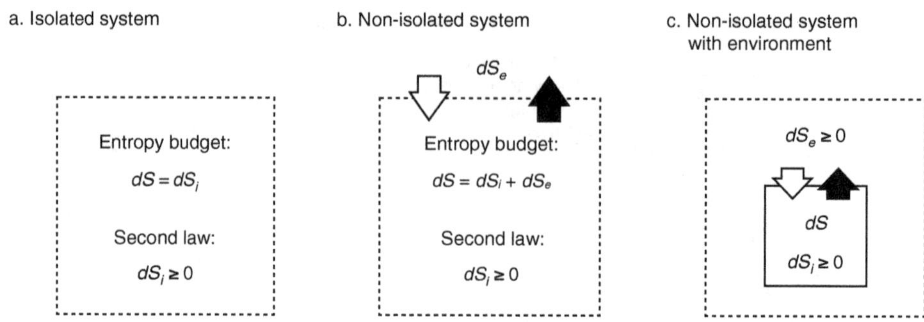

Figure 3.3 Illustration of the second law for (a) an isolated system, and (b) a non-isolated (closed or open) system. When a non-isolated system is considered together with its environment (c), then we deal again with an isolated system. The diagrams refer to eqn. 3.6.

non-isolated system jointly with its environment again as an isolated system, as shown in Fig. 3.3c.

When these changes in entropy are considered through time, we use the symbol $\sigma = dS_i/dt$ to describe the rate of entropy increase, or entropy production, due to internal processes. When processes produce entropy, they alter the spread of energy within the system and cause an irreversible change to the system. Hence, the rate of entropy production is a measure for the irreversibility of a process. Irreversibility refers to the fact that this process occurs spontaneously and its effect in terms of increasing the entropy cannot be undone unless work is performed on the system. We refer to the contribution by an individual process to the total entropy production as the entropy production associated with that process. We will see later that it closely relates to the rate of dissipation, so that entropy production can be seen as a measure for the intensity by which energy is being converted. The net entropy exchange through time associated with energy or mass exchange across the system boundary is denoted by $J_s = dS_e/dt$, which represents the sum of all entropy fluxes across the system boundary. When using these rates, we can extend eqn. 3.6 and obtain the entropy budget of the system:

$$\frac{dS}{dt} = \sigma + J_s \tag{3.7}$$

Note that the sign convention for the net entropy exchange is such that it is negative when the entropy export of the system is greater than the entropy import, that is, $J_s < 0$.

In the context of the entropy budget, the second law of thermodynamics requires that $\sigma \geq 0$. With this requirement, we can easily see that the entropy budget of an isolated system yields the original formulation of the second law expressed by eqn. 3.5. In the isolated system, $J_s = 0$, so that $dS/dt = dS_i/dt = \sigma \geq 0$. Hence, the entropy of an isolated system cannot decrease. It will eventually result in the final, steady state of thermodynamic equilibrium which is associated with $\sigma = 0$.

The case is different, however, when we consider a non-isolated system. Such a system can maintain a state away from thermodynamic equilibrium with a lower entropy without violating the second law. It requires the presence of net entropy exchange with the surroundings of the system, so that more entropy is exported to the surroundings than is produced within or imported into the system. In fact, when a system is driven by an entropy exchange of a rate J_s, then it will eventually reach a steady state in which its entropy does not change in time, $dS/dt = 0$. In this steady state, the entropy production within the system is balanced with the net entropy exchange with the surroundings, $\sigma = -J_s$. During transient periods, deviations are possible in either direction as the entropy of the system can change in time.

It is through entropy exchange across the system's boundary that thermodynamic systems can maintain an entropy that is lower than the corresponding value at thermodynamic equilibrium. This lower entropy is perfectly in accordance with the second law, as it only happens locally. If we consider the system in combination with its environment, as shown in Fig. 3.3c, we obtain again an isolated system. The entropy of the combined system and the environment then consists of the entropy S of our original system and the entropy of the environment, S_{env}. The change of entropy in time is given by

$$\frac{dS}{dt} + \frac{dS_{env}}{dt} = (\sigma + J_s) - (J_s) = \sigma \geq 0 \tag{3.8}$$

In total, this is just what we would expect from an isolated system which obeys the second law.

Entropy is not only exchanged by heat fluxes, but also by radiative and mass fluxes (cf. Section 2.4). Energy exchanged by radiation is associated with radiative entropy which characterizes the distribution of radiative energy across wavelengths. As radiative energy is being converted, it is subject to the second law as well. Budgets for radiative entropy balance the change of radiative entropy of the system with the entropy production by radiative processes within the system and the entropy that is being exchanged by radiative fluxes across the system boundary. Entropy exchange by radiation represents by far the largest form of entropy exchange so that it plays a monumental role for the Earth system. Likewise, the second law applies to conversions in which mass changes its phase or is chemically transformed, processes that we refer to as mass conversions in the following. Mass that is exchanged across a system boundary carries a molar entropy associated with the molecular binding energies of the substances that are being exchanged. For instance, a system that exchanges liquid water and water vapor with its surroundings may exchange the same mass, but these substances involve different entropies as water molecules have a much higher molar entropy in vapor form than when they are bound in a liquid state. The same applies to the mass exchange fluxes of chemical species that have a different range of molar energy levels and hence have different molar entropies. Again, the second law applies to mass conversions inside a system, so that these conversions produce entropy. The formulation of the entropy budget associated with mass conversions of a system is straightforward: it needs to balance changes in entropy within the system with entropy production by mass conversions and the entropy exchange across the system boundary by mass exchange. An important extension to this budget formulation is that mass conversions typically also exchange heat with the surroundings and can alter the pressure. The entropy exchange due to the differences in heat and pressure need to be taken into account as well. Just think of the vast amounts of heat that are needed to evaporate water, or the heat that is being released by exothermic reactions. Hence, a system that exchanges mass of different entropies typically also exchanges thermal energy which needs to be taken into account in the overall entropy budget.

To summarize, the second law states that any process taking place within a system maintains or increases the entropy of the system. The application of the second law requires the formulation of entropy budgets that balance the internal changes of entropy of a system with the entropy produced by internal processes and the entropy that is being exchanged across the system boundary. Entropy budgets are not just formulated for thermal energy, but are also required for radiative fluxes and mass exchange.

3.4 Thermodynamic equilibrium and disequilibrium

The manifestation of the second law and its relation to states of thermodynamic equilibrium and disequilibrium is demonstrated by a very simple system that is shown in Fig. 3.4. In the following, we formulate the energy and entropy budgets for this system. We also relate the dynamics inside the system to the second law and to the thermodynamic state of the system.

The system we consider consists of two reservoirs of thermal energy, A and B. The two reservoirs are in contact with each other, so that heat can flow from one reservoir to another. We further look at this system in two different setups: a first setup in which there is no exchange with the surroundings, hence forming an isolated system, and a second setup in which heat is added to one reservoir while it is removed from the other reservoir. As heat is exchanged between the system and its surroundings in this latter setup, it represents a non-isolated system.

We start the formal description of the system by the energy balances of the two reservoirs which are characterized by their respective temperatures, T_a and T_b. The

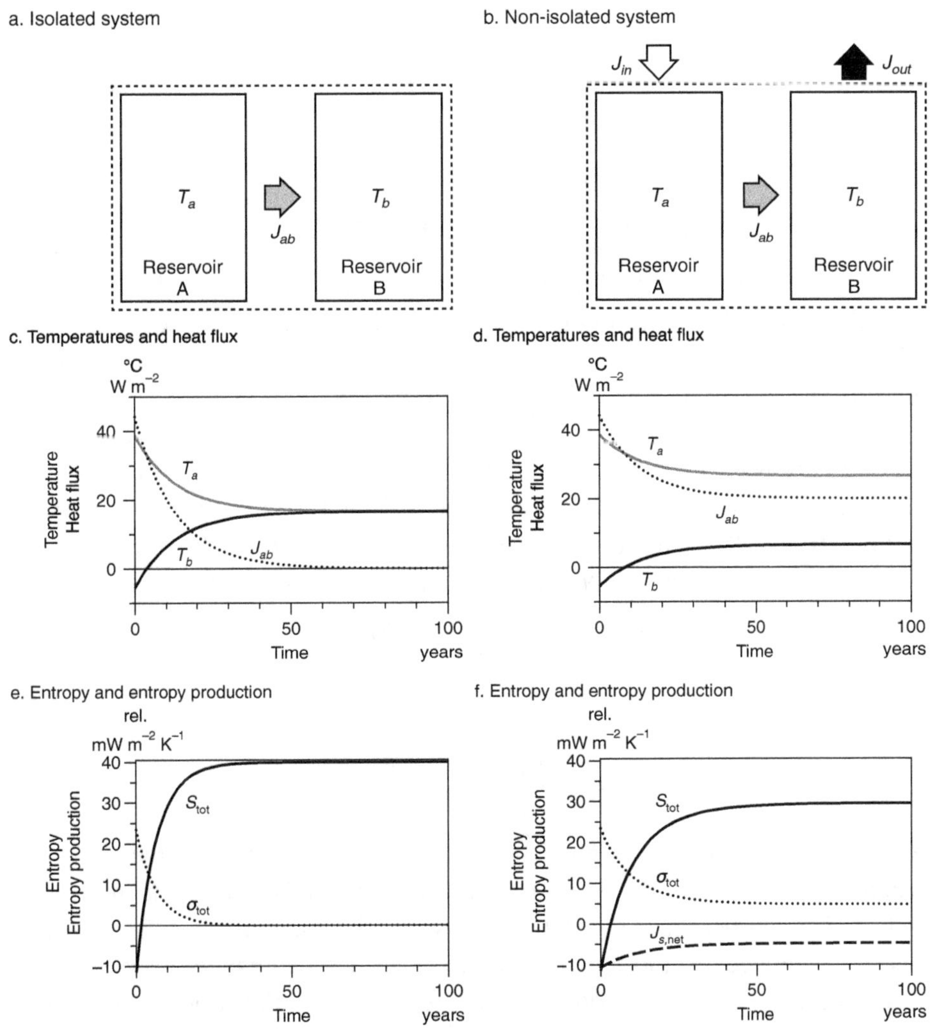

Figure 3.4 Two heat reservoirs form a simple (a) isolated and (b) non-isolated thermodynamic system. The reservoirs initially have different temperatures, T_a and T_b, and are connected by a heat flux, J_{ab}. The non-isolated system is driven by the heating of a heat flux, J_{in}, into reservoir A and the cooling by a heat flux J_{out} out of reservoir B. The transient changes of the system are shown for (c) and (d) temperature and heat flux as well as (e) and (f) entropy production σ_{tot}, net entropy exchange $J_{s,\text{net}}$, and the total entropy S_{tot} (scaled with respect to its maximum value). Values of $J_{\text{in}} = 20$ W m^{-2}; $c = 2 \cdot 10^8$ J m^{-2} K^{-1}; and $k = 1$ W m^{-2} K^{-1} were used. The initial conditions were set to $T_a(0) = 38.7$ °C; $T_b(0) = -5.5$ °C. The dashed boxes delineate the system boundaries. After Kleidon (2009c).

energy balances are given by:

$$c\frac{dT_a}{dt} = J_{\text{in}} - k(T_a - T_b) \tag{3.9}$$

and

$$c\frac{dT_b}{dt} = -J_{\text{out}} + k(T_a - T_b) \tag{3.10}$$

where c is the heat capacity of the two reservoirs (which is assumed to be equal for simplicity), and k is the heat conductivity. The heat flux J_{in} is added to reservoir A, the heat flux $J_{ab} = k(T_a - T_b)$ exchanges heat between the reservoirs, and J_{out} represents the flux of heat that is removed from reservoir B.

The two energy balances are complemented by the entropy balances of the two reservoirs. For this, we need to relate the entropy fluxes to the heat fluxes and the temperatures at which the heat exchange takes place. Using Clausius' expression of thermal entropy, $dS = dQ/T$ (eqn. 2.2), we note that entropy fluxes are expressed by $(1/T)\ dQ/dt$. The entropy budgets of the two systems are then given by

$$\frac{dS_a}{dt} = \frac{J_{\text{in}}}{T_a} - \frac{k(T_a - T_b)}{T_a} + \sigma_a \tag{3.11}$$

and

$$\frac{dS_b}{dt} = -\frac{J_{\text{out}}}{T_b} + \frac{k(T_a - T_b)}{T_b} + \sigma_b \tag{3.12}$$

The first terms on the right hand side describe the entropy exchange with the surroundings. The addition of heat by the flux J_{in} adds entropy to reservoir A, which is at a temperature T_a. Here we assume that the entropy of the heat flux J_{in} is characterized by the temperature T_a. The heat exchange between the two reservoirs removes entropy from reservoir A by J_{ab}/T_a. Equivalently, the removal of heat from reservoir B removes entropy at a rate J_{out}/T_b, while the heat exchange between the two reservoirs adds entropy at a rate J_{ab}/T_b. The last terms in the equations describes the entropy production, σ_a and σ_b, in the two reservoirs. The overall entropy exchange of the two reservoirs with the surroundings is given by

$$J_{s,\text{net}} = \frac{J_{\text{in}}}{T_a} - \frac{J_{\text{out}}}{T_b} \tag{3.13}$$

which we refer to as the net entropy exchange of the system.

The second term on the right hand sides of eqns. 3.11 and 3.12 describe the changes in entropy that take place due to the internal redistribution of heat between the two reservoirs. The heat flux, J_{ab}, removes heat from reservoir A with a temperature T_a and adds it to reservoir B, which is at a different temperature, T_b. The mixing of the heat transported from reservoir A with the heat of reservoir B results in entropy production. We obtain the rate of entropy production of the system from the entropy budget of the whole system, which is given by the sum of the budgets of the two reservoirs. Using the eqns. 3.11 to 3.15, we obtain for the change of entropy, S_{tot}, of the whole system

$$\frac{dS_{tot}}{dt} = \sigma_{tot} + J_{s,net} \tag{3.14}$$

This change in entropy is composed of the total entropy production, $\sigma_{tot} = \sigma_a + \sigma_b$, within the system and the net entropy exchange of the system with its surroundings, $J_{s,net}$. As we consider the redistribution of thermal energy as the only process within the system, the total entropy production is due to the mixing of thermal energies between the two reservoirs. We then obtain this rate of total entropy production, σ_{tot}, as

$$\sigma_{tot} = k(T_a - T_b)\left(\frac{1}{T_b} - \frac{1}{T_a}\right) \tag{3.15}$$

This expression relates the entropy production to the heat flux, J_{ab}, and the difference of inverse temperatures, which is the simplified version of the more general form of entropy production due to heat conduction (e.g., Kondepudi and Prigogine 1998).

To illustrate how the dynamics of these energy and entropy budgets play out, we assume that the heat exchange is near a steady state, so that $J_{in} = J_{out}$. With this assumption, we look at the temporal dynamics of the system in terms of the temperatures of each reservoir and the associated entropies, which is shown in Fig. 3.4 for the cases of an isolated system (with $J_{in} = 0$) and for a non-isolated system ($J_{in} > 0$). Because we assume that $J_{in} = J_{out}$ and so no net heat is added to the system, the overall energy remains constant during these temporal dynamics and the mean temperature of the whole system does not change.

We first look at the isolated system (Fig. 3.4, left column). Initially, the two reservoirs have different temperatures, that is, they are initially in disequilibrium with each other. The difference in temperatures causes a heat flux to develop, which depletes the temperature difference as we would expect. This decrease in the temperature difference is associated with an increase of entropy in the whole system. This increase of entropy is entirely due to the redistribution of heat within the system and the resulting entropy production, as we set the boundary fluxes to zero so that the heat content inside the system does not change. The maximum in entropy is reached when the heat is distributed uniformly within the system, as we would expect in a state of thermodynamic equilibrium (as in the simple example in Section 2.4.1).

The non-isolated system initially shows a similar response (Fig. 3.4, right column). The temperature difference first declines, and so does the heat flux between the reservoirs, and the entropy of the system increases. Yet, in contrast to the isolated system, this system reaches a steady state which is characterized by a non-zero flux and a temperature difference between the reservoirs. The entropy production of the system is greater than zero. The figure also shows that in this steady state, the entropy production within the system is balanced by the net entropy exchange of the system with its surroundings. Through this net entropy exchange, the system maintains a state with an entropy that is below the state of thermodynamic equilibrium of the equivalent, isolated system shown in the left column in Fig. 3.4. This steady state of thermodynamic disequilibrium is reached when the flux between the reservoirs balances the influx, so that $J_{\text{in}} = J_{ab} = k(T_a - T_b)$. The temperature difference is then given by $(T_a - T_b) = J_{\text{in}}/k$, which relates directly to the extent of thermodynamic disequilibrium that is maintained within the system. From this we can directly see that the extent to which thermodynamic disequilibrium is being maintained relates directly to the magnitude of exchange fluxes across the system boundary. The properties of the system can also modulate this disequilibrium, which in our simple case is reflected by the conductivity k in the denominator.

This simple example illustrates two critical points that we will keep encountering in later chapters of the book: first, the redistribution of heat within the system, which in our simple system represents the "dynamics" of the system, can directly be seen as a consequence of the second law. After all, the heat flux is directed in such a way as to increase the entropy of the system. As we will see later on, this applies to essentially all Earth system processes as well, although the relation to the second law may not be as clear. Note, however, that in this example the heat flux of the system is entirely determined in steady state by the exchange fluxes at the system boundary, so that the second law does not need to be explicitly invoked to determine the state of the system. Second, the example illustrates how important the exchange fluxes at the boundary of a system are for maintaining disequilibrium and the dynamics within a thermodynamic system. It is through the exchange of heat at different temperatures that the simple system is able to maintain a state of thermodynamic disequilibrium and the dynamics of a heat flux that redistributes heat within the system. It is this exchange that prevents the depletion of the gradients and the system to reach a uniform state of thermodynamic equilibrium at which nothing would happen at the macroscopic level. Hence, the entropy exchange of a system explains why dynamics and gradients can be sustained in a thermodynamic system that is in a steady state.

3.5 Free energy and thermodynamic potentials

The dynamics of the simple system shown in Fig. 3.4 can also be expressed in terms of free energy, which is a system-level aspect of a thermodynamic system that makes it interesting. It closely relates to disequilibrium and the dynamics that are possible from the energy stored in the system and that result from the second law. Free energy in thermodynamics is specifically defined as Helmholtz or Gibbs free energy that we deal with further on. We want to introduce the concept of free energy here more generally as the amount of energy within a system that is capable of performing work. In the engineering literature, this concept is often referred to as the exergy of a system (e.g., Bejan 2002). As we will see in the following, free energy relates to the differences in thermal energy between the two reservoirs, but its magnitude is typically much smaller because it also involves the differences in entropy and the distance to thermodynamic equilibrium. Note that the presence of this free energy does not necessarily mean that it actually is converted into work, as it may also simply be dissipated by an irreversible process such as diffusion.

To describe the free energy of the system shown in Fig. 3.4, let us look again at the factors that shape the dynamics in the system. The term "dynamics" refers here to the maintenance of the heat flux between the reservoirs, as this is what makes the system interesting. This heat flux would not be expected if the system was in a state of thermodynamic equilibrium. Its driver is the second law of thermodynamics, as the heat flux acts to level out the differences in temperature and increase the entropy of the system. It is thus insightful to express the dynamics that take place inside a system in terms of its distance to the state of thermodynamic equilibrium.

The distance of the system to thermodynamic equilibrium ΔS_{diseq}, is measured by the difference between the entropy of the system at thermodynamic equilibrium S_{eq}, and the actual entropy of the system S:

$$\Delta S_{diseq} = S_{eq} - S \tag{3.16}$$

The entropy S_{eq} is calculated from the thermal energy content of the whole system, while S is computed from the sum of the entropies of the two reservoirs, which was referred to as S_{tot} in Section 3.4. This measure of disequilibrium is shown in Fig. 3.5. Initially, when the difference in thermal energy is greatest in the system, ΔS_{diseq} is greatest as well. We can clearly see that the depletion of ΔS_{diseq} in time relates closely to the temperature difference and the heat flux between the two reservoirs by comparing this change to the changes in Fig. 3.4. In the isolated system, ΔS_{diseq} becomes zero in the final state of thermodynamic equilibrium as the two reservoirs are in equilibrium. For the non-isolated system, ΔS_{diseq} reaches a non-zero steady-state value that reflects the disequilibrium of the system that is being maintained by the exchange fluxes of the system.

The disequilibrium ΔS_{diseq} directly relates to the free energy in the system. Recall that entropy is the conjugate variable to temperature, which taken together describe thermal energy. We can thus assign an energy to this disequilibrium.

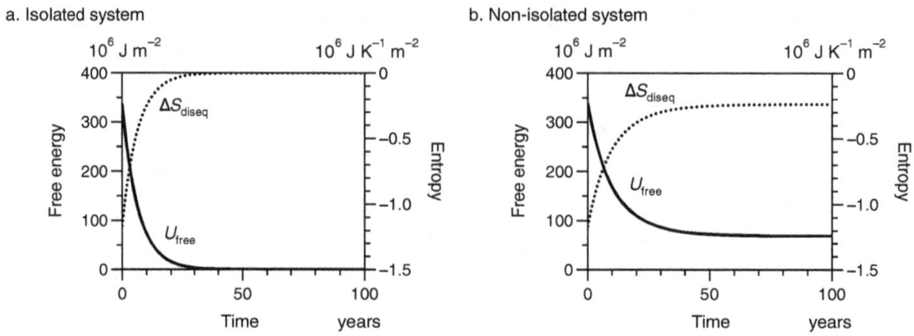

Figure 3.5 Free energy, U_{free}, and disequilibrium, ΔS_{diseq}, for the simple systems considered in Section 3.4. Based on Kleidon (2010).

Since the thermal energy of the whole system remains unchanged, this free energy, U_{free}, is simply described by

$$U_{free} = T\Delta S_{diseq} \tag{3.17}$$

where T is the mean temperature of the system. The free energy of the system is much smaller than the difference in heat content between the two reservoirs (Fig. 3.5), because it also considers the entropies that are associated with the heat contents. It can be viewed as the fraction of the energy in the system to which no entropy is attached, so that it can be converted into another form without violating the second law. The free energy follows the same dynamics as the disequilibrium, ΔS_{diseq}, because the mean temperature of the system is fixed in the example we consider here. With the heat flux depleting the temperature gradient in time, it dissipates the free energy in the system. When the system reaches the final state of thermodynamic equilibrium, there is no free energy in the system. Free energy is thus not a conserved property within the system. The non-isolated system maintains some free energy as its state of disequilibrium is being maintained by its net entropy exchange at the system boundary.

This use of free energy is somewhat different from the common use of free energy in thermodynamics, yet it describes the same idea. In thermodynamics, free energy are forms of so-called thermodynamic potentials of a system. Thermodynamic potentials represent combinations of different energy forms. The most common thermodynamic potentials are the internal energy U, defined by

$$dU = TdS - pdV \tag{3.18}$$

the enthalpy H,

$$H = U + pV \tag{3.19}$$

and the Helmholtz free energy F,

$$F = U - TS \tag{3.20}$$

The definition of enthalpy and Helmholtz free energy are such that a change in these is given by $dH = dU + pdV + Vdp = TdS + Vdp$ and $dF = dU - TdS - SdT = -SdT - pdV$. Another commonly used thermodynamic potential is the Gibbs free energy of a system, particularly when dealing with phase transitions and chemical reactions. The Gibbs free energy is based on the Helmholtz free energy, but also considers the contribution by the uncompensated heat pV:

$$G = F + pV = U - TS + pV \tag{3.21}$$

The use of these combined forms of energy is shown later in the book when we deal with phase transitions and chemical reactions.

Free energy is a critical concept for energy conversions. It represents the fraction of energy of a system that can be used to perform work, cause dynamics, and hence be converted into other forms of energy. We will encounter this term at many more occasions in the following parts of this book.

3.6 Dynamics as energy conversion

Thermodynamic equilibrium, disequilibrium, and free energy apply to other forms of energy as well. This extension forms the critical basis to place the range of Earth system processes into the broader picture of how the whole Earth functions as a thermodynamic system. Just as temperature differences reflect thermodynamic disequilibrium and relate to thermal energy, there are differences in other variables that relate to disequilibrium and other forms of energy. The basics of how disequilibrium and the corresponding difference in a variable is created, maintained, and depleted is analogous to thermal energy.

In this book, we refer to this depletion of disequilibrium as dynamics and postulate that it is this form of dynamics that describes the dynamics of the Earth system and represent the manifestation of the second law. As disequilibrium is reflected in a difference in the associated, intensive variable, and fluxes generally act to deplete this difference, these result in dissipative processes within the system. We then characterize the activity of a process by the magnitude of its dissipation. Some examples for differences related to disequilibrium and different forms of energy are illustrated in Fig. 3.6. A system which contains pressure differences, for instance, reflects a disequilibrium in the distribution of uncompensated heat. The pressure difference in the system is associated with forces which result in a mass flux. This mass flux aims to deplete the pressure difference until there is a uniform distribution of pressure

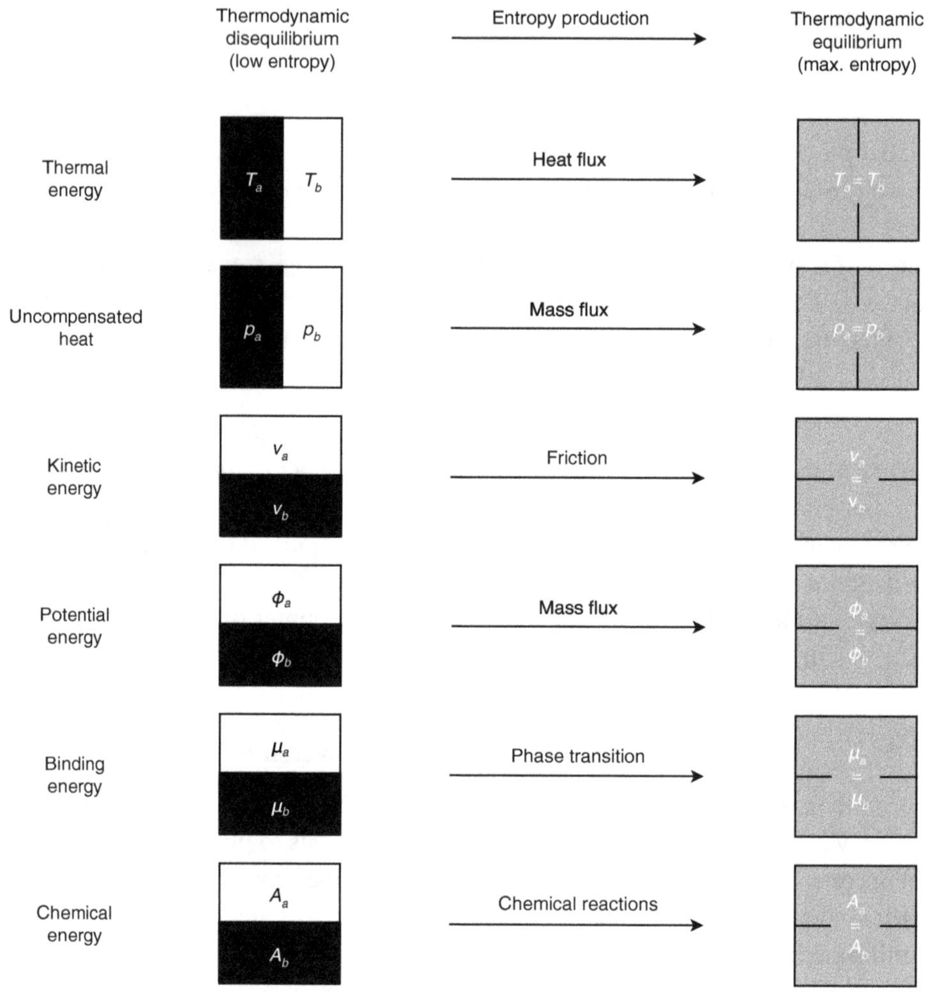

Figure 3.6 Examples of differences that reflect disequilibrium for different forms of energy, the kind of fluxes that deplete these differences, and their resulting equilibrium states. The symbols refer to the intensive variables: T_a, T_b – temperature; p_a, p_b – pressure; v_a, v_b – velocity; ϕ_a, ϕ_b – geopotential; μ_a, μ_b – chemical potential; A_a, A_b – chemical affinity. After Kleidon (2010).

within the system. Likewise, a system with velocity differences results in a momentum flux, better known as a friction force, that aims to compensate this difference in velocity. Further examples are shown for differences in geopotential that are leveled out by mass fluxes, differences in chemical potential that are depleted by mass fluxes or transformations such as phase transitions, and differences in chemical affinity that are depleted by chemical reactions. In all systems, the dynamics that are represented by mass fluxes, forces, and momentum fluxes are driven by a gradient that is depleted by the resulting dynamics. This general scheme follows the same direction as it was described earlier for thermal energy with its temperature differences. Yet, while the second law of thermodynamics provides a clear direction for thermal energy, the application of the second law to non-thermal forms of energy is not quite as straightforward, mainly because different forms of energy are connected, making the state of equilibrium not as simple as it is shown in Fig. 3.6.

The link between non-thermal forms of energy and thermal entropy is established by noting that the depletion of the differences shown in Fig. 3.6 reduce the associated form of energy. This reduction in energy by associated dynamics within the system is compensated for by an increase of thermal energy, e.g., by frictional heating or heat release by a chemical reaction, which in turn increases the entropy of the system except for some idealized cases such as an ideal pendulum. The connection to thermal entropy is thus made when we consider a non-thermal form of energy together with the thermal energy of the system. To illustrate this point, let us consider an isolated system for simplicity. For this system, we consider one non-thermal form of energy, U_{other}, together with its thermal energy, U_{te} (and for a gas, we would need to account for the uncompensated heat as well as it is linked to the thermal energy, as described in Section 2.3.3.), so that the total energy of the system that we consider here is described by $U_{\text{tot}} = U_{\text{te}} + U_{\text{other}}$. Since energy is being conserved and thus $dU_{\text{tot}} = 0$, any change in dU_{other} is compensated for by a change in thermal energy dU_{te}, with $dU_{\text{other}} = -dU_{\text{te}}$. This consideration also applies to more extensive, cascading conversions of non-thermal energy. The change in thermal energy relates to a change in entropy by Clausius' expression, so that the change in entropy of the system, dS, is given by:

$$dS = -\frac{dU_{\text{other}}}{T} \tag{3.22}$$

In other words, when U_{other} is reduced, the entropy of the system is increased. When we consider this expression in time, we can express this depletion of U_{other} as a contribution to the entropy production of a system:

$$\frac{dS}{dt} = \sigma = -\frac{dU_{\text{other}}}{T} = J_{td} \cdot F_{td} \tag{3.23}$$

The last term in this equation relates this entropy production to a typical form in non-equilibrium thermodynamics, which expresses entropy production as a product of thermodynamic fluxes J_{td}, and thermodynamic forces F_{td}. The thermodynamic flux J_{td}, is generally expressed as a function of the driving gradient that relates to one of the variables shown in Fig. 3.6. The heat flux in the example in Section 3.4, for instance, was expressed in terms of the temperature difference of the form $k\ (T_a - T_b)$. The thermodynamic force, F_{td}, is of a more abstract form. Using the example of the heat flux, the thermodynamic force corresponds to $1/T_b - 1/T_a = (T_a - T_b)/(T_a T_b)$. The product of these two expressions then yields the entropy production as given in eqn. 3.15. As the fluxes related to the examples shown in Fig. 3.6 are typically expressed in the form of a gradient in the intensive variable, we can also express the dynamics in terms of thermodynamic fluxes and forces.

What this formulation then tells us is that the second law requires non-thermal forms of free energy in an isolated system to be reduced in time, so that $dS \geq 0$. As S increases, ΔS_{diseq} decreases, so that we can also formulate the second law as stating that the free energy in a system is reduced and reaches a minimum in a state of thermodynamic equilibrium. The state of thermodynamic equilibrium and of maximum entropy then corresponds to a state in which the other forms of energy, U_{other}, are at a minimum . This minimum in U_{other} may not necessarily imply that the energy content is zero because it depends on the reference value of U_{other} and on potential constraints imposed by mass or momentum balances. It may also be that a system does not reach this minimum state because the process that could deplete the free energy is not available.

The application of the second law to a non-thermal form of energy and the resulting minimization of energy is illustrated in the simple example of kinetic energy shown in Fig. 3.7. As described before, kinetic energy is expressed by velocity and momentum as the conjugate pair of variables to which the conservation of momentum applies. The system is set up in a similar way as the example with the two reservoirs of thermal energy, but we now consider two reservoirs with different velocities, v_a and v_b which exchange momentum by a flux, J_{mom}. An example for such a system would, for instance, be a set of two connected wheels that rotate at different speeds. As in the example for thermal energy, we consider an isolated system and a non-isolated system that is driven by an input of momentum across the system boundary.

The starting point to describe the temporal evolution of the system are the momentum balances of the two reservoirs, which ensure the conservation of momentum. For simplicity, we assume that the reservoirs contain equal mass m, and have same temperatures. We write the momentum balances as

$$m\frac{dv_a}{dt} = -J_{\text{mom}} + F_{\text{acc}} \tag{3.24}$$

and

$$m\frac{dv_b}{dt} = J_{\text{mom}} - F_{\text{fric}} \tag{3.25}$$

where F_{acc} is an accelerating force, that is, a momentum flux, which is zero in the isolated system, but non-zero in the non-isolated system. The momentum flux between the reservoirs depends on the difference in velocities, $v_a - v_b$, but we do

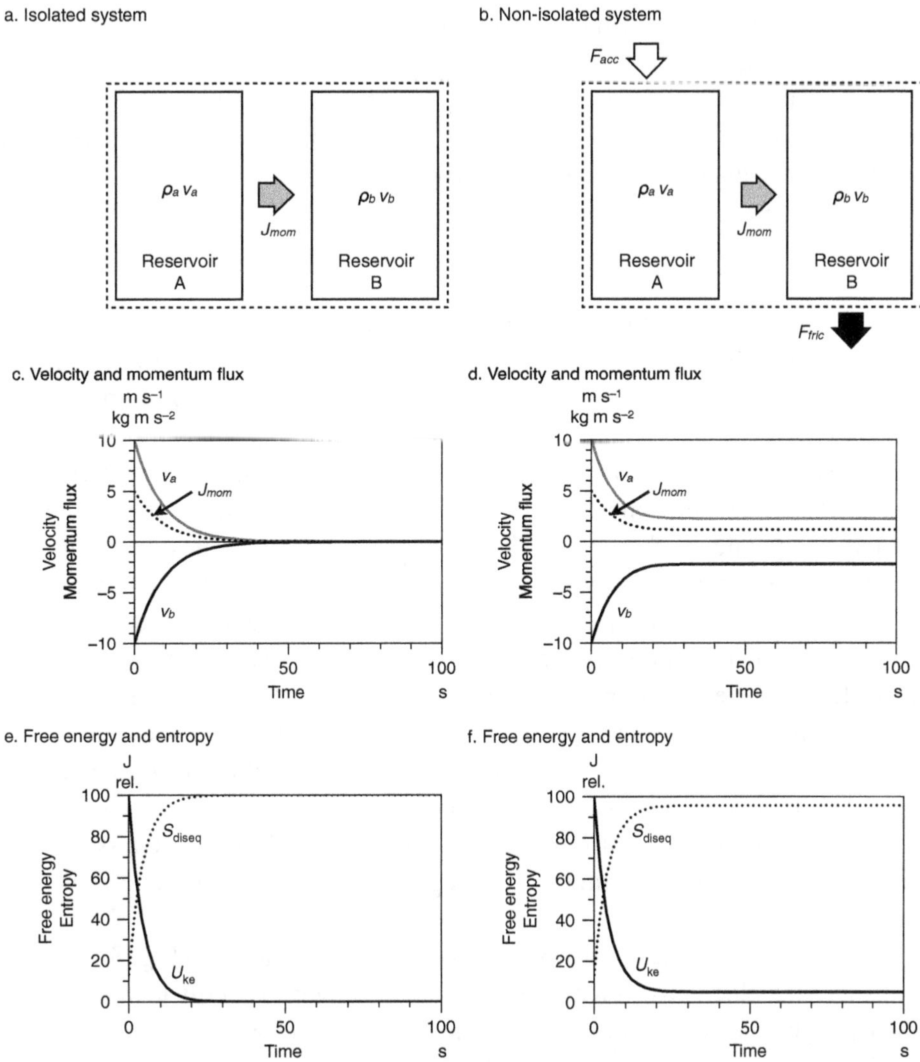

Figure 3.7 Two momentum reservoirs similar to the model in Fig. 3.4, but for kinetic energy, with (a) being an isolated system while (b) is an open system driven by external input in the form of an accelerating force F_{acc}. The reservoirs have the same mass and initially opposing velocities, v_a and v_b, of the same magnitude and are connected by a momentum flux, J_{mom}, between the reservoirs. The transient changes of the system are shown for (c) and (d) velocity and momentum flux as well as (e) and (f) free energy U_{ke} and disequilibrium S_{diseq}. Values of $m = 1$ kg, $J_{\text{mom}} = k(v_a - v_b)$ with $k = 0.05$, and $G = 1$ W were used. The dashed boxes delineate the system boundaries.

not need to specify its functional form in this example because it is fixed in steady state to $F_{acc} = J_{mom} = F_{fric}$.

As momentum is exchanged by J_{mom}, which is accompanied with frictional heating, the kinetic energy as well as the thermal energy of the system change. The total kinetic energy of the system is given by

$$U_{ke} = \frac{m}{2}v_a^2 + \frac{m}{2}v_b^2 \tag{3.26}$$

while the total energy of the system is conserved, so that $U_{tot} = U_{te} + U_{ke}$ is constant. If we consider that the accelerating force F_{acc} originates from the depletion of potential energy, then we would need to account for this form of energy in the total as well.

As already explained earlier in more general terms, the maximum in entropy is reached when the kinetic energy of the system is at a minimum. This minimum can be obtained analytically from eqn. 3.26, because the two velocities are not independent, but rather constrained by the conservation of momentum. When we express this constraint as $2mv = mv_a + mv_b$ with a fixed value of the total momentum $2mv$, then the state of minimum kinetic energy, and thus of maximum entropy, can be derived from the derivative of U_{ke} with respect to one of the velocities and yields as a solution, $v = v_a = v_b$. Hence, in the isolated system considered here, the manifestation of the second law is reflected in the momentum flux J_{mom} which acts to deplete differences in velocity. The resulting state of equilibrium is one in which the velocity is uniformly distributed inside the system. This is of course what we would expect as the momentum flux that depletes the velocity difference is typically described as friction.

The temporal dynamics of this simple system are displayed in Fig. 3.7, showing the changes in velocity as well as the associated changes in energy towards the state of thermodynamic equilibrium. Equivalent to the case of a temperature difference described earlier, it requires exchange fluxes to maintain a state of disequilibrium in the form of velocity gradients. These exchange fluxes add free energy to the system by creating and maintaining a momentum gradient between the reservoirs that is then depleted by processes that follow the second law inside the system.

In several cases, equilibrium states of Earth system processes involve more than two forms of energy so that the considerations need to involve more processes. The associated equilibrium state may then maintain simultaneous differences in variables. For instance, the state of hydrostatic equilibrium in the atmosphere represents a form of equilibrium in which the vertical gradient in geopotential is compensated for by a vertical pressure gradient of opposite sign. Further examples that involve multiple forms of energy are phase transitions and chemical reactions. Nevertheless, also for these forms, the state of maximum entropy is represented by a state of minimum free energy, which is typically formulated in terms of the Gibbs free energy in the system (e.g., Wallace and Hobbs 1977; Kondepudi and Prigogine 1998).

In summary, what we have illustrated here is that the second law and states of maximum entropy can be applied to forms of energy beyond thermal energy and relate to a state of minimum free energy. Hence, the second law of thermodynamics provides a general direction not just for heat fluxes, but for all kinds of dissipative Earth system processes. To maintain these processes and their associated driving gradients requires exchange fluxes with some other form of energy, typically thermal energy. Hence, these exchange fluxes play a critical role as the "glue" between Earth system processes. This aspect will be inspected in greater detail in the following as it provides the basis for a holistic, thermodynamic view of how the whole Earth system operates.

3.7 Thermodynamic laws in the Earth system

At the end of this chapter, we summarize the implications of the first and second law of thermodynamics for the functioning of the Earth system. For every form of energy and Earth system processes that act on these, these two laws apply. The first law ensures the accounting of energy changes, flows, and conversions, while the second law sets the direction for processes to produce entropy and to evolve towards states of thermodynamic equilibrium. This entropy production either results directly from the redistribution of heat, or indirectly from the dissipation of free energy.

When placing these conversions in the context of the Earth system, processes moderating the exchange fluxes form a critical link between the forms of energy that they act upon. In the absence of exchange fluxes, processes would proceed to dissipate free energy, produce entropy, and result in an evolution of the system towards a state of thermodynamic equilibrium in which no dynamics at the macroscopic scale would take place. It is only through exchange fluxes that different forms of energy are maintained in states of thermodynamic disequilibrium. Exchange fluxes ultimately relate to the planetary forcing, which mostly is accomplished by the radiative exchange between Earth and space, and which is successively converted to different forms of energy.

This successive conversion of energy from the planetary, radiative forcing to different forms of energy is illustrated in Fig. 3.8. This figure is, of course, highly simplified. It shows three forms of energy, labeled as box A, B, and C, that we account for separately, hence each forming a subsystem within the greater context of the Earth system. The planetary exchange by the fluxes J_{in} and J_{out} result in an uneven heating and cooling even if they are in a steady state of $J_{in} = J_{out}$ and thereby maintain a disequilibrium in the distribution of thermal energy (box A).

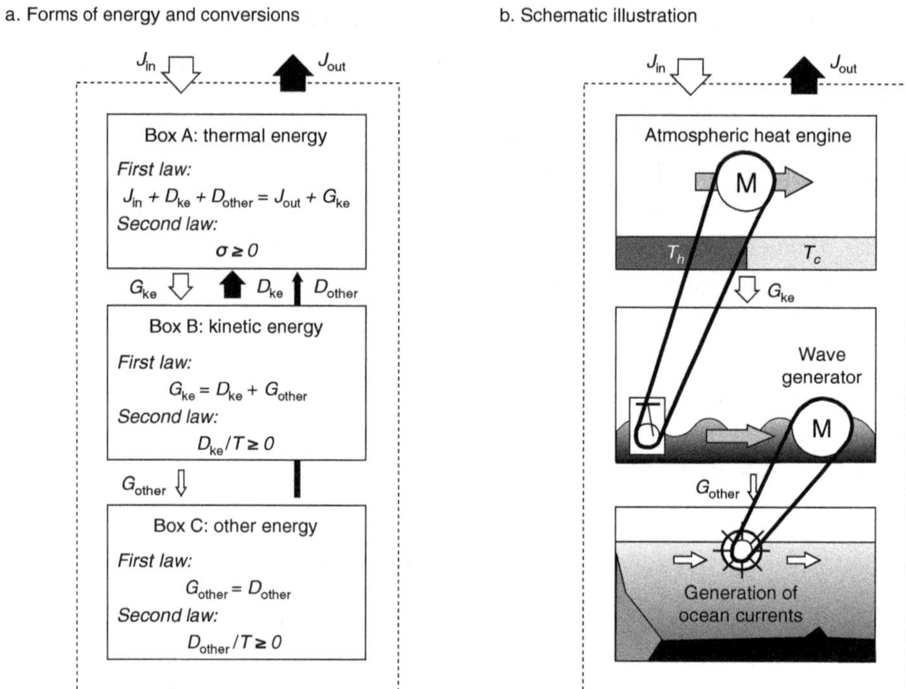

Figure 3.8 Illustration of how the first and second laws apply to different forms of energy that are coupled within the context of the Earth system. The fluxes at the top of J_{in} and J_{out} represent the radiative planetary forcing. Panel (a) shows three forms of energy by the three boxes A, B, and C that are linked by the respective generation rates, G_{ke} and G_{other}, as well as their dissipation rates, D_{ke} and D_{other}. The first and second law of thermodynamics set fundamental constraints and the direction for the dynamics by Earth system processes that convert these forms of energy. Panel (b) shows a schematic illustration of such conversions in the Earth system; a radiative forcing drives a heat engine that drives motion and further conversions into ocean waves and currents, as described in the text. The conversions of energy act as motors and are marked by "M."

Examples for this uneven heating by planetary exchange is the surface heating by the absorption of solar radiation while the atmosphere is cooled by emission, the uneven heating between tropical and polar regions due to different inclination of the surfaces to the incident solar radiative flux, and the uneven heating during day and night. Some fraction of the energy associated with this disequilibrium is converted into kinetic energy at a rate G_{ke}, driving the dynamics that are associated with motion and kinetic energy (box B). In the Earth system, this kinetic energy is mostly related to the atmospheric circulation. Some of the energy of box B may be converted further into another form of energy at a rate G_{other} into box C. This can, for instance, represent the lifting of dust particles into the atmosphere. The energy of box C would then correspond to the potential energy of the dust particles. Another example for box C is the generation of ocean waves. Here, the energy would correspond to the potential and kinetic energy of waves. This energy could be converted even further into the kinetic energy associated with the wind-driven oceanic circulation. Yet, the point of this consideration is to emphasize that each box is, directly or indirectly, fed by the planetary exchange, the critical connections are maintained by G_{ke} and G_{other}, and each conversion process is subjected to the first and second law of thermodynamics.

To illustrate the application of the first and second laws to these conversions, we consider the conversions in a steady state. In this steady state, the planetary exchange is balanced, with $J_{in} = J_{out}$, and the conversions are all balanced as well, with generation balanced by dissipation and further conversion, $G_{ke} = D_{ke} + G_{other}$ and $G_{other} = D_{other}$. Such a steady state, at least in a mean state averaged over a sufficiently long time, is inevitable, because otherwise there would be a steady increase (or decrease) of energy within the system. These balances all reflect the first law. The second law applies to the system as a whole, but also to each of the boxes, as we can describe each box as a separate system. It then states that whatever processes take place inside the box, they follow the second law and either produce entropy directly (in the case of thermal energy, as indicated by $\sigma \geq 0$ in Fig. 3.8), or by dissipating free energy ($D_{ke} \geq 0$ and $D_{other} \geq 0$ in Fig. 3.8), which ultimately produces thermal energy and thereby entropy. By setting these conditions, the first and second law formulate a fundamental framework for the interactions between the different forms of energy within the Earth system and thereby for the interactions between Earth system processes.

There is more to be learned from the laws of thermodynamics. When the first and second law are combined, they yield fundamental limits on the energy conversion rates (i.e., G_{ke} and G_{other} in Fig. 3.8). From these limits, transport characteristics can be inferred from the associated optimum flux (i.e., the conductivity k in the example of Section 3.4), and we can infer the thermodynamic consequences of these conversions for the dissipative nature of the whole system. These aspects will be considered in the next chapter.

4

Thermodynamic limits

4.1 Limits to energy conversions

Energy conversions play a central role in driving the dynamics of Earth system processes. In the last chapter, we have seen how energy conversions are governed by the first and second law of thermodynamics. Energy conversions and the associated dissipative processes are sustained in natural systems by the entropy exchange across the system's boundary and are constrained by the second law. The subsequent generation of free energy out of this entropy exchange maintains the disequilibrium associated with a variety of variables and is reflected in the dynamics of the Earth system.

In this chapter, we take the laws of thermodynamics a step further. We first show how these laws yield fundamental conversion limits when combined, which then set the dissipative "speed limits" for the dynamics of a system. It sets the limit on the rate by which work can be performed through time, that is, the power that is associated with the conversion of thermal energy into another form. When applied to the Earth system, it is important to note that the work is performed inside the system, so that the consequences of the work need to be taken into account in the formulation of thermodynamic limits.

To illustrate thermodynamic limits qualitatively, consider the energy conversion associated with the heat engine shown in Fig. 4.1a. A heat engine is an abstract device that converts thermal energy into mechanical work and can represent a steam engine, a turbine, or atmospheric convection. The heat engine is driven by a heat flux J_{in}, from a hot reservoir, expels a certain fraction into a so-called waste-heat flux J_{out}, to a cold reservoir, and converts the other fraction into work at a rate G, with work performed through time representing the power of the engine. The heat flux J_{out} is referred to as a waste heat flux because its energy is not converted into free energy. The entropy exchange of the heat engine only consists of the heat fluxes between the reservoirs with their respective temperatures. This entropy exchange

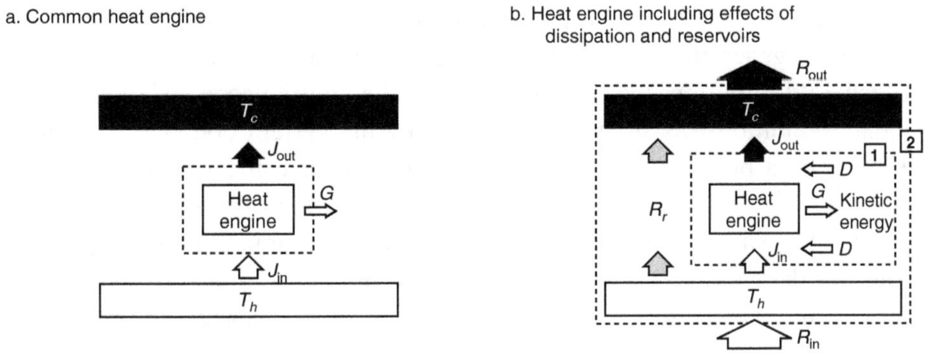

Figure 4.1 A heat engine that is driven by a heat flux J_{in} from a warm reservoir with temperature T_h, produces work at a rate G (i.e., power), and a heat flux J_{out} to a cold reservoir T_c. The second law yields a limit on how much power G can maximally be produced by the heat engine when (a) only the engine is considered, and (b) the dissipative heating that results from the generated work is considered (with system boundary labeled "1") and the engine is considered jointly with the fluxes that determine the temperatures of the reservoirs (with system boundary labeled "2").

decreases the greater the value of G, because for the same entropy import associated with J_{in}, less is exported by J_{out}. The second law sets the ultimate limit to G, because the heat engine needs to at least export as much entropy as it is imported to maintain a non-negative entropy exchange. Hence, there is a fundamental limit to how much work can be performed, or, more generally, how much energy can be converted from the heat flux J_{in} into another form at rate G. This limit is known as the Carnot limit. It sets the fundamental conversion limit of the heat engine, and, as we will see further on, depends on the boundary conditions of the system, but not on the details that make up the engine.

When we consider a heat engine in the context of the Earth system, it is important to note that it does not operate in isolation, but affects the system. Work is not performed outside the system, as is the case for a car engine, but rather takes place within the system. The work to generate another form of energy will eventually result in its dissipation, that is, its conversion back into thermal energy, and this dissipative heating needs to be taken into account. This is shown in Fig. 4.1b by the system boundary labeled "1." When this dissipative heating is accounted for, it yields a slightly different limit, which we refer to as the Carnot limit of a dissipative heat engine.

We also need to consider the consequences of the engine in terms of how it alters its boundary conditions that affect the operation of the engine. When, for instance, a temperature difference is used to generate motion in the atmosphere, the moving mass transports heat, resulting in a heat flux. As it transports heat in the direction from hot to cold, heat transport depletes the driving temperature difference that maintains the generation of motion. So when we aim to estimate the limit associated with generating motion, we need to consider the consequences of motion on the boundary conditions, as shown in Fig. 4.1b by the system boundary labeled "2." This requires a broader definition of how we delineate the system boundary to not simply include the heat engine, but also the effects of the engine, particularly regarding the temperature difference that drives the engine. To do so, we need to consider the energy balances that determine the temperatures of the reservoirs as well as the energy fluxes that are considered in the balances. It results in an extension of the Carnot limit which includes the consideration of the magnitude of the heat flux as well as the consequences on the driving temperature difference. This limit will be referred to as the maximum power limit. This maximum power limit applies not just to heat engines, but also to other forms of energy conversions and is well known in a range of applications (Odum and Pinkerton 1955). This limit constrains further conversions of energy into other, non-thermal forms of energy. The maximum power limit is then set into the context of proposed thermodynamic extremum principles, specifically those of maximum and minimum entropy production. The chapter closes by setting thermodynamic limits into the context of the Earth system.

4.2 The Carnot limit

The Carnot limit expresses the maximum work that can be derived by a heat engine from a heating source. This limit follows directly from the application of the first and second law of thermodynamics and as such is independent of the details of the engine that we consider. To derive the limit, we only need to use the conditions at the boundary of the heat engine, as shown in Fig. 4.1a. In contrast to the common derivation of this limit in thermodynamics textbooks, the limit is derived here from the energy and entropy balance in steady state. It results in the same expression, but the derivation is much simpler and more general.

We consider an engine that is heated by a flux J_{in} from a hot reservoir with temperature T_h, expels heat to a cold reservoir with a temperature T_c at a rate J_{out}, and yields power $G = dW/dt$, that is, work per time. We assume that the temperatures of the reservoirs do not change in time, either because the reservoirs are large, or because they have been heated and cooled and have reached a steady state. The first law ensures energy conservation within the heat engine, which in a steady state is expressed as:

$$J_{in} = J_{out} + G \tag{4.1}$$

The entropy balance of the engine is given by the entropy production within the engine σ, and the net entropy exchange due to the two heat fluxes at their respective temperatures:

$$\sigma + \left(\frac{J_{\text{in}}}{T_h} - \frac{J_{\text{out}}}{T_c}\right) = 0 \tag{4.2}$$

The second law requires that $\sigma \geq 0$. The net entropy exchange is determined from the heating J_{in} of the engine, which imports entropy at a rate J_{in}/T_h, and the cooling J_{out}, which exports entropy at a rate J_{out}/T_c. Note that the power G does not enter the entropy balance because it expresses a generation rate of free energy, and, as described in the previous chapter, has no entropy attached to it. When this power produces some form of free energy, the energy is eventually dissipated into heat, which generates entropy at a rate $\sigma_{\text{diss}} = D/T$ outside the engine, with T being the temperature at which the free energy is dissipated.

To derive a limit on the power G, we use the first law (eqn. 4.1) to express J_{out} as a function of J_{in} and G, and combine it with the entropy balance (eqn. 4.2). We then obtain an expression for the power that can be derived from the entropy budget as:

$$G = J_{\text{in}} \frac{T_h - T_c}{T_h} - \sigma\, T_c \tag{4.3}$$

The upper limit for power is given by the case in which there is no internal entropy production inside the engine, $\sigma = 0$, which yields the Carnot limit, G_{Carnot}:

$$G_{\text{Carnot}} = J_{\text{in}} \frac{T_h - T_c}{T_h} \tag{4.4}$$

We can also express this limit in terms of an efficiency by relating the power to the heat flux that enters the engine:

$$\eta_{\text{Carnot}} = \frac{G_{\text{Carnot}}}{J_{\text{in}}} = \frac{T_h - T_c}{T_h} \tag{4.5}$$

This efficiency is referred to as the Carnot efficiency of the heat engine. It is independent of the heat flux through the engine. Note that it only depends on the temperatures at the boundary of the engine, and not on the design and material properties of the engine itself. How this limit can be reached is, however, nontrivial and a different question. To illustrate this efficiency, we can evaluate the limit using typical conditions of the Earth's surface–atmosphere system, with a warm surface temperature of $T_h = 288$ K and a cold atmospheric radiative temperature of $T_c = 255$ K. This yields a temperature difference of about 33 K, and a Carnot efficiency of $\eta_{\text{Carnot}} = 33/288 = 11.5\%$.

To illustrate further how the entropy balance sets the limit to the power of the engine, the partitioning of energy fluxes as well as the entropy production terms

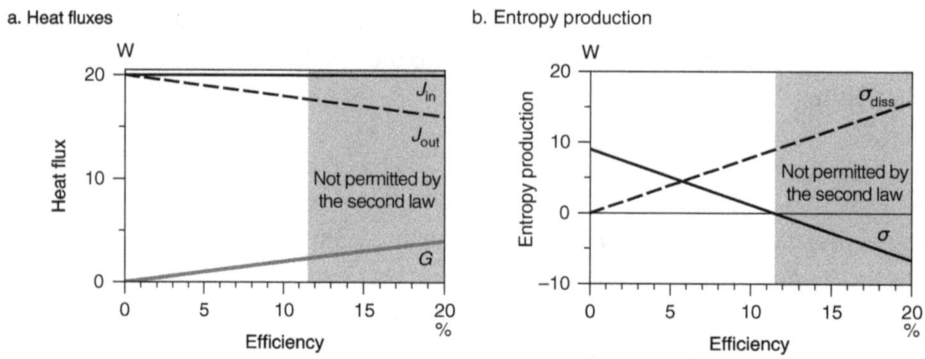

Figure 4.2 Heat fluxes, J_{in} and J_{out}, power, G, (a) and (b) entropy production, σ and σ_{diss}, of a heat engine as shown in Fig. 4.1a as a function of the conversion efficiency $\eta = G/J_{in}$. The two terms of entropy production shown in (b) represent the internal entropy production, σ, within the heat engine and the entropy production, σ_{diss}, that results from the dissipation of the free energy outside the engine. For the figure, a heat flux of $J_{in} = 20$ W was used with temperatures of $T_h = 288$ K and $T_c = 255$ K. The Carnot limit for these values is $(T_h - T_c)/T_h = 11.5\%$. The Carnot limit sets the left boundary of the area shaded in grey, which marks conditions that violate the second law because it would require internal entropy production inside the engine to be negative.

are shown in Fig. 4.2. Recall that the conditions at the boundary of the heat engine were prescribed, in terms of the heat flux that enters the engine J_{in}, as well as the temperatures of the reservoirs, T_h and T_c. The only flexibility that is not set by the boundary conditions is how much of the heat flux that enters the engine J_{in}, is converted into power G or leaves the engine to the cold reservoir, given by J_{out}. At the one limit in which the heat engine produces no power ($G = 0$), there must be some internal entropy production inside the engine to meet the entropy balance ($\sigma = J_{in}(T_h - T_c)/(T_h T_c)$, which can be inferred directly from eqn. 4.3). The more power the engine generates, the lower this rate of internal entropy production is going to be. The Carnot limit is then given by the other extreme, which is set by the requirement of the second law that $\sigma = 0$.

The free energy generated by the engine will, of course, eventually be dissipated as well, with $G = D$ in the steady state. This dissipation takes place somewhere else outside the heat engine. Yet, if this place is linked to the same heat reservoirs, then, at most, the entropy that is being produced by this dissipation is given by $\sigma_{diss} = D/T_c$. This expression can be derived simply from the entropy balance in steady state given by eqn. 4.2 and by $D = G_{Carnot}$, and represents the share by which the internal entropy production is reduced when $G > 0$. In other words, this is a zero sum game because the total entropy production in steady state is fixed by the boundary conditions as specified by the heat flux J_{in}, and the temperatures of the reservoirs, T_h and T_c. The degree to which the heat engine generates power then merely affects the partitioning of the total entropy production into the internal entropy production inside the heat engine and the external entropy production related to the dissipation of the free energy that the heat engine generated.

These considerations are still a somewhat abstract view of a heat engine. We only dealt with the first and second laws as applied to the system we called the heat engine, combined these, and derived the limit to the power that the engine could generate at best operation. We have not dealt with the specific steps involved in an engine that generate this power yet. These steps are described by a so-called thermodynamic cycle, a concept that is described next.

4.3 The Carnot cycle

When we think of real heat engines, we typically have a concrete image of the different steps of the engine that result in work being performed. With a concrete picture of these steps, the Carnot limit is no longer an abstract construct. In the following, we consider such a heat engine (shown in Fig. 4.3) and analyze the different steps that are involved in running this engine and that produce mechanical work. As we will see, these steps in the ideal case also result in the Carnot limit and represent its common derivation in textbooks. After working through these steps for the heat engine, we then compare this cycle to how natural systems perform this work.

The heat engine shown in Fig. 4.3 involves four steps that add and remove heat and involve the expansion and compression of a gas inside the engine. In the first step, heat is added to the engine and the air inside expands. In the second step, the

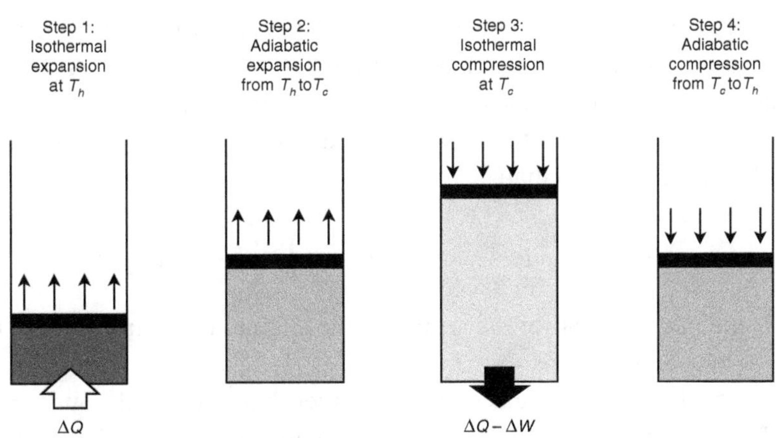

Figure 4.3 Sketch of the different steps that form a Carnot cycle operating between a hot temperature T_h, and a cold temperature T_c.

air expands further without the addition of heat. In step 3, heat is removed with a compression of the air, while in step 4, the air is compressed further without the removal of heat. This sequence of steps and the associated assumptions as detailed further on is referred to as the Carnot cycle. Note that the process must be cyclic so that in a steady state averaged over the cycle, the mean state of the system does not change in time. More generally, any cyclic sequence of such steps is referred to as a thermodynamic cycle. The steps are summarized in Table 4.1 and are shown graphically in Fig. 4.4a and b.

The steps result in characteristic changes in the variables that are associated with the two forms of energy that are involved, thermal energy and uncompensated heat. The variables involved are temperature T, entropy S, pressure p, and volume V, of the gas. These variables change in a cyclic fashion, so that at the end of the four steps, they obtain the same values as at the beginning of the first step. The variables thus form a cycle, as shown in Fig. 4.4. In the following, we describe the changes in these variables that occur during the four steps, with capital letters as indices marking the four points shown in Fig. 4.4.

In *Step 1* of the Carnot cycle, heat is added to the engine isothermally, which means that it is assumed that the temperature does not change. This step connects the points A and B in Fig. 4.4. As the temperature does not change, we denote the temperature during this step as the hot temperature $T_h = T_A = T_B$. The addition of heat results in an increase of entropy, with $\Delta S = S_B - S_A = S_h - S_l$. The latter variables, S_h and S_l, are introduced because these stay constant during steps 2 and 4. We can thus express the added heat as

$$\Delta Q_1 = \int_{S_A}^{S_B} TdS = T_h \Delta S = T_h\,(S_h - S_l) = \Delta Q \tag{4.6}$$

The changes in uncompensated heat during this step are assumed to be governed by the ideal gas law. As there is no change in temperature, the ideal gas law requires that the product pV remains constant as well. Yet, as the air expands isothermally, its volume increases, so that the pressure needs to decrease. This expansion takes up the added heat, ΔQ_1, so that:

$$\Delta Q_1 = \int_{V_A}^{V_B} pdV = RT_h \ln\left(\frac{V_A}{V_B}\right) = RT_h \ln\left(\frac{p_B}{p_A}\right) \tag{4.7}$$

where R is the gas constant for the air inside the engine. In the integration, the ideal gas law, $pV = RT$, was used.

In *step 2* of the cycle, the gas expands further, but without the further addition of heat. This conversion is called adiabatic expansion and refers to an expansion without a change in internal energy. As the heat content does not change, this step is isentropic, that is, it occurs without a change in entropy. The decrease

Table 4.1 *Four steps of a heat engine to produce work as shown in Fig. 4.3.*

Step	Description	Change in $T\ S$	Change in $p\ V$	Link to Fig. 4.4a and b
1	Expansion of the gas by the addition of heat, $\Delta Q > 0$.	Isothermal addition of heat at a temperature $T_A = T_B = T_h$ with $\Delta T = 0$. Entropy increases from S_l to S_h, with $\Delta S = S_h - S_l = \Delta Q/T_h$.	Isothermal expansion of the gas. Since the expansion is isothermal and the temperature does not change, the ideal gas law requires that the product of pressure and volume $p\ V$, remains constant. Hence, $p_A\ V_A = p_B\ V_B$.	Connects points A and B.
2	Adiabatic expansion of the gas with no heat exchange ($\Delta Q = 0$).	Isentropic at constant entropy S_h with $\Delta S = 0$. The temperature decreases from $T_B = T_h$ to $T_C = T_c$ with $\Delta T = T_c - T_h < 0$.	Since no heat is exchanged, the first law jointly with the ideal gas law require changes in pressure and volume of a form in which the product $p\ V^{c_p/c_v}$ is constant. Hence, $p_B\ V_B^{c_p/c_v} = p_C\ V_C^{c_p/c_v}$. The adiabatic condition can also be expressed by a constant potential temperature, given by $\theta_h = T_h = T_B = T_C(p_C/p_B)^{R/c_p}$.	Connects points B and C.
3	Isothermal compression of the gas with a removal of heat, $-(\Delta Q - \Delta W)$.	Isothermal at temperature $T_C = T_D = T_c$ with $\Delta T = 0$. The entropy decreases from S_h to S_l with $\Delta S = S_l - S_h = -(\Delta Q - \Delta W)/T_c$.	Isothermal compression at constant temperature T_c. The ideal gas law requires that the product of pressure and volume $p\ V$, remains constant, so that $p_C\ V_C = p_D\ V_D$.	Connects points C and D.
4	Adiabatic compression of the gas with no heat exchange ($\Delta Q = 0$).	Isentropic at constant entropy S_l with $\Delta S = 0$. The temperature increases from T_c to T_h with $\Delta T = T_h - T_c > 0$.	Since no heat is exchanged, the first law jointly with the ideal gas law require changes in pressure and volume of a form in which the product $p\ V^{c_p/c_v}$ is constant. Hence, $p_D\ V_D^{c_p/c_v} = p_A\ V_A^{c_p/c_v}$. The adiabatic condition can also be expressed by a constant potential temperature, given by $\theta_l = T_h = T_A = T_D(p_A/p_D)^{R/c_p}$.	Connects points D and A.

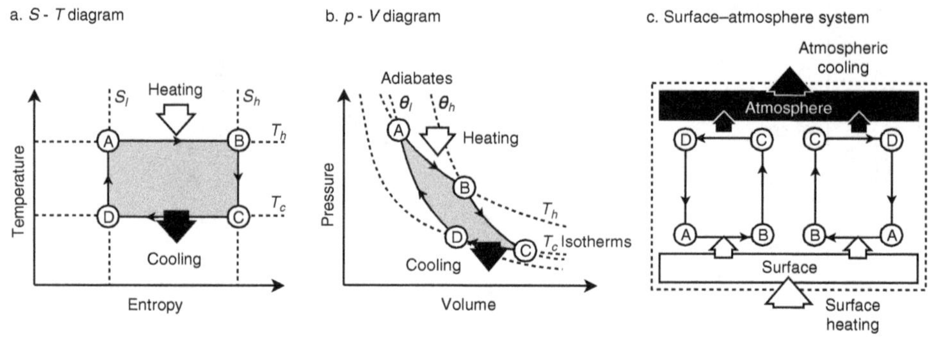

Figure 4.4 The different steps of the Carnot cycle in (a) $T - S$ coordinates, (b) $p - V$ coordinates, and (c) illustrated in relation to convection between the surface and the atmosphere. The letters of the points and the symbols refer to the different steps of the Carnot cycle, as summarized in Table 4.1. Not drawn to scale.

in entropy that results from the cooling in this step is compensated for by the increased contribution of the greater volume to entropy. Hence, the entropy at the end of this step remains at the same value as at the beginning of the step, so that $S_C = S_B = S_h$. The temperature, however, decreases from $T_B = T_h$ to $T_C = T_c$ as air is expanded. Energy conservation (the first law) requires that the energy change due to the decrease in thermal energy $c_v\, dT$, is compensated for by a change in uncompensated heat due to the expansion of the gas $p\, dV$, so that $dQ = c_v\, dT + p\, dV = 0$. The ideal gas law states for these changes in p, V, and T that $d(p\, V) = p\, dV + V\, dp = R\, dT$. Taken together, these yield the relationship

$$\frac{dp}{p} = -\frac{c_v + R}{c_v}\frac{dV}{V} \tag{4.8}$$

When integrated, this yields, with $c_p = c_v + R$, the well-known adiabatic condition

$$p\, V^{c_p/c_v} = \text{const.} \tag{4.9}$$

Using the ideal gas law, this condition can also be formulated in terms of pressure and temperature, which yields the expression for the so-called potential temperature θ, given by

$$\theta = T_0 \left(\frac{p_0}{p}\right)^{R/c_p} \tag{4.10}$$

where the index "0" refers to a reference level. The potential temperature describes the temperature of the air at a pressure p that it would attain when it is brought back to the reference pressure p_0. It is commonly used in meteorology as it has the advantage that it does not change during the adiabatic expansion of step 2. The changes during step 2 are thus governed by

$$p_B \, V_B^{c_p/c_v} = p_C \, V_C^{c_p/c_v} \tag{4.11}$$

or, alternatively formulating it in terms of the potential temperature with the temperature of the hot reservoir as a reference level,

$$T_B = T_h = \theta_h = T_C \left(\frac{p_C}{p_B}\right)^{R/c_p} \tag{4.12}$$

The temperature at the end of this step T_C, then represents the temperature T_B, yet its temperature is lower because it is corrected for the lower pressure at point C. The change in heat content during this step is then simply related to the change in pressure by the conservation of energy, given by:

$$c_v(T_h - T_c) = \int_{V_B}^{V_C} pdV \tag{4.13}$$

In *step 3*, heat is removed from the engine during the compression of the gas, which is again assumed to occur without a change in temperature, as in step 1. Hence, $T_c = T_C = T_D$, and the change in heat, ΔQ_3, that is removed from the engine is given by

$$\Delta Q_3 = \int_{S_C}^{S_D} T \, dS = T_c \, \Delta S = T_c \, (S_l - S_h) = \Delta Q - \Delta W \tag{4.14}$$

where, as in the general case we considered in the previous section, the added heat from step 1 minus the work done, $\Delta Q_3 = \Delta Q - \Delta W$, is being removed from the engine. The changes in pressure and volume during step 3 are determined as in step 1, and yield

$$\Delta Q_3 = \int_{V_C}^{V_D} p \, dV = RT_c \ln\left(\frac{p_D}{p_C}\right) \tag{4.15}$$

Finally, in *step 4*, the gas is further compressed adiabatically without the removal of heat ($\Delta Q = 0$) or change in entropy ($\Delta S = 0$). It is almost the opposite of step 2, except that it takes place at a lower entropy S_l, since heat was removed from the engine in the previous step. The temperature, however, increases from $T_D = T_c$ to $T_A = T_h$. This increase in thermal energy is compensated for by an associated decrease in uncompensated heat. The compensation is achieved by an adiabatic compression from T_c to T_h. The change in pressure and volume are determined by the same relationships as in step 2. They are thus given by

$$p_D \, V_D^{c_p/c_v} = p_A \, V_A^{c_p/c_v} \tag{4.16}$$

or, alternatively,

$$T_A = T_h = \theta_l = T_D \left(\frac{p_D}{p_A} \right)^{R/c_p} \tag{4.17}$$

where again the temperature of the hot reservoir is used as a reference level (noting that $T_h = T_A = T_B$). The associated change in pressure is then again given by the change in heat content

$$-c_v(T_h - T_c) = \int_{V_D}^{V_A} pdV \tag{4.18}$$

When we consider the cycle as a whole, the total entropy exchanged is determined from the difference of the entropy added to the system during step 1, $\Delta Q/T_h$, and the removal of entropy during step 3, $(\Delta Q - \Delta W)/T_c$. When integrated over the whole cycle, the entropy of the system does not change, and the addition and removal of entropy balance each other. By equating these entropy changes, we obtain the net entropy exchange of the system

$$\frac{\Delta Q}{T_h} - \frac{\Delta Q - \Delta W}{T_c} = 0 \tag{4.19}$$

This condition can be solved for ΔW, from which we obtain the Carnot limit,

$$\Delta W = \Delta Q \, \frac{T_h - T_c}{T_h} \tag{4.20}$$

When the cycle is analyzed regarding the total change in pressure over the cycle, we note that the contributions by the work done during the adiabatic changes in steps 2 and 4 cancel, so that the total change is given by

$$\begin{aligned} \int pdV &= \int_{V_A}^{V_B} pdV + \int_{V_B}^{V_C} pdV + \int_{V_C}^{V_D} pdV + \int_{V_D}^{V_A} pdV \\ &= RT_h \ln \left(\frac{p_B}{p_A} \right) + RT_c \ln \left(\frac{p_D}{p_C} \right) \\ &= \Delta W \end{aligned} \tag{4.21}$$

so that the total work performed by the cycle is ΔW as derived by the Carnot limit given in eqn. 4.20. This total work corresponds to the grey area shown in Fig. 4.4b.

To sum up, the work performed by the cycle results from the four steps that exchange entropy in terms of the thermal energy, and the associated changes in uncompensated heat that perform the mechanical work of the engine. The key aspect that allows for the work to be done is the heat exchange at different temperatures: heat is added when the gas is warm (in step 1) and removed when it is cold (in step 3). These changes are associated with expansion and compression of the gas taking place at different pressure, which then performs the work of the engine.

The work done by the engine ΔW, can also be related to the heat added ΔQ. By using eqns. 4.7 and 4.21, we again obtain the expression for the Carnot efficiency:

$$\eta = \frac{\Delta W}{\Delta Q} = \frac{\Delta(pV)_{tot}}{\Delta(pV)_1} = \frac{T_h - T_c}{T_h} \tag{4.22}$$

Hence, the particular set of steps considered here and the assumptions made for the conditions that take place during each step results in the same expression for the maximum efficiency derived earlier in eqn. 4.5. Yet, we should also note that the derivation in Section 4.2 is much simpler and more suitable for our analysis of Earth systems that are driven by heat fluxes and does not depend on the particular steps of the engine that is being considered.

When we consider the addition and removal of heat through time Δt, with $J_{in} = \Delta Q/\Delta t$, $G = \Delta W/\Delta t$, and $J_{out} = (\Delta Q - \Delta W)/\Delta t$, then we get a direct correspondence with the more general derivation of Section 4.2. The exchanges of entropy during steps 1 and 3 then represents exchanges of entropy identical to the entropy export term in eqn. 4.2 in the more general case. The exchange of heat and the processes in the cycle would need to take place at a slow rate to ensure the isothermal conditions assumed during heat exchange in steps 1 and 3, and to avoid frictional losses. Otherwise, these would result in irreversibility and entropy production within the system, which would inevitably reduce the maximum efficiency by which the engine can convert thermal energy into other forms of energy. This is the main reason why the Carnot limit is typically not reached by real heat engines.

Such steps of a thermodynamic cycle are applicable to processes in the Earth system as well. Figure 4.4c shows an atmospheric convection cell, which acts as a heat engine just as the one considered earlier. Surface heating by absorption of radiation adds heat to the engine, near surface air expands and rises adiabatically, cools at a height due to emission of radiation, removing heat from the engine, and the air is compressed adiabatically during descent. Hence, the same steps apply to this atmospheric heat engine, although it should be noted that the steps do not act in a strict temporal sequence inside one piston, but rather occur simultaneously in a spatially organized way. Another important difference to a real heat engine is that the way that updrafts and downdrafts are organized within the atmosphere are not fixed by a mechanical enclosing, but can freely self-organize in space and time. We will see in the following chapter that this spatial organization plays a very important role. It allows the atmospheric heat engine to reduce internal friction so that it can, in fact, operate near the Carnot limit.

This separation of the heat engine into several steps can also be extended to other forms of energy. The starting point is, again, the conjugate variables that are associated with the form of energy that is being converted. When these variables are plotted against each other, just as it was done for temperature and entropy (T, S) and for pressure and volume (p, V), they yield the amount of energy being exchanged and the work that is done. This amount of energy is determined from the integral over the closed cycle of these variables, and the power is given by how often this cycle is performed in time. The integration applied to other pairs of conjugate variables then yields the work done by (or on) this form of energy. However, as the derivation of the Carnot limit from the entropy budget is simpler and more general, we use the approach described in Section 4.2 to derive energy conversion limits.

4.4 The dissipative heat engine

The derivation of the Carnot limit does not account for what happens to the work done by the engine. This omission is fine for a heat engine when the generated free energy is dissipated outside the system. In the Earth system, however, any generated power is also dissipated within the system, and this needs to be accounted for as a separate dissipative heating term. This accounting is easily done and yields a slightly modified Carnot limit, a limit that has been referred to as the Carnot limit of a dissipative heat engine in atmospheric science (Renno and Ingersoll 1996; Bister and Emanuel 1998; Bister et al. 2011).

To include dissipation of the generated free energy in the derivation of the Carnot limit, we move the description of our system from the one outlined in Fig. 4.1a to the system boundary labeled "1" in Fig. 4.1b. As before, we consider a steady state in which $J_{in} = J_{out}$ and in which the generated power balances dissipative heating, so that $G = D$. In principle, this dissipative heating can take place anywhere in the system, wherever friction takes place. To evaluate its effects on the Carnot limit, we consider two extreme cases that do not alter the energy balance of the system, but affect the heat engine. We thus distinguish the heat fluxes from and to the heat reservoirs from the heat fluxes in and out of the heat engine. In the first case, the dissipative heating takes place near the cold sink of the engine, which is represented by the upper arrow labeled "D" in Fig. 4.1b. Dissipative heating would then simply add to the waste heat flux, so that $J_{out} + D$ is added to the cold reservoir. The inclusion of the dissipative heating leads to balanced heating and cooling of the heat engine with $J_{in} = J_{out} + D$. Yet, the dissipative heating would not affect the heat flux into the engine, so that the efficiency of the heat engine is unchanged. In the second case, the dissipative heating takes place near the hot reservoir, as indicated by the lower arrow labeled "D" in Fig. 4.1b. This heating adds to the heat flux into the engine, so that the engine is heated by $J_{\text{in}} + D$ and the steady state of the heat engine is represented by $J_{\text{in}} + D = J_{\text{out}}$.

For this setup of a dissipative heat engine, the first and second law are described by

$$J_{\text{in}} + D = J_{\text{out}} + G \tag{4.23}$$

and

$$0 = \frac{J_{\text{in}}}{T_h} - \frac{J_{\text{out}}}{T_c} + \frac{D}{T_h} \tag{4.24}$$

Noting that $J_{\text{in}} = J_{\text{out}}$ and $G = D$ in steady state, the Carnot limit of the dissipative heat engine, G_{diss}, is directly derived from the entropy budget and is given by

$$G_{\text{diss}} = J_{\text{in}} \frac{T_h - T_c}{T_c} \tag{4.25}$$

Note that this expression is almost identical with the Carnot limit (eqn. 4.4), except that it uses the temperature of the cold reservoir T_c, instead of the hot reservoir T_h, in the denominator. Because $T_c < T_h$, this limit is higher than the Carnot limit by a factor of $G_{\text{diss}}/G_{\text{Carnot}} = T_h/T_c > 1$, so that the efficiency, η_{diss}, of the dissipative heat engine is higher than the Carnot efficiency

$$\eta_{\text{diss}} = \frac{T_h - T_c}{T_c} > \eta_{\text{Carnot}} \tag{4.26}$$

Using conditions of the Earth's surface–atmosphere system, this efficiency is $\eta_{\text{diss}} = 33/255 = 12.9\%$, which is about 13% greater than the Carnot efficiency of 11.5%.

When we look at the entropy balance of the system, we can easily see that the limit of the dissipative heat engine is entirely consistent with the second law of thermodynamics. The entropy production by dissipation, $\sigma_{\text{diss}} = D/T_h$ is given by

$$\sigma_{\text{diss}} = J_{\text{in}} \frac{T_h - T_c}{T_h T_c} = \frac{G_{\text{Carnot}}}{T_c} = \frac{G_{\text{diss}}}{T_h} \tag{4.27}$$

The first expression on the right hand side is the typical expression for the entropy production that we encountered before and is set by the boundary condition of the system. The second expression divides this entropy production into the Carnot limit of the heat engine G_{Carnot}, and assumes that the generated free energy is dissipated at T_c, while the last expression divides the terms differently into the limit of the dissipative heat engine G_{diss}, which dissipates the energy at T_h. Hence, both limits are entirely consistent with the entropy balance, yet the latter case of the dissipative heat engine generates somewhat more power.

4.5 The maximum power limit

So far, we have evaluated the thermodynamic limit to the work a heat engine can perform with fixed boundary conditions. The derivations of the previous sections prescribed the heat flux and the temperature difference to which the heat engine is subjected to. However, heat engines in the Earth system do not operate in isolation, but alter the boundary conditions. When kinetic energy is generated out of radiative heating differences, the resulting motion transports the heat that constitutes the heat flux into the heat engine, but the heat flux also acts to deplete the driving temperature difference. When we want to apply the heat engine concept to Earth system processes, this effect on the driving difference needs to be included in the treatment to obtain an adequate description of how heat engines operate inside the Earth system. This can easily be accomplished by slightly extending the boundary of our system to include the processes that shape the temperature difference in our considerations.

To include this effect on the boundary conditions, we consider a slightly different setup that is shown by the system boundary labeled "2" in Fig. 4.1b. We again consider a heat engine that is placed between two heat reservoirs, but we also include the energy balances that describe the heat reservoirs and thus the temperature difference, $T_h - T_c$. The warmer reservoir is heated by a continuous heat flux R_{in}, for instance by absorption of solar radiation, while the cold reservoir is cooled by a heat flux R_{out}, e.g., by emission of radiation. We also consider another process that transfers energy at a rate R_r between the two reservoirs that does not involve the heat engine and that is not associated with work being performed. This can, for instance, represent diffusive transport of heat or, as we will see in the following chapters, radiative exchange between the reservoirs. The reservoirs are considered to be in a steady state, in which their heating and cooling rates balance so that $R_{\text{in}} = R_{\text{out}}$ and in which we can neglect temporal changes in temperature.

To evaluate the limit in energy conversion in this setup, we start by formulating the energy balances of the reservoirs. The balance for the hot reservoir in steady state is described by

$$0 = R_{\text{in}} - R_r - J_{\text{in}} \tag{4.28}$$

and

$$0 = R_r + J_{\text{out}} - R_{\text{out}} \tag{4.29}$$

for the cold reservoir, where J_{in} is the heat flux that enters the heat engine (with $J_{\text{in}} = J_{\text{out}}$). The energy exchange by R_r is expressed here for simplicity as a linear function of the temperature difference between the reservoirs, with $R_r = k_r(T_h - T_c)$

and k_r being an effective conductivity. Such a linear expression is typically adequate for describing a diffusive process and also serves as a first order approximation to radiative exchange. With this expression for R_r, the energy balances yield an expression for the temperature difference, $T_h - T_c$, between the reservoirs which depends on the heating rate R_{in}, and also on the heat flux, J_{in}, through the heat engine

$$T_h - T_c = \frac{R_{in} - J_{in}}{k_r} \tag{4.30}$$

Note how an increasing heat flux J_{in} through the heat engine results in a reduction of the temperature difference between the reservoirs. In fact, the temperature difference vanishes when the heat flux J_{in} approaches the external heating rate of the hot reservoir, $J_{in} = R_{in}$. Since the Carnot limit depends on both the heat flux and the temperature difference, there is a particular value of the heat flux at which the power of the heat engine is at a maximum. This maximum is derived by combining the expression of the Carnot limit of the dissipative heat engine (eqn. 4.25) with the expression of the temperature difference obtained from the energy balances (eqn. 4.30):

$$G = J_{in} \frac{T_h - T_c}{T_c} = J_{in} \frac{R_{in} - J_{in}}{k_r T_c} \tag{4.31}$$

Alternatively, the Carnot limit given by eqn. 4.4 could be used as well (depending on where the dissipation takes place within the system), which would result in a different temperature T_h in the denominator and would yield a slightly lower limit. The maximum of this expression is derived by $\partial G/\partial J_{in} = 0$. This maximization yields an optimum value for J_{in} of:

$$J_{opt} = \frac{R_{in}}{2} \tag{4.32}$$

which we refer to as an optimum heat flux J_{opt}. The associated maximum in power, G_{max}, is given by:

$$G_{max} = \frac{R_{in}^2}{4k_r T_c} \tag{4.33}$$

and the associated temperature difference is given by

$$\Delta T = T_h - T_c = \frac{R_{in}}{2k_r} \tag{4.34}$$

At the limit of maximum power, the temperature difference is reduced to half its maximum value when $J_{in} = 0$. This results in a lower temperature of the hot reservoir, which is the result of convective cooling, an overall lower thermal energy content, and a lower entropy of the whole system.

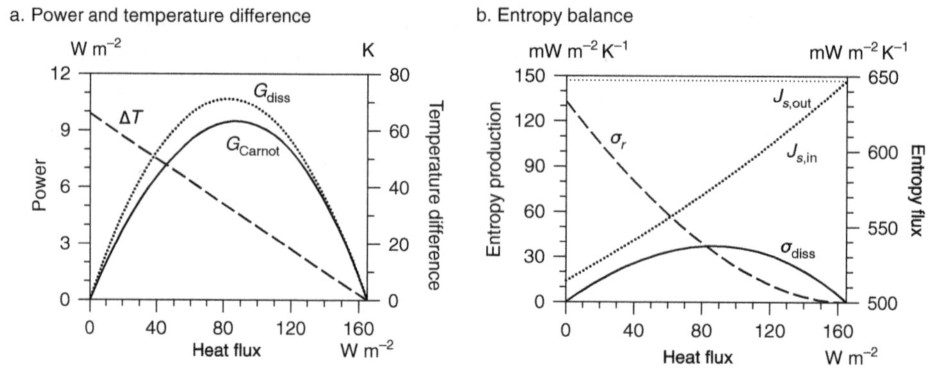

Figure 4.5 The sensitivity of (a) the Carnot limit (G_{Carnot}, solid line and G_{diss}, dotted line) and the temperature difference between the reservoirs, $\Delta T = T_h - T_c$, and (b) of entropy production and entropy exchange to the heat flux, J_{in}, that flows through the heat engine. The components of the entropy budget are given with $J_{s,\text{in}}$ and $J_{s,\text{out}}$ being the entropy fluxes in and out of the system, σ_{diss} being the entropy production by dissipation of the free energy generated by the engine, and σ_r being the entropy production by the competing heat flux R_r. Conditions similar to the global mean atmosphere were used, with $R_{\text{in}} = 165$ W m^{-2}, $k_r = 2.5$ W m^{-2} K^{-1}, and $T_a = 255$ K.

Note that the maximum power limit does not represent a maximum in efficiency, as expressed by the Carnot efficiency. This maximum in efficiency is represented by the case in which the temperature difference is the greatest, which is the case when $J_{\text{in}} = 0$. Since ΔT is reduced to half its maximum value, the Carnot efficiency at maximum power is also reduced to half its maximum value.

The maximum power limit results from a strong interaction between the heat flux through the heat engine and the temperature difference between the reservoirs. As shown in Fig. 4.5a, the power first increases linearly for small values of J_{in}, with the slope given by the Carnot efficiency. This increase in J_{in} is accomplished by a corresponding decrease in the flux R_r, because the sum is fixed at R_{in} by the energy balance constraint. Since R_r is also driven by the temperature difference with a fixed conductivity k_r, a decrease in R_r can only be accomplished by a reduced temperature difference ΔT. This reduction in the temperature difference then reduces the efficiency, and thus reduces the slope by which power increases with an increasing heat flux up to the maximum. This line of reasoning applies to both kinds of Carnot limits discussed earlier, in the common form as well as the limit of a dissipative heat engine (both are shown in Fig. 4.5a for comparison). The maximum in power thus ultimately originates from the presence of another, unavoidable heat flux that is described by R_r that is also driven by the temperature difference. If this alternative flux were absent, it would not result in a maximum power limit. The steady state would then require $R_{\text{in}} = J_{\text{in}}$ and we would need additional information to specify the temperature difference.

This sensitivity to the heat flux entering the heat engine strongly alters the entropy budget of the system. The components of this budget are shown in Fig. 4.5b. The steady-state entropy budget consists of three components: (i) the net entropy exchange, consisting of an influx of entropy, $J_{s,\text{in}} = R_{\text{in}}/T_h$, associated with the heating rate of the hot reservoir, and the entropy export, $J_{s,\text{out}} = R_{\text{out}}/T_c$, associated with the cooling rate of the cold reservoir, (ii) the entropy production, σ_{diss}, associated with the dissipation of free energy generated by the heat engine, and (iii) the entropy production of the other heat flux R_r. In steady state, these components balance, so that the net entropy exchange balances the entropy production within the system, $J_{s,\text{out}} - J_{s,\text{in}} = \sigma_{\text{diss}} + \sigma_r$. The sensitivity of the entropy budget to J_{in} shows two major trends: the first trend reflects the shift in the heat fluxes from R_r to J_{in} that results in a decrease in σ_r and an increase in σ_{diss}. The second trend relates to a reduced temperature of the hot reservoir T_h. As the hot reservoir is cooled more effectively in the presence of J_{in}, resulting in a reduced temperature difference ΔT and a lower temperature T_h, the entropy influx into the system increases with greater values of J_{in}. As a consequence, the net entropy export decreases with greater values of J_{in}, and thus the sum $\sigma_{\text{diss}} + \sigma_r$ is reduced as well. These two trends taken together result in a maximum in the entropy production of σ_{diss} that is closely related to the maximum power state. This direct correspondence in the maxima can be seen in the entropy production associated with the performance of work, as described by eqn. 4.27. Hence, the performance of work within the system alters the thermodynamic state of the system, as well as its entropy exchange. This is a very important aspect as it shows how energy conversions and interactions result in system-wide consequences. This aspect is explored further at the end of this book in Chapter 12.

There are two general aspects that result in the maximum power limit. The first aspect is that in the derivation of the limit, another dissipative process (the flux R_r) was included that unavoidably also depletes the temperature difference. The second aspect is that the interaction of the heat flux through the engine with the driving temperature gradient is included in the derivation. The maximum power limit thus emerges from a broader systems perspective that does not just focus on the energy conversion process, but also includes a dissipative loss process as well as the consequences of the energy conversion on the driving difference of the heat engine. These two aspects are rather general; hence, the maximum power limit should be rather general and should be applicable to many forms of energy conversions within the Earth system.

4.6 Maximum power limits for non-thermal energy conversions

The maximum power limit of the heat engine just described forms the first step in the Earth system that converts heating differences into forms of free energy that are then converted further. For these further conversions of non-thermal energy, forms of the maximum power limit should apply as well. What makes the maximum power limit rather general and applicable to energy conversions beyond heat is that at its center is a general trade-off between a flux and a gradient, where a greater flux results in a weaker gradient. This trade-off is applicable to all forms of energy that were described in Chapter 2, because this trade-off relates to the set of conjugate variables that describes the different forms of energy (also shown in Table 2.1). Converting a certain form of energy is associated with a flux in the extensive variable, which is driven by a gradient in the intensive variable. The conversion of thermal energy, for instance, involves heat fluxes that are driven by temperature gradients. Kinetic energy conversions involve momentum fluxes, better known as forces, that depend on velocity gradients, while potential energy conversions involve mass fluxes that are driven by geopotential gradients. The effect of the fluxes on the driving gradients is then a consequence of the associated conservation laws. The momentum flux that converts kinetic energy is subjected to momentum conservation, while a mass flux is subjected to mass conservation. When a system maintains fluxes and energy conversions in a steady state, the associated conservation laws demand that the driving gradient is reduced when a higher flux is being maintained. Because of this trade-off, the maximum power limit can apply to other energy conversions beyond thermal energy. Strictly speaking, this is then no longer a thermodynamic limit as it does not involve considerations of entropy directly. Nevertheless, as this conversion relates to energy that was originally generated by some form of thermodynamic engine and since it involves thermal dissipation at some stage as well, it indirectly relates to thermodynamic limits. In the following, two examples of a maximum power limit are shown for rather different forms of energy conversions involving electric energy and kinetic energy.

The maximum power limit has long been known in electrical engineering, where it is known as the maximum power transfer theorem (e.g., Odum and Pinkerton 1955). It applies to an electric circuit in which power is transferred from an electrical power source, such as a generator, to a device (a "load") that is attached to it. The power that the device can draw from the power source is less than what is generated by the source because the power source has an internal resistance that dissipates some of the generated power (see Fig. 4.6a). The power generated by the power source G, is thus converted into dissipative heating by the internal resistance, D_i, and converted into another form by the load at a rate G_{load}.

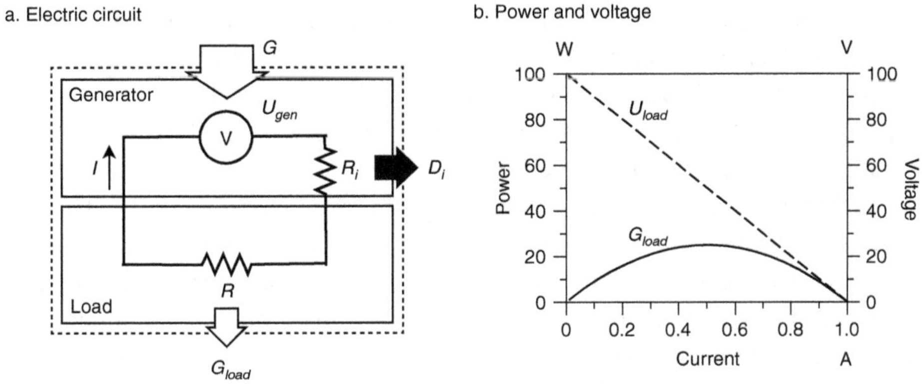

Figure 4.6 The maximum power limit in an electric circuit. (a) A simple electric circuit consisting of a voltage source (a generator with a voltage U_{gen}) with an internal resistance R_i, and a load with resistance R that is attached to the source and transfers the power G_{load}, to another process. (b) Sensitivity of the voltage drop across the load U_{load}, and the transferred power G_{load}, as a function of the current in the circuit I. The sensitivity was created by varying the resistance R, of the load. The maximum power is transferred when the resistance of the load matches the internal resistance of the generator ($R = R_i$). Values of $U_{\mathrm{gen}} = 100$ V and $R_i = 100\ \Omega$ were used to obtain the plot.

To derive the maximum rate by which G is converted into G_{load}, we consider a generator that is made up of a constant voltage source, U_{gen} with an internal resistance R_i, and a load with a resistance R (Fig. 4.6a). The power transferred to the load is a function of the resistance R_{load}. This power is described by the expression for dissipation of a resistance, $G_{\mathrm{load}} = D = I\Delta U$, where ΔU is the voltage drop across the resistance. To express it in terms of the resistance R, we note that the voltage drop is given by Ohm's law, $\Delta U = RI$. The current I is inferred by applying Ohm's law to the whole circuit, $U_{\mathrm{gen}} = (R_i + R)I$, which yields $I = U_{\mathrm{gen}}/(R_i + R)$. Taken together, these expressions show that the power transferred to the load is $G_{\mathrm{load}} = R/(R_i + R)^2 U_{\mathrm{gen}}^2$. The expression reaches a maximum value, $\partial G_{\mathrm{load}}/\partial R = 0$, when $R = R_i$, that is, when the resistance of the load matches the internal resistance of the generator (Fig. 4.6b). This is a well-known result (Odum and Pinkerton 1955).

The example given here mirrors very closely what we have found in the derivation of the maximum power limit earlier. The correspondence is easy to note when we look at the conjugate variables that describe electric energy and heat. The conjugate variables of electric energy are the charge Q and the voltage U, while for heat these are entropy S and temperature T. The current I of the circuit is a flux of charge, that is, $I = dQ/dt$, while the heat engine is driven by a heat flux J. The voltage difference that causes the current corresponds to the temperature difference that causes the heat flux. To compare how the maximum power states were accomplished, it is easier to see when the heat flux through the heat engine is expressed in terms of a temperature difference and a conductivity k (i.e., $J = k(T_s - T_a)$), with conductances being the inverses of resistances. With this formulation, the maximum power limit for the heat engine corresponds to the state in which $k = k_r$, where k_r is the effective conductivity of the competing process. This result is equivalent to the case of the electrical network. Technically speaking, a direct equivalence can be made when an electric circuit with a constant current source is considered in which the resistances are attached in parallel. A maximum power state for such a circuit is also achieved when the resistances, or the conductances, are equal. While we will not deal with electric energy further in this book, the image of an electric circuit and its maximum power limit can be handy to conceptualize processes and their limits to energy conversions.

We can also identify a maximum power limit in the conversion of other, non-thermal forms of energy that are more relevant to the dynamics of Earth system processes. Fig. 4.7 shows an example of a simple system in which kinetic energy is converted to some other form (taken from Miller et al. (2011a) and Gans et al. (2012)). The system is characterized by two reservoirs of momentum, with one of the reservoirs being maintained in disequilibrium by an accelerating force F_{acc}. For simplicity, let us assume that the other reservoir is at rest. We can think of this

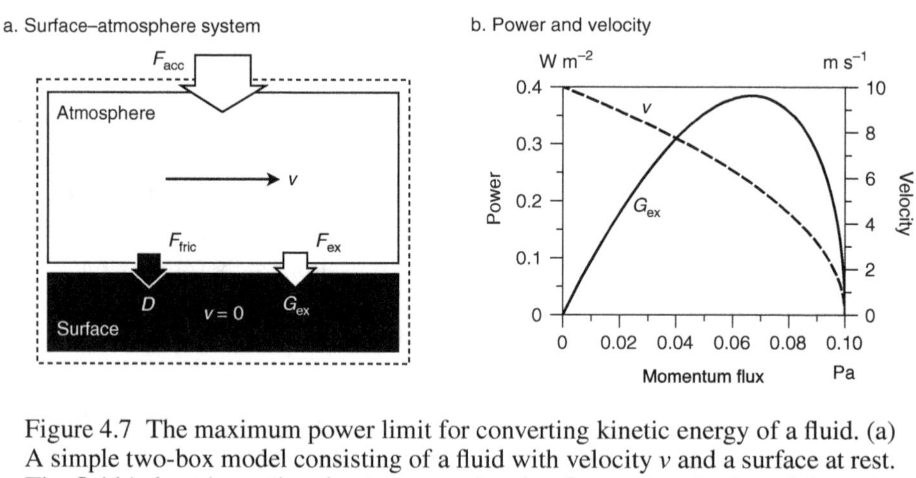

Figure 4.7 The maximum power limit for converting kinetic energy of a fluid. (a) A simple two-box model consisting of a fluid with velocity v and a surface at rest. The fluid is kept in motion due to an accelerating force, F_{acc}, is slowed down by friction F_{fric}, and an additional force F_{ex} extracts kinetic energy from the flow to convert it into another form. These forces are associated with the generation of kinetic energy at a rate G_{ke}, frictional dissipation D, and conversion into another form of energy G_{ex}. (b) Sensitivity of the velocity of the fluid v, and the extracted power G_{ex}, to the momentum flux, or force F_{ex}. Values of $F_{acc} = 0.1$ N m^{-2} (or Pa) and $k = 10^{-3}$ kg m^{-3} were used to generate the plot.

system as the atmospheric boundary layer being one of the reservoirs, and the surface being the other. Friction depletes some of the momentum difference between the reservoirs. Turbulent friction is typically expressed as a quadratic function of wind speed, so that the friction force is expressed as $F_{fric} = k\,v^2$ with k being a frictional parameter (which is essentially a drag coefficient, as described in greater detail in Section 7.3). Another force, F_{ex}, is present to convert some of the free energy of the system into other form. This can be a natural process, such as the lifting of dust particles into the atmosphere, or this force can be associated with the generation of wind power by a wind farm. The starting point for deriving the maximum power limit in this case is the momentum balance of the upper reservoir in steady state:

$$F_{acc} - F_{fric} - F_{ex} = 0 \tag{4.35}$$

Using the assumed relationship of F_{fric}, we can use the momentum balance to obtain an expression for the velocity:

$$v = \left(\frac{F_{acc} - F_{ex}}{k}\right)^{1/2} \tag{4.36}$$

The power associated with converting kinetic energy into another form G_{ex}, has then the form

$$G_{ex} = F_{ex} v = F_{ex}\left(\frac{F_{acc} - F_{ex}}{k}\right)^{1/2} \tag{4.37}$$

This expression has a well-defined maximum (cf. Fig. 4.7b), which is obtained by $\partial G_{ex}/\partial F_{ex} = 0$. The analytic expressions for the optimum values of the force $F_{ex,opt}$, the associated velocity v_{opt}, and the maximum power $G_{ex,max}$, are given by

$$F_{ex,opt} = \frac{2}{3} F_{acc} \tag{4.38}$$

$$v_{opt} = \left(\frac{F_{acc}}{3k}\right)^{1/2} \tag{4.39}$$

$$G_{ex,max} = \frac{2}{3^{3/2}} \frac{F_{acc}^{3/2}}{k^{1/2}} \approx 0.38\, G_0 \tag{4.40}$$

where G_0 is the original power that enters the system in the absence of F_{ex} and with eqn. 4.36 given by $G_0 = k^{-1/2} \cdot F_{acc}^{3/2}$. The maximum in power results from the same trade-off between flux and difference as in the case of a heat engine. The flux corresponds to the force, which is a flux of momentum, while the difference is represented by the difference in velocity between the two reservoirs. Hence, only a fraction of the kinetic energy can be converted into another form. Yet the example also shows that because of the difference in the functional form of the flux as the momentum flux F_{fric} depends quadratically on the gradient, the resulting optimum flux $F_{ex,opt}$ is not simply half of the forcing flux, F_{acc}. As we will see in Chapter 11 towards the end of this book, this example relates to how much kinetic energy can be converted by wind turbines at large scales and used as renewable energy.

Both examples illustrate that maximum power limits can apply to various forms of energy conversions. A critical component that results in the maximum and that prevents all of the energy to be converted relates to the fact that there are "competing" or "alternative" processes involved in the energy conversion process that results in a fraction of the energy being dissipated into thermal energy. For the conversion of the electric energy of the generator, this competing process is the dissipation at the internal resistance of the generator, while in the example for kinetic energy, it is the frictional dissipation. In both cases, this dissipative loss can be reduced, but not fully avoided. This limits the ability to convert energy unavoidably and quite substantially. If this "competing" process were not present, then the steady state of the system would be fixed by the flux through the boundary condition as this flux determines the flux within the system. Then, there would be no trade-off between flux and force, and the rate of energy conversion within the system would be determined by the boundary condition. Yet, such alternative processes are a very common feature for Earth system processes; hence the maximum power limit should apply to a wide range of processes that convert energy.

4.7 Maximum entropy production (MEP) and other thermodynamic principles

The maximum power limit relates very closely to the proposed principle of maximum entropy production (MEP). The MEP principle states that thermodynamic non-equilibrium systems with sufficient degrees of freedom adapt steady states in which their rate of entropy production is maximized subject to relevant constraints. The roots of MEP date back to the 1970s to the work of Paltridge, who applied this formulation to the Earth's atmospheric heat transport (1975; 1978; 1979). It has recently gained renewed interest (Ozawa et al. 2003; Kleidon and Lorenz 2005; Martyushev and Seleznev 2006; Kleidon et al. 2010; Dewar et al. 2014), partly due to some theoretical advances (Dewar 2003, 2005a,b, 2010), and partly due to some rather successful applications to Earth and planetary systems (e.g., Ozawa and Ohmura 1997; Ozawa et al. 2001; Lorenz et al. 2001; Kleidon et al. 2003; Wang et al. 2004; Kleidon et al. 2006; Wang et al. 2007; Ou 2001, 2006, 2007). The MEP principle has also received substantial criticisms, mostly due to its ambiguity as to which entropy production ought to be maximized (Volk and Pauluis 2010) as there are different gradients that result in entropy production when dissipated (as in Fig. 3.6), and also regarding the mechanisms which would accomplish this maximization (Goody 2007). These critiques can be addressed by being more specific about the processes involved in generating entropy, and by formulating the dynamics in terms of feedbacks that shed light on the mechanisms by which such a maximization could be accomplished.

The close linkage between the maximum power limit and a state of MEP can be seen in Fig. 4.5 by comparing the maximum power with the maximum entropy production associated with the dissipation of the free energy generated by the heat engine. Alternatively, we can look at eqn. 4.27, which directly relates entropy production to the power of the heat engine. The difference between a state of MEP and the maximum power limit only relates to the temperature in the denominator and can hardly be noticed. The same trade-off is at play between a force and a flux that results in a maximum power state and in states of maximum entropy production. Hence, there is little difference in the outcome between a state of MEP and the state associated with the maximum power limit.

The perspective taken in this book is that even if the overall system maximizes entropy production to the extent possible – and we will get back to such an evaluation towards the end of the book –, the focus on power and associated conversions of energy is advantageous, because it is a more familiar concept and is more specific about the dynamics, the interactions, and feedbacks within the system. Furthermore, the formulation of the maximum power limit allows us to separate a thermodynamic limit from the question whether a system would evolve to and operate at this limit. The MEP principle combines these two aspects – it formulates a limit to the entropy production of a process with the hypothesis that the system would operate at this limit. The separation between the thermodynamic limit of a system and the hypothesis that the system evolves to and eventually operates at this limit may seem somewhat formal. The advantage of this separation is that thermodynamic limits are very well established, while the hypothesis that systems operate at their thermodynamic limit certainly does not apply to all systems so that the conditions when this hypothesis is applicable need to be carefully formulated. It is a rather general and highly relevant evolutionary hypothesis that was already formulated by Alfred Lotka (1922a, b). We will evaluate this hypothesis in the next chapter that deals with a general formulation of dynamics and feedbacks.

The focus on the limits of energy conversions rather than entropy production also allows us to evaluate the mechanisms that are involved in greater detail. Entropy production can result either immediately from dissipation, e.g., by diffusion, or through the combination of free energy generation and its subsequent dissipation. By relating the generation, dissipation, as well as the free energy involved to the processes that operate within a system, we gain a mechanistic understanding of where and why entropy is being produced. This more detailed view helps us to identify to which process the thermodynamic limit should be applied. In the example in Fig. 4.5 for instance, it requires mechanical power to generate motion and convective heat transport, while the competing process, thermal diffusion or radiative transfer, does not involve motion and thus does not require mechanical power. This justifies the application of the maximum power limit to convection. The focus on thermodynamic limits on energy conversions is not in contradiction to the proposed MEP principle, but rather a more refined view on the detailed processes in a system that are involved in shaping the thermodynamic state of a system. In the following chapters, applications of MEP are sometimes referred to as support for a process to operate at maximum power, because of this focus and the close relationship between these two limits.

The description of thermodynamic extremum principles would be incomplete without mentioning at least two other thermodynamic limits or principles: Prigogine's theorem of minimum entropy production (MinEP) (Prigogine 1947) and the thermodynamic limit of finite-time thermodynamics (Novikov 1958; Curzon and Ahlborn 1975; Andresen et al. 1977).

Prigogine's theorem states that a system in its steady state is characterized by a minimum in entropy production compared to the system's transient behavior. Typical examples that are given to illustrate this theorem is heat diffusion (see e.g., Kondepudi and Prigogine 1998). This trend can also be seen in the examples of the last chapter, for instance shown in Fig. 3.4. It shows how the entropy production declines from its initial state to a steady-state value. Note that the MinEP theorem is not contrary to maximum power or MEP, because it describes transient behavior, while maximum power and MEP apply to the system in a steady state. In a steady state, however, the thermodynamic rate of entropy production is already determined by the entropy exchange at the system's boundary, so that the minimization of entropy production does not provide additional information.

Another statement of minimum entropy production can be found in the engineering literature (Bejan 1996). The minimization of entropy production is described, for instance, as an engineering method for power plants and other devices to minimize dissipative losses so that the power output of the device is maximized. The Carnot limit can be seen as the ideal outcome of such a design, because it represents the case in which there is no internal entropy production (i.e., $\sigma = 0$ in eqn. 4.3). In other words, the minimization of entropy production internal to the heat engine merely uses a different perspective on which aspect is optimized, and thus also focuses on maximization of power in an engineering context.

Closely related to engineering is the field of so-called finite time thermodynamics and the associated Curzon–Ahlborn limit (Novikov 1958; Curzon and Ahlborn 1975; Andresen et al. 1977, 1984), where the term "finite time" refers to the Carnot cycle being passed in a finite amount of time. This limit is similar to the Carnot limit, except that it also takes into account dissipative losses that are associated with conductive heat transfer across the boundary of the system. The engine is then referred to as an endoreversible engine. Since entropy production is associated with conductive heat transfer, the resulting thermodynamic limit is less than the Carnot limit, with the maximum efficiency given by $\eta = 1 - \sqrt{T_c/T_h}$. Finite-time thermodynamics is also often motivated by the common interpretation of the Carnot limit as taking infinitely long time to complete a thermodynamic cycle to avoid irreversible losses. However, as is shown at the beginning of this chapter, the Carnot limit can be derived in general form directly from the first and second laws in steady-state conditions without a specific cycle of operation in mind. The natural systems that are examined in this book are not necessarily subjected to conductive heat transfer at the system boundary (for instance, when water vapor condenses inside the atmosphere to generate moist convection); hence they will certainly be subjected to the Carnot limit, but not necessarily to the limit of finite time thermodynamics.

4.8 Thermodynamic limits in the Earth system

In this chapter, we described a sequence of thermodynamic limits that originate directly from the first and second laws of thermodynamics. The resulting limits – from the classical Carnot limit to the dissipative heat engine to the maximum power limit – only differed by how the system was being defined and which effects were included in the derivation, as was illustrated in Fig. 4.1. The maximum power limit was demonstrated for a heat engine by the combination of the Carnot limit with the inclusion of the energy balance constraint. The other examples in Section 4.6 inferred similar limits for other forms of energy that followed the same template. The derivation of these limits was done in a highly simplified way for steady-state conditions. As we will see in later chapters, the actual state may depend on other factors, such as temporal variations; yet, the main mechanism that results in a maximum power limit is unaffected by such refinements.

What is at the core of these limits is that a flux of some physical property is driven by a gradient, yet the flux also depletes the driving gradient. This is what we would expect from the second law of thermodynamics. The associated rate of energy conversion is represented by the combination of the flux and the gradient, as these represent a pair of conjugate variables of the form of energy that is being converted (cf. Table 2.1). Conservation laws related to the extensive variable then link the magnitude of the flux to the depletion of the driving gradient. This general template in combination with a competing process that depletes the gradient without generating a form of free energy then yields the maximum power limit.

Hence, the maximum power limit is a very general thermodynamic limit that results from the combination of the second law with conservation laws. Since the limit is set by the depletion of the driving gradient, it reflects a strong interaction between the process and its driver and thus alters the thermodynamic state of the system.

In the following chapters, we will use this general template and apply it to several Earth system processes. To fill the general template, we need to fill in the necessary details of the processes that are being considered. These details include a description of the forcing and the boundary conditions, a minimum description of the energy conversion processes that yield different forms of energy and describe the associated dynamics, and, rather importantly, the feedbacks of the energy conversion on the forcing and the boundary conditions of the process that is considered. This feedback is central in setting the maximum power limit, as discussed earlier in general terms.

The consequence of this feedback is rather profound and ultimately manifests itself in the thermodynamic state of the planetary system. Recall the chain of energy conversions that we discussed at the end of the last chapter in Fig. 3.8, which link the dynamics of a process all the way to the planetary forcing. When we now add the maximum power limit to this chain of conversions, it implies that when this process operates at this maximum power limit, it interacts strongly with its boundary conditions, and the effects would feed back all the way to the planetary boundary conditions. The planetary boundary conditions of Earth are not fixed, but are affected rather strongly by the internal dynamics. Clouds, ice, and other internal processes affect how much of the incident solar radiation is being absorbed by the Earth system and thus affects the influx into the system that acts as the main driver for energy conversions. Heat redistribution by planetary motion and greenhouse gases in the atmosphere affect the temperature with which the absorbed radiative energy is being exported to space. Both aspects taken together imply that the planetary boundary conditions of the Earth system exhibit a great extent of flexibility in terms of the planetary exchange of radiative energy and entropy. Through these chains of energy conversions and the associated feedbacks, the internal dynamics of the Earth system strongly affect this planetary exchange. The coupled dynamics of the system thereby affect the thermodynamic state of the planet. We will get back to this aspect towards the end of the book after we have dealt with the main processes of the Earth system.

So far, we have derived here thermodynamic limits on energy conversions but we have not dealt with the issue if, why, and how processes actually would achieve this limit. This issue relates closely to the differentiation that we discussed earlier in the context of the proposed MEP principle where we separated the thermodynamic limit from the hypothesis that processes evolve towards their limit. Before we apply thermodynamic limits to Earth system processes, we will look at such evolutionary dynamics in relation to the maximum power limit in the next chapter.

5

Dynamics, structures, and maximization

5.1 Energy conversions, maximization, and evolution

So far, we formulated Earth system processes in terms of the different forms of energy that these involve and identified fundamental limits to the conversion of one form of energy into another. We have not yet discussed why processes should evolve to thermodynamic limits and how this evolution should take place. The goal of this chapter is to explore exactly these questions of why, how, and to what extent systems should be expected to evolve towards their thermodynamic limits.

At the core of these questions are the dynamics that take place within a system. The term dynamics is used here in a general sense on a thermodynamic system in disequilibrium that maintains fluxes that are directed towards depleting this thermodynamic disequilibrium. The extent of dynamics can thus be measured by how much free energy is being generated and dissipated. The motivation of this definition of dynamics is that a system in thermodynamic equilibrium would not show macroscopic dynamics. In contrast, a system with free energy generation can sustain fluxes and macroscopic changes taking place inside the system. As these fluxes and changes involve energy, mass, momentum, and other variables that are represented in the conjugated pairs of variables described in the earlier chapters, these dynamics relate to conversions of different forms of energy. We can thus view dynamics as the consequence of how a certain form of energy is generated and dissipated. The term evolutionary dynamics is then used here to characterize dynamics that go beyond energy, mass, and momentum balances and are specifically characterized by a change in free energy generation, dissipation, and the thermodynamic state of the system.

A general template for the conversions involved in a particular form of energy is shown in Fig. 5.1. At the top of the diagram is a driving gradient, for instance, a temperature difference, from which a certain form of free energy is generated. This form of free energy is then represented by a gradient in another variable.

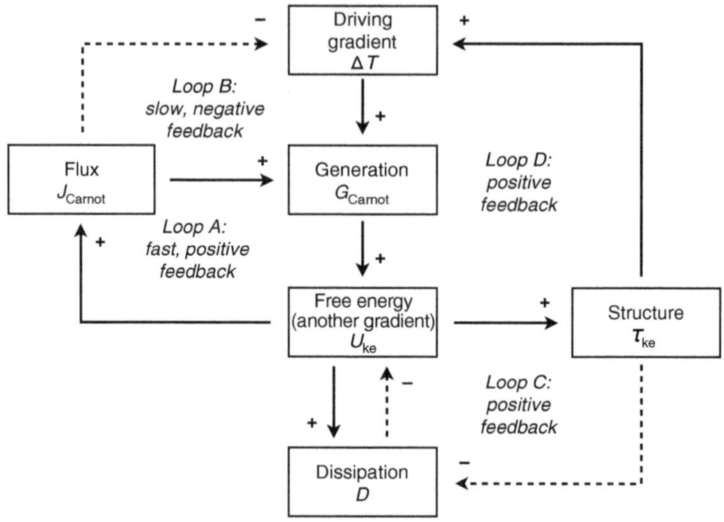

Figure 5.1 A feedback diagram to illustrate the feedbacks involved in shaping the maximum power limit, using the variables of the example described in this chapter. The relation to structure development in the system and its role in modulating dissipation is also shown. After Kleidon et al. (2013).

For kinetic energy, this gradient is represented by velocity differences within the fluid and its surroundings. Ultimately, this form of free energy is dissipated into thermal energy, with a greater amount of free energy being typically associated with a greater dissipation rate. For simplicity, conversions into other forms of energy during dissipation are not included here. An example for this sequence is the heat engine described in the previous chapter. Differential heating of a system generates a temperature difference, from which work can be performed and another form of energy such as kinetic energy can be generated. Frictional dissipation would then convert this energy into thermal energy, but without re-establishing the original difference.

The question why this energy conversion sequence should evolve to its thermodynamic limit relates to the effects associated with the energy conversions. The free energy in the system maintains the flux that is involved in the generation rate of this free energy. Kinetic energy, for instance, directly maintains the heat transport by the fluid. This heat transport has two effects: it affects the generation rate, because greater motion results in a greater heat flux and power, and it acts to deplete the driving gradient, because a greater heat flux results in a greater depletion of the temperature difference. We have encountered both of these effects before in the derivation of the maximum power limit: the Carnot limit depends on the heat flux through the engine, so that the work performed by the engine to generate free energy depends on the heat flux. The enhanced depletion of the driving gradient by the heat flux was the reason why we obtained the maximum power limit from the Carnot limit. To understand the dynamics towards the maximum power limit, we look at these two effects from a somewhat different perspective in that we formulate these effects in terms of feedbacks.

Feedbacks describe how the state of a system reacts to a perturbation and whether this reaction amplifies or dampens the system's response to the perturbation. Feedbacks are commonly used to describe the response of the climate system to change (e.g., Hansen et al. 1984). A feedback loop is formed by a sequence of effects. For example, one feedback loop (Loop A in Fig. 5.1) is formed by the following sequence: a greater generation rate results in a greater amount of free energy, which results in a greater flux, which should then result in a greater generation rate as the Carnot limit depends on the flux. Each effect is characterized by a positive relationship between the variables, so that a plus sign is attached to the arrows. Technically, the sign of the influence depends on the sign of the partial derivative of one variable to another, an aspect we deal with in greater detail later. This example for a feedback loop results in an amplification of the initial perturbation, thus forming a positive feedback. If a response results in a dampening of the perturbation, it forms a negative feedback loop. Feedbacks relate to the maximum power limit because the limit can be seen as the result of two contrary feedbacks, labeled by loops A and B in Fig. 5.1: the first feedback (loop A) is a feedback that enhances the dynamics within the system, which is achieved by enhancing the power associated with the driving process and which we refer to as a "power-enhancing" feedback. The second feedback (loop B) shown in Fig. 5.1 deals with the depletion of the driving difference by the flux which causes the dynamics to slow down and forms a "gradient-depletion" feedback. This feedback is a negative feedback and limits the power within the system and thus the dynamics. As we will see, these two feedbacks taken together can be seen as a prerequisite for the evolutionary dynamics towards the maximum power limit.

While feedbacks describe the dynamical changes within a system, they do not answer the question how a system is actually organized to achieve the maximum power limit. This organization is related to the spatial organization of the flow in the form of structures, such as convection cells, turbulent eddies, or network structures. These structures are characterized by non-uniform spatial and temporal variations of the flow. They can be viewed as "dissipative structures," a term introduced by Prigogine (1962) to refer to those structures that are characterized by strong thermodynamic disequilibrium (see also Kondepudi and Prigogine (1998) for examples, and also alternative approaches, e.g., Haken (1975)). Here, we explore the role of such structures in how they affect the sequence of energy conversion shown in Fig. 5.1. It is argued here that it is through these structures that systems evolve to and maintain a state at their thermodynamic limit. The role of these structures is twofold: first, by structurally organizing the flow, they reduce frictional losses within the system (loop C), rather than enhancing the power directly (which is represented by loop A), and second, they can locally enhance the driving gradient (loop D), particularly at the system boundary and thus the exchange fluxes at the boundary. Both effects alter the overall energy conversion sequence of the whole system. This hypothesis of "maximization by structure formation" is illustrated with a few examples of common structures found in the Earth system at the end of this chapter.

We thus have two different views on how systems are able to evolve to their thermodynamic limit. The first view is based on feedbacks and how these shape the dynamics towards a maximum power state. The second view describes the spatial organization within systems and how these are related to the ability of the system to enhance power. These two views do not contradict, but rather complement each other. The link is established by feedbacks that combine structure formation with the dynamics within the system.

In the following, we first use the maximum power example of a heat engine from the previous chapter to describe the concept of feedbacks and how these are formulated in mathematical terms. The two feedbacks related to the maximum power limit are derived analytically. We then describe the other view of maximization by structure formation and illustrate how structures alter the ability of a system to derive power. We close with a summary of the different types of structures within the Earth system, how these relate to energy conversions of different types, and how these likely make the whole Earth system more dissipative.

5.2 Dynamics of an evolving heat engine

The application of feedbacks to energy conversions is illustrated using the example from the last chapter of the maximum power limit associated with convective motion. For this application, we need descriptions for the functional relationships between the different boxes shown in Fig. 5.1 that link the driving gradient with the generation rate, the amount of free energy, the resulting heat flux, and the effects of the resulting heat flux. Feedbacks are then described in mathematical terms as the product of partial derivatives of these relationships, with each derivative representing how one variable changes due to a change in another variable. If the product of the derivatives is positive, then this forms a positive feedback as it acts to enhance the change, while a product that is negative is associated with a negative feedback. It is this sign of the derivative that is indicated at the arrows shown in Fig. 5.1. This mathematical formulation also allows us to quantify the relative magnitudes of different feedbacks within a system and how these change as the system evolves.

We use the functional relationships of the heat engine from the last chapter. For the generation rate, we use the expression of the Carnot limit given by eqn. 4.4,

$$G_{\text{Carnot}} = J_{\text{in}} \frac{T_h - T_c}{T_h} \tag{5.1}$$

This power results in the generation of kinetic energy U_{ke}. The balance of kinetic energy U_{ke}, is then determined in steady state from the power and the frictional dissipation D, within the system,

$$\frac{dU_{\text{ke}}}{dt} = G_{\text{Carnot}} - D = 0 \tag{5.2}$$

where the frictional dissipation D is taken to be proportional to the kinetic energy

$$D = \frac{U_{\text{ke}}}{\tau_{\text{ke}}} \tag{5.3}$$

with a characteristic time scale τ_{ke}. This time scale depends on material properties, particularly the viscosity of the fluid (we deal with a more common formulation for frictional dissipation in Section 7.3), but also on the arrangement of the flow within the system, an aspect that we deal with later this chapter. In steady state, the amount of kinetic energy in the system is then given by

$$U_{\text{ke}} = G_{\text{Carnot}} \cdot \tau_{\text{ke}} \tag{5.4}$$

The relationship of the kinetic energy in the system to the heat flux J_{in}, is obtained by noting that the kinetic energy is associated with a characteristic velocity v, so that it can be expressed by

$$U_{\text{ke}} = \frac{\rho V}{2} v^2 \tag{5.5}$$

where ρ is the density of the fluid and V the volume of the system. This velocity determines the rate by which heat is being transported between the heat reservoirs by convection, and thus shapes the convective heat flux J_{in}. This connection is reflected in the following expression for the convective heat flux:

$$J_{\text{in}} = c_p \rho v \Delta T = c_p \left(\frac{2\rho U_{\text{ke}}}{V} \right)^{1/2} \Delta T \tag{5.6}$$

where c_p is the specific heat capacity of the fluid and v is determined from the kinetic energy of the system (eqn. 5.5). The motivation for expressing the heat flux J_{in}, in terms of the velocity v is to explicitly account for the fact that this heat flux is intimately related to the strength of motion within the system, although the actual relationship in a real system may take a somewhat different form. The driving gradient ΔT, is derived from the energy balances in steady state as in Section 4.5 and is given by eqn. 4.30

$$\Delta T = T_h - T_c = \frac{R_{in} - J_{in}}{k_r} \tag{5.7}$$

We thus have four functional equations that express the relationships necessary to describe the two feedback loops A and B in Fig. 5.1. These relationships are given by $G_{Carnot} = f_a(J_{in}, \Delta T)$ (eqn. 5.1, box "Generation" in Fig. 5.1), $U_{ke} = f_b(G_{Carnot}, \tau_{ke})$ (eqn. 5.4, box "Free energy"), $J_{in} = f_c(U_{ke}, \Delta T)$ (eqn. 5.6, box "Flux"), and $\Delta T = f_d(J_{in})$ (eqn. 5.7, box "Driving difference"). The forcing is described by the temperature difference ΔT, while a primary, internal parameter of the system that is not determined is the characteristic time scale τ_{ke}.

5.3 Dynamic feedbacks and maximum power

With this set of functional relationships, we can now evaluate by how much each of the variables change when a perturbation in one of the variables occurs, and thus describe the feedback loops A and B in Fig. 5.1. To do so, we consider a small perturbation in the kinetic energy ΔU_{ke}, for instance due to random motion of molecules within the fluid. This perturbation results in a perturbation of the heat flux ΔJ_{in}, which is given by $\Delta J_{in} = (\partial J_{in}/\partial U_{ke})\ \Delta U_{ke}$. Because the derivative is positive ($\partial J_{in}/\partial U_{ke} > 0$ for $\Delta T > 0$), an increase in U_{ke} results in an increase in J_{in}, so that this arrow is marked with the "+" sign in Fig. 5.1. This increase in the heat flux by ΔJ_{in} then results in an increase in the generation rate by ΔG_{Carnot}, which is given by $\Delta G_{Carnot} = (\partial G_{Carnot}/\partial J_{in})\ \Delta J_{in}$. This derivative is again positive and it denotes that the generation rate increases when the heat flux increases. This increase in the generation rate ΔG_{Carnot}, then results in an alteration of the initial change by $(\Delta U_{ke})_A = (\partial U_{ke}/\partial G_{Carnot})\ \Delta G_{Carnot}$. Taken together, the chain of changes results in the feedback loop A, with the magnitude being described by the product of the partial derivatives,

$$(\Delta U_{ke})_A = \frac{\partial U_{ke}}{\partial G_{Carnot}} \cdot \frac{\partial G_{Carnot}}{\partial J_{in}} \cdot \frac{\partial J_{in}}{\partial U_{ke}} \cdot \Delta U_{ke} \tag{5.8}$$

Each partial derivative in this expression corresponds to one of the arrows in feedback loop A in Fig. 5.1. Since the product of all derivatives is positive, the initial perturbation ΔU_{ke} is enhanced, so that this constitutes a positive feedback that we earlier referred to as the "power-enhancing" feedback.

We can formulate the effect of the feedback in a slightly different way by expressing the total change in kinetic energy, $(\Delta U_{ke})_{total}$ as the sum of the initial perturbation and the modification due to the feedback:

$$(\Delta U_{\rm ke})_{\rm total} = \Delta U_{\rm ke} + (\Delta U_{\rm ke})_A = (1 + f_A) \cdot \Delta U_{\rm ke} \tag{5.9}$$

where f_A is a feedback factor that is composed of the product of partial derivatives in eqn. 5.8. Since $f_A > 0$, this is a positive feedback and the total response is greater than the initial perturbation.

The feedback loop B in Fig. 5.1 is constructed in an analogous way. The perturbation $\Delta U_{\rm ke}$ causes a change in flux $\Delta J_{\rm in}$, as in loop A, but we now consider the effect of a changed flux on the driving difference. This effect is described by the perturbation of the temperature difference, $\Delta(\Delta T) = (\partial \Delta T/\partial J_{\rm in})\ \Delta J_{\rm in}$, with the double Δ symbol describing a change in the temperature difference, ΔT. Using eqn. 5.7 for calculating the partial derivative of ΔT, we note that this derivative is negative, meaning that the temperature difference is reduced in the presence of a greater flux. The altered temperature difference then affects the generation rate, which results in a change of $\Delta G_{\rm Carnot} = (\partial G_{\rm Carnot}/\partial \Delta T)\ \Delta(\Delta T)$. The Carnot limit (eqn. 4.4) shows that this partial derivative is positive, so that a greater temperature difference results in a greater generation rate, while a reduced temperature difference results in a reduced generation rate. The last link in feedback loop B is established by the effect of an altered generation rate on the kinetic energy of the system. This link is, as in loop A, described by $(\Delta U_{\rm ke})_B = (\partial U_{\rm ke}/\partial G_{\rm Carnot})\ \Delta G_{\rm Carnot}$. In total, this chain of effects associated with loop B results in a modification of the initial perturbation $\Delta U_{\rm ke}$ by

$$(\Delta U_{\rm ke})_B = \frac{\partial U_{\rm ke}}{\partial G_{\rm Carnot}} \cdot \frac{\partial G_{\rm Carnot}}{\partial \Delta T} \cdot \frac{\partial \Delta T}{\partial J_{\rm in}} \cdot \frac{\partial J_{\rm in}}{\partial U_{\rm ke}} \cdot \Delta U_{\rm ke} \tag{5.10}$$

Or, expressing the total change due to this feedback, we can write it as

$$(\Delta U_{\rm ke})_{\rm total} = \Delta U_{\rm ke} + (\Delta U_{\rm ke})_B = (1 + f_B) \cdot \Delta U_{\rm ke} \tag{5.11}$$

where f_B is the feedback factor associated with loop B, composed of the product of partial derivatives in eqn. 5.10. Because the derivatives are positive except for the third derivative $(\partial \Delta T/\partial J_{\rm in})$, the feedback factor is negative $f_B < 0$. It thus forms a negative feedback that we referred to earlier as the "gradient-depletion" feedback.

The two feedbacks do not operate in isolation, but jointly in shaping the response of the system to a perturbation in $U_{\rm ke}$. We note that the alterations of the initial perturbation is additive, so that the effects of the combination of the two feedbacks is given by

$$(\Delta U_{\rm ke})_{\rm total} = (1 + f_A + f_B) \cdot \Delta U_{\rm ke} \tag{5.12}$$

The two feedback factors, f_A and f_B, are given by the products of partial derivatives expressed earlier. Using the minimum formulation of the system dynamics given in Section 5.2, these factors are given by

$$f_A = \frac{\partial U_{\rm ke}}{\partial G_{\rm Carnot}} \cdot \frac{\partial G_{\rm Carnot}}{\partial J_{\rm in}} \cdot \frac{\partial J_{\rm in}}{\partial U_{\rm ke}}$$
$$= \frac{c_p \tau_{\rm ke}}{T_h} \left(\frac{\rho}{2VU_{\rm ke}} \right)^{1/2} \Delta T^2 \tag{5.13}$$

and

$$f_B = \frac{\partial U_{\rm ke}}{\partial G_{\rm Carnot}} \cdot \frac{\partial G_{\rm Carnot}}{\partial \Delta T} \cdot \frac{\partial \Delta T}{\partial J_{\rm in}} \cdot \frac{\partial J_{\rm in}}{\partial U_{\rm ke}}$$
$$= -\frac{c_p^2 \rho \tau_{\rm ke}}{k_r T_h V} \Delta T^2 \tag{5.14}$$

These expressions show the influence of some physical parameters of the system, but also some of the system variables that change with time. Most notably, both factors depend on the magnitude of the driving gradient squared ΔT^2, and the time scale $\tau_{\rm ke}$ that is related to how frictional dissipation takes place within the system. The factors differ in their functional form in that f_A entirely depends on properties associated with heat transport and dynamics, such as heat capacity, temperatures, and the kinetic energy within the system, while f_B does not depend on the kinetic energy, but on the other energy transport mechanism within the system through its dependence on k_r. With ΔT and $U_{\rm ke}$ being central system variables that change in time, what these expressions state is that the strengths of the feedbacks change with the evolutionary dynamics in time. As a next step, we consider these temporal dynamics in more detail.

5.4 Evolutionary dynamics and power

Let us now consider the evolutionary dynamics of the system just described to see how the feedbacks shape the dynamics. The description of the feedbacks was subjected to steady-state conditions of the energy balances that describe the driving gradient ΔT (with the energy balance of the hot reservoir given by $R_{\rm in} = R_r + J_{\rm in}$), and of the kinetic energy balance with $G_{\rm Carnot} = D$. We thus focus, in the following, on dynamics that go beyond the temporal dynamics towards a steady state that were used in the last chapters and consider feedbacks on changes of variables within the system in their steady state.

We consider the system initially at rest with an external forcing, described by $R_{\rm in}$ (as in Fig. 4.1b), that maintains a driving gradient, $\Delta T > 0$. The initial state is thus represented by $U_{\rm ke} = 0$, $J_{\rm in} = 0$, and $G_{\rm Carnot} = 0$. Note that this state is a steady state with respect to the energy balances and with respect to the kinetic energy balance. The heating of the hot reservoir is completely compensated for in steady state by the flux of energy, $R_r = k_r \Delta T$, which transports energy without involving macroscopic motion such as conduction or radiative exchange. The energy balance is then given by $R_{in} = R_r$, while the kinetic energy balance is given by $G_{Carnot} = D = 0$. In this state, the temperature difference has the greatest value with $\Delta T_{init} = R_{in}/k_r$.

When a perturbation in the kinetic energy ΔU_{ke}, of the system occurs, this results in a non-vanishing heat flux, $J_{in} > 0$, which then results in the generation of kinetic energy, $G_{Carnot} > 0$. The associated feedback factor f_A is extremely large because of its inverse dependence on U_{ke}, which initially is zero. This initial instability is likely due to the simplicity of the formulation used here. Since f_B is well bounded for $U_{ke} = 0$, its value is going to be substantially smaller. In total, feedback A will thus dominate the behavior of the system, and the perturbation is going to grow rapidly. The steady-state kinetic energy balance involves terms that are greater than zero, so that $G_{Carnot} = D > 0$. Since $J_{in} > 0$, the temperature difference is somewhat reduced from its initial value, $\Delta T < \Delta T_{init}$.

As the kinetic energy grows in time, the heat flux, J_{in}, increases as well, which further increases G_{Carnot}. With a greater value of J_{in}, the temperature difference ΔT is being reduced. Both the increase in U_{ke} and the decrease in ΔT result in a reduced value of f_A, so that this feedback loses its strength with greater kinetic energy in the system. The value of f_B also decreases because of the decrease in ΔT, but because it does not depend on U_{ke}, its decline with greater kinetic energy is less than that of f_A. The kinetic energy balance is maintained at states with increasingly greater values of G_{Carnot} although the growth rate of the kinetic energy is successively reduced as feedback A loses its strength. Since J_{in} still increases with greater values of U_{ke}, the temperature difference between the reservoirs decreases further.

Eventually, the point is reached at which the two feedbacks compensate each other. At this point, an increase in J_{in} no longer results in an increase in G_{Carnot}, because the reduction in ΔT weighs equally as strong as the increase in J_{in} in the expression for G_{Carnot}. This state then forms a "steady state" with respect to the evolutionary dynamics, in which G_{Carnot} no longer grows. To see its relation with the maximum power limit, we next evaluate this state in greater detail.

5.5 Feedbacks, maximum power, and dynamic stability

To evaluate the state in which the two feedback factors are equal, we first revisit the two expressions for the factors, given by eqns. 5.13 and 5.14. We note that they share two of the partial derivatives, $\partial U_{ke}/\partial G_{Carnot}$ and $\partial J_{in}/\partial U_{ke}$. The difference between the factors thus only relates to the terms $\partial G_{Carnot}/\partial J_{in}$ and $\partial G_{Carnot}/\partial \Delta T \cdot \partial \Delta T/\partial J_{in}$. Hence, the difference in the strength of the two feedbacks relates to the difference in these terms. We can thus express the feedback factor f_B in terms of f_A, and obtain

$$f_B = \frac{(\partial G_{\mathrm{diff}}/\partial \Delta T)(\partial \Delta T/\partial J_{\mathrm{in}})}{\partial G_{\mathrm{Carnot}}/\partial J_{\mathrm{in}}} f_A = -\frac{J_{\mathrm{in}}}{k_r \Delta T} \cdot f_A \tag{5.15}$$

By expressing it in this form, we can see directly the relationship to the evolutionary steady state described at the end of the last section. The increase in G_{Carnot} is halted when the feedbacks A and B are of equal magnitude. This is the case when the term $J_{\mathrm{in}}/(k_r \Delta T) = 1$, which requires that $J_{\mathrm{in}} = k_r \Delta T = R_r$ because of eqns. 4.28 and 4.30 (see also Fig. 4.1b). Since the energy balance requires $R_{\mathrm{in}} = R_r + J_{\mathrm{in}}$, this results in $J_{\mathrm{in}} = R_r = R_{\mathrm{in}}/2$. This is exactly the expression that we obtained from the maximization of power in the last chapter (cf. eqn. 4.32). The maximum power limit is thus equivalent to a state in which the two feedbacks A and B are of equal magnitude, but of opposite sign. The temperature difference ΔT, at this limit is given by $\Delta T_{\mathrm{opt}} = R_{\mathrm{in}}/(2k_r)$, half of the initial value in the absence of a heat flux, $\Delta T_{\mathrm{opt}} = \Delta T_{\mathrm{init}}/2$.

Once a system reaches its maximum power state, it is dynamically stable and responds to a perturbation by a negative feedback. This negative feedback can be related back to eqn. 5.15. When the perturbation results in a smaller heat flux $J_{\mathrm{in}} < J_{\mathrm{opt}}$ or a greater value of $\Delta T > \Delta T_{\mathrm{opt}}$, feedback B is weaker than feedback A, so that $f_B < -f_A$. The combined effect of the feedbacks is then such that it amplifies the dynamics (cf. eqn. 5.12), resulting in a greater value of U_{ke}, a greater flux J_{in}, and a reduced gradient ΔT. In the other case, when a perturbation results in a greater heat flux, $J_{\mathrm{in}} > J_{\mathrm{opt}}$, or a smaller gradient, $\Delta T < \Delta T_{\mathrm{opt}}$, feedback B is stronger than feedback A, so that $f_B > -f_A$. This then results in dynamical changes that reduce the value of U_{ke}, lead to a weaker heat flux, J_{in} and a greater gradient ΔT. The maximum power limit as an evolutionary steady state can thus be seen as the outcome of the balancing of these two feedbacks.

Yet, the evolution to the maximum power limit is not a necessary outcome as it involves structural changes and adjustments in how the system organizes its flow, among other factors. To identify this connection, we note that the heat flux J_{in} in eqn. 5.15 is not an independent variable, but depends on the kinetic energy within the system. Using eqn. 5.6 for J_{in} in eqn. 5.15, we can write f_B as a function of U_{ke}, rather than J_{in}, as

$$f_B = -\frac{c_p}{k_r}\left(\frac{2\rho U_{\mathrm{ke}}}{V}\right)^{1/2} \cdot f_A \tag{5.16}$$

The balancing of the feedbacks thus depends on the particular value of U_{ke}. This value is not only determined by the power, but equally by the time scale τ_{ke} that is related to the dissipation of kinetic energy. This time scale is not about how much is dissipated, which is after all set by how much kinetic energy is generated. It is rather a property that relates to how dissipation occurs in the system. By combining the expressions of Section 5.2, we can in fact express the kinetic energy of the system by

$$U_{\mathrm{ke}} = \frac{2\rho c_p^2}{V} \frac{\Delta T^4}{T_h^2} \cdot \tau_{\mathrm{ke}}^2 \tag{5.17}$$

This expression combines the forcing of the system, in terms of ΔT, and places the value of τ_{ke} at a prominent place. It can be related to the kinetic energy associated with maximum power, so that we can infer an optimum time scale, $\tau_{\mathrm{ke,opt}}$, that corresponds to the system being at a state of maximum power

$$\tau_{\mathrm{ke,opt}} = \frac{2k_r^3 V T_h}{\rho c_p^2} R_{\mathrm{in}}^{-2} \tag{5.18}$$

Note how this time scale only includes physical properties of the system as well as the forcing; it does not include any frictional parameters like viscosity. What this means is that friction would need to be arranged in a way that is fairly independent of viscosity so that the kinetic energy of the system can adapt a value by which generation and dissipation is not characterized by the material properties, but by the thermodynamic limit. This does not necessarily need to be the case, which would then result in suboptimal states of less than maximized power. Whether friction in a system is determined by its material properties or by the thermodynamic limits relates to how the flow, and thus its frictional dissipation, is organized within the system. It is through structure formation that a system can reach different amounts of kinetic energy with the same rate of frictional dissipation, an aspect that we explore next.

5.6 Structural organization of flow

To better understand how a system can achieve the limit of maximum power, a certain flexibility is needed regarding how friction is distributed within the system. We refer to this flexibility as degrees of freedom within a system to arrange its dissipative properties. This flexibility depends on the size of the system, because the larger the system, the greater the role that internal friction plays compared to friction at the system boundary. In the example we explore in the following, this refers to the flexibility regarding the relationship between kinetic energy and frictional dissipation that allows for a range of different values for τ_{ke}. When this range includes the optimum value, $\tau_{\mathrm{ke,opt}}$, then a system is able to evolve to the maximum power limit.

How such a flexibility can be accomplished by arranging the flow is shown in Fig. 5.2, following Kleidon et al. (2014a). This figure shows a qualitative

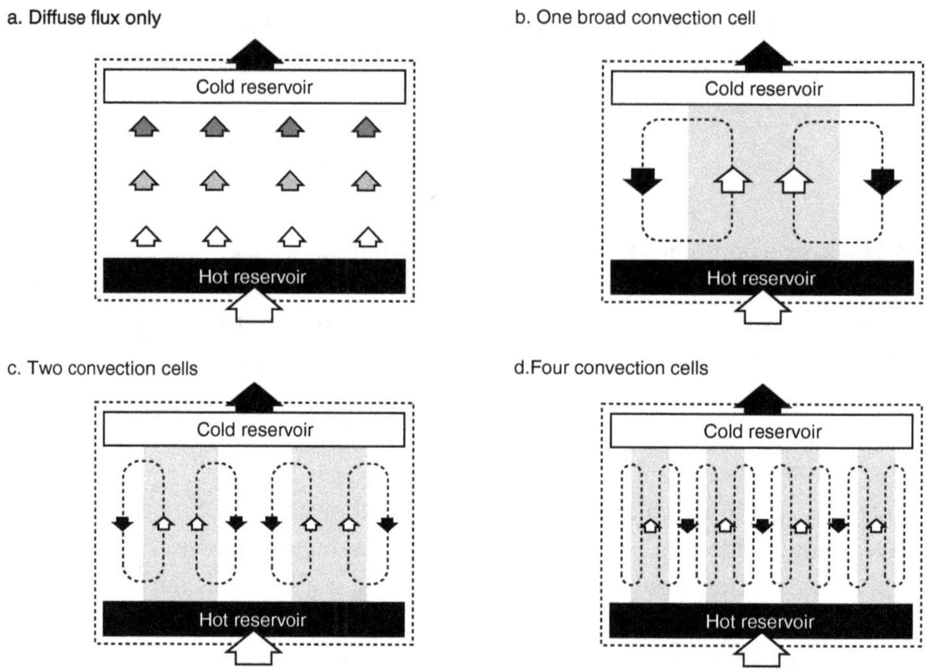

Figure 5.2 Illustration of different configurations of flow. The areas shaded grey illustrate regions of upward motion.

illustration of how flow can be arranged in different ways, showing the example of no flow in Fig. 5.2a – which was the initial state considered in Section 5.4 – and three examples with different numbers of convection cells within the system in Fig. 5.2b–d. The purpose of the following description is to qualitatively show that different relationships between U_{ke} and D due to different values of τ_{ke} are possible (cf. eqn. 5.3), providing different steady states for a given power. As these different relationships relate to different ways of spatially organizing the flow, these are referred to here as "structural organization" of the flow. It is an aspect that is very close to Prigogine's notion of dissipative structures, since this structural organization also affects dissipation and the thermodynamic state of the system.

The different forms of how flow is organized in the examples result in different ways of how frictional dissipation occurs in the system. Frictional dissipation relates to a friction force that acts to deplete momentum gradients, much like the example used in Section 4.6. The friction force, also referred to as shear stress, depends on spatial gradients in the velocity field, with the proportionality given by the viscosity, a material property of the fluid. Frictional dissipation is then given by the product of friction force, that is, the momentum flux, and the velocity difference as the conjugated pair of variables. This background is sufficient to understand why the examples shown in Fig. 5.2 differ in terms of their values of τ_{ke}. We consider these systems in similar overall steady states given by $G = D$, so that the rate of frictional dissipation is fixed. Since $G = D = 0$ in Fig. 5.2a, we do not consider this case further. What differs in the other three examples is the spatial velocity gradients between the updrafts and the downdrafts, which are shown as the black and white arrows. The distance between the updraft and the downdraft in Fig. 5.2b is much greater than in Fig. 5.2d, where the up- and downdrafts occur much closer to each other. Since we consider setups in which the overall rate of dissipation D, is fixed, and frictional dissipation depends on the ratio of velocity differences to the distance over which they occur, the organization of flow shown in Fig. 5.2b could maintain a larger velocity difference because of the greater spatial separation. The configuration in Fig. 5.2b could then maintain a greater kinetic energy than the other configurations shown in Fig. 5.2c or d. Hence, the flow organization shown in Fig. 5.2b should have a greater value of τ_{ke}, while the organizations shown in Fig. 5.2c or d are associated with lower values of τ_{ke}. When the flow rate is greater, as in Fig. 5.2b, then there is also a greater velocity gradient near the hot (and cold) reservoir, which would allow for greater turbulent exchange at the system's boundary. While this description is rather qualitative, it nevertheless illustrates that by spatial organization of the flow, different values of τ_{ke} are represented, a condition necessary for the system to reach its thermodynamic limit of maximum power.

An aspect that is directly related to the spatial organization of the flow is that the driving gradient is being altered. In the absence of convection, the heating of the hot reservoir in steady state is balanced by the flux R_r. As this flux does not include material transport, the transport of energy depends on a uniform temperature gradient throughout the system, as shown by profile (a) in Fig. 5.3. In the presence of convection, heat is transported by convective motion in the interior, so that it does not require a temperature gradient within the system to drive a heat flux. Instead, the temperature gradients are pushed towards the system boundary, where they are also steeper, creating a "boundary layer," thus allowing for a greater heat exchange between the reservoirs and the fluid. This phenomenon is well observed, e.g., in Rayleigh–Bénard convection cells, and thermodynamically explained (e.g., Schneider and Kay 1994b). In the Earth system, such profiles are also well observed, particularly in the atmospheric boundary layer that develops near the surface (e.g., Oke 1987; Stull 1989). The atmospheric boundary layer shows a similar profile as shown in Fig. 5.3, except that it is the potential temperature that is constant within the system (recall that it is this property that is conserved during steps 2 and 4 of the Carnot cycle, Section 4.3). There may also not necessarily be a boundary layer at the top as there is no rigid surface, although there may be an entrainment zone.

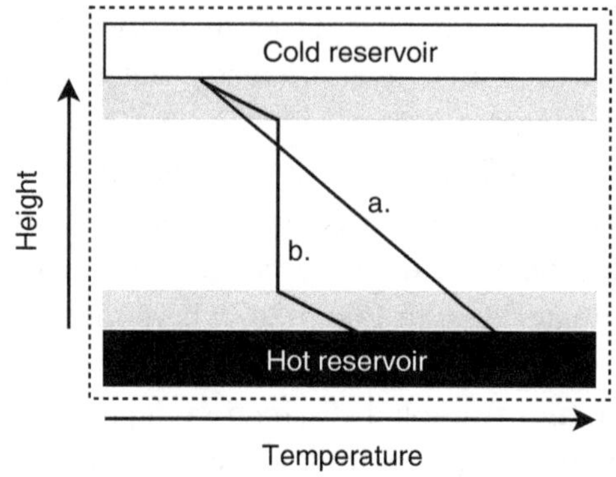

Figure 5.3 Illustration of the effect of convective motion on the temperature profile for a system as shown in Fig. 5.2. The profile (a) results from purely diffusive heat transfer, while the profile (b) includes the effect of convection. The areas shaded grey show boundary layers with steepened temperature gradients.

What these profiles indicate is that the formation of structured flow is associated with a modification of internal dissipation that affects the value of τ_{ke}, and it alters the location and magnitude of the driving gradients at the system boundary. These effects are shown in Fig. 5.1 by the loops C and D. The formulation of these feedback loops is not mathematical as in Section 5.3. The box "structure" in this figure stands as a surrogate for the spatial organization of the flow that is associated with the free energy within the system. With more kinetic energy in the system, the organization of the flow can reduce the internal frictional dissipation within the fluid, leading in a greater value of τ_{ke}. This would then constitute a positive feedback by structure formation (loop C). Associated with this structure formation is then the steepening of the driving gradient, as shown by profile (b) in Fig. 5.3. This steepening of the gradients results in a greater generation rate and more free energy, thus forming a positive feedback as well (loop D). Ultimately, these feedbacks are intimately related to the variables of feedback loops A and B, and are thus constrained by the thermodynamic limit of maximum power.

Even though this line of reasoning is nowhere near a rigorous proof, it nevertheless suggests that the thermodynamic limit sets a constraint on the extent of structure formation in the system, and that an evolution towards the maximum power limit needs to involve a structural organization of the flow.

5.7 Maximization by structures

There are, of course, many other forms of structures within the Earth system. We may define structures here as organized heterogeneity reflected in the spatiotemporal variations of the state variables that affect the ability to generate power within the system that feeds their formation (see also Kleidon et al. 2013). In addition to the flow structures that are involved in convective motion, there are wave-like structures represented by, for instance, ocean waves, sand dunes, and vegetation patterns (Fig. 5.4), or tree-like structures found in river networks and vascular systems of plants and animals, as shown in the introduction (Fig. 1.4). These structures occur across a wide range of processes and scales, indicating that there is a general underlying mechanism and functional significance that generates and maintains such structures. The functional significance should then be reflected in the thermodynamic state and the associated rates of energy conversions of such structures.

The maintenance of structures is closely related to the fluxes that generate and maintain these, and thus are related to energy conversions. Wave-like patterns are often observed at the interface between two substances across which momentum is transported. Winds near the fluid surface of the ocean generate waves and water flow, water and air flow causes sediment transport and the formation of ripples

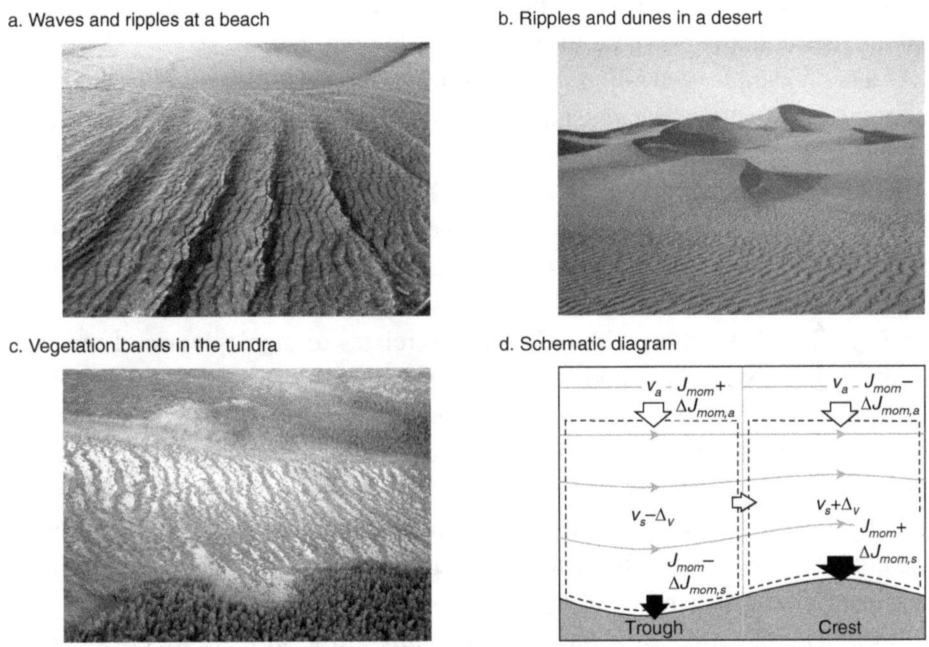

Figure 5.4 Examples of wave-like patterns in (a) water, in the form of surface waves and ripples at the seafloor, (b) ripples and sand dunes in the desert, and (c) vegetation bands in the wet tundra in western Siberia. The schematic diagram in (d) shows two boxes (dotted lines) of uneven extent. This example is used in the text to describe how such wave-like patterns relate to the local concentration of driving gradients and the associated alteration of energy transfer and dissipation. Photos by the author.

in sand. They result in conjunction with momentum transport across the interface and thus are linked to the conversion of the kinetic energy of the driving fluid into frictional dissipation and the work done to generate wave-like structures. Tree-like structures effectively collect and transport water from a surface or a volume to a single outlet, thereby reducing dissipative losses as the water flow depletes a driving gradient. The explanations and underlying dynamics of such patterns have long been explored (e.g., Cross and Hohenberg 1993).

At the end of this chapter, the goal is to relate such structures to a general hypothesis that structure formation represents the means of a system to reach its thermodynamic limit, thereby depleting the driving gradient at the maximum possible rate. This would then imply that the presence of such structures indicate that the associated process and energy conversion operates near – or even at – the thermodynamic limit. Hence, these structures may have a profound functional significance in shaping the thermodynamic state of Earth systems. To support this notion, these structures are evaluated in a system's perspective that includes the dynamics of the driving difference, and their effects are related to the energy conversions within the system and the conditions at the system boundary. In the following, merely a qualitative account is provided to support this hypothesis.

As a first example, wave-like structures are evaluated with respect to their effects on energy conversions. We consider a setup of two boxes that represent the trough and the crest region of an uneven, wave-like surface at rest (Fig. 5.4d). A fluid at the top of the boxes (the "driving fluid") flows with a steady-state velocity v_a, which is associated with a downward momentum flux J_{mom}. It is assumed that the surface is at rest, and that the momentum flux results in a near-surface flow velocity v_s in the absence of a wave-like surface with a uniform, downward momentum flux J_{mom}. The dissipation within the system relates to the depletion of the kinetic energy of the driving fluid, and is given by $D_{\text{tot}} = J_{\text{mom}} \cdot v_a = D_a + D_s$. This total dissipation consists of a contribution D_a, by the dissipation between the driving fluid and the near-surface flow velocity, and a contribution D_s, by the dissipation between the near-surface flow velocity and the surface at rest. When the surface is even, we would expect that this dissipation occurs uniformly across the surface. This uniform case is then associated with a uniform near-surface flow velocity, and a dissipation near the surface of $D_s = J_{\text{mom}} \cdot v_s$.

To evaluate the effect of a wave-like structure at the surface, we consider some variation in surface topography associated with variations in potential energy, resulting in troughs and crests. This surface topography results in a variation of the near-surface flow velocity and thus also in alterations in the vertical momentum fluxes. These alterations are shown in Fig. 5.4d. In the trough box, the near-surface flow velocity is somewhat reduced by $-\Delta v$ (with $\Delta v \geq 0$), since the near-surface mass flux is spread across a wider vertical extent. In the crest box, this flow is compressed, so that the velocity would be enhanced. This enhancement is taken here to be opposite of the trough box, so that it is given by $+\Delta v$. Since momentum fluxes depend on velocity gradients, this alteration of the flow velocity affects the momentum fluxes. At the top boundary of the trough box, the momentum flux from the driving fluid is somewhat enhanced by $\Delta J_{mom,a} \geq 0$, because the velocity difference is now given by $(v_a - v_s + \Delta v)$. The opposite is the case at the top boundary of the crest box, where the momentum flux is reduced by $-\Delta J_{mom,a}$ because of the reduced velocity difference. At the surface, this situation is reversed. The lower flow velocity in the trough box results in a lower velocity difference between the flow and the surface, so that the momentum flux should be reduced by some value $-\Delta J_{mom,s} \leq 0$, while the momentum flux across the surface in the crest box is enhanced. This uneven momentum flux across the top and bottom of the boxes is compensated for by a net momentum exchange between the two boxes. Also note that these alterations still conserve mass and momentum, so that the total momentum flux from the driving fluid to the surface is, on average, given by J_{mom}.

While these alterations to the near-surface flow cancel out with respect to their effect on the average momentum flux, they enhance the dissipation of kinetic energy of the driving fluid. To show this, we separate the dissipation into the two components described earlier. The first component is given by $D_a = 1/2 \cdot [(J_{mom} + \Delta J_{mom,a}) \cdot (v_a - v_s + \Delta v) + (J_{mom} - \Delta J_{mom,a}) \cdot (v_a - v_s - \Delta v)] = J_{mom} \cdot (v_a - v_s) + \Delta J_{mom,a} \cdot \Delta v$, where we assumed for simplicity that both boxes cover the same spatial extent to obtain the mean dissipation rate across both boxes. Equivalently, surface dissipation is given by $D_s = 1/2 \cdot [(J_{mom} - \Delta J_{mom,s}) \cdot (v_s - \Delta v) + (J_{mom} + \Delta J_{mom,s}) \cdot (v_s + \Delta v)] = J_{mom} \cdot v_s + \Delta J_{mom,s} \cdot \Delta v$. While the first terms on the right hand side correspond to the mean dissipation rate in the absence of heterogeneity, the second terms do not cancel out and yield an additional positive contribution that is directly related to the imposed variation in the momentum transport. Hence, these variations act to enhance the overall dissipation of the system even though the mean momentum flux remains unchanged. This enhanced dissipation would then feed back to the driving fluid, resulting in a lower flow velocity v_a and thus a lower driving difference with potentially further changes in the atmosphere.

To relate these effects to the ability of a process to reach its thermodynamic limit, we next look at these effects in the context of the feedbacks shown in Fig. 5.1. We need to relate the links in the feedback diagram to the processes that generate and maintain the wave-like pattern at the surface. As a start, the driving gradient of the process is the velocity difference between the fluid and the surface, as already mentioned before. The generation of the wave-like pattern at the surface relates to the surface stress that is associated with the frictional dissipation of the fluid motion. A fraction of the kinetic energy is, however, not dissipated into heat, but used to perform work at the surface. This work comes at the expense of the kinetic energy of the fluid, so that it would be included in the term D_s described earlier. This work is needed to generate and maintain the heterogeneity that is reflected by the different heights of the surface and the associated potential energy. The box "generation" thus relates to the work performed to generate the wave profile at the surface, which is related to the dissipation rate at the surface. The "free energy" is the potential energy associated with the heterogeneity in elevational difference, which also serves as a measure for the "structure" that we describe. The two feedback loops A and B are then identified in a straightforward manner: feedback loop A is described by the effect of a greater magnitude of a wave-like pattern on a greater variation in surface stress, which in turn would allow greater surface dissipation, a fraction of which would contribute to a greater generation rate. Feedback loop B then describes the overall effect of these dynamics to enhance the dissipation rate of the whole system, which would need to reduce the driving gradient. It would thus seem that at least some of the wave-like structures could be described by the same type of dynamic feedbacks as in convection. Wave-like patterns may thus indicate that these structures reflect an enhancement or even a maximization of the energy conversions associated with their maintenance.

There are two examples, sand dunes and striped vegetation patterns, for which similar considerations have been formulated. For sand dunes, it is well recognized that the wind regime plays a pivotal role in shaping the different forms found in nature (Werner 1999; Kok et al. 2012; Charru et al. 2013). Rubin and Hunter (1987) formulated the hypothesis that the formation of sand dunes act to maximize gross bedform-normal transport of sand, and supported this hypothesis with experimental results. As sand transport involves the work of lifting the grains, and this work is derived from the kinetic energy of the near-surface air flow, the transport rate and the dissipation rate are closely related to each other. It would thus seem that the hypothesis that dunes maximize the transport rate of sediments describes a very similar outcome as that these structures operate near their limit of performing work out of the driving fluid.

The processes involving the formation of striped vegetation patterns are similar, although it involves biotic activity so this is not simply an abiotic phenomena as in the formation of waves and dunes. The formation of such patterns have been reproduced by a relatively simple, spatially explicit reaction–diffusion type of differential equation which couples the surface flow of water with the diffusive spread of a vegetation (Klausmeier 1999; Lejeune and Tlidi 1999). These patterns have been explained as an outcome of a strong, local-scale and non-linear coupling between plant productivity and infiltration in water-limited regions for which the lateral redistribution of water forms a key limiting constraint that limits the generation of new biomass. Lejeune et al. (2002) explained that the formation of such vegetation patterns can be seen as a local adaptation that maximizes biomass and thus growth, given the water constraint imposed by the precipitation rate of the region, while Schymanski et al. (2010) related this interpretation to thermodynamic limits. In arctic regions, similar dynamics are found that result in pattern formation (as shown in Fig. 5.4c), although the constraint has been attributed to the limitation imposed by nutrients rather than water availability (Rietkerk et al. 2004).

A considerably more complex type of structure is represented by the tree-like structures that can be found in river networks and vascular networks in plants and animals. Tree-like structures have been described as means to minimize the frictional dissipation of fluid flow within the network (West et al. 1997, 1999) or by related energetic concepts such as minimum energy expenditure (Howard 1990; Rodriguez-Iturbe and Rinaldo 1997; Rinaldo et al. 1998) or others (Bejan 2007). This effect can be illustrated quite easily by the two examples shown in Fig. 5.5. Note that while river networks and vascular networks look very similar, they differ in important aspects that are explained further here.

To understand the effect of river networks on the frictional dissipation of water flow along a topographic gradient, we consider the drainage of water of a certain area that is fed by precipitation as shown in Fig. 5.5a. The difference in topographic heights within this area result in the water at the surface having different values of potential energy, which can be converted into kinetic energy associated with water flow. The kinetic energy associated with water flow then depends on how much potential energy can be converted, and also on the intensity of frictional dissipation as the water drains the region. This frictional dissipation depends on the magnitude of water flow, and also on the contact area between the flowing water and the solid underlying surface. This contact area forms the interface over which momentum is exchanged between the water flow and the solid surface at rest and thus plays an important role in shaping the magnitude of frictional dissipation. It

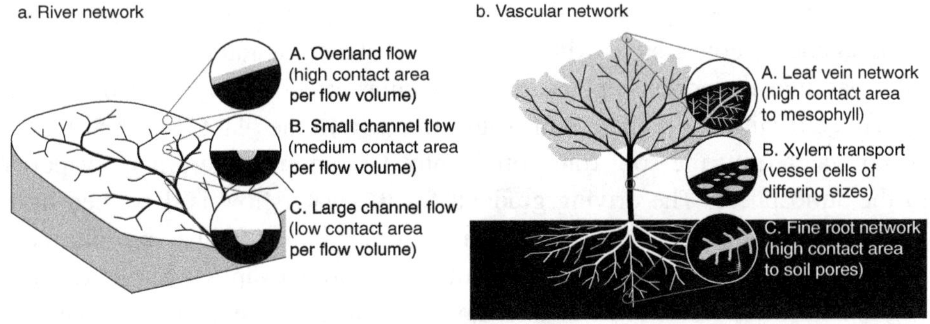

Figure 5.5 Illustration of two types of structures that affect frictional losses of water flow. (a) A river network that channels runoff from the land surface into an outlet. (b) A tree as a vascular network that conducts soil water to the atmosphere.

thus makes a difference if water flows relatively uniformly across a surface or if it flows in channels. When water flows over the surface, it has a relatively large contact with the surface, as shown by inlet A in Fig. 5.5a, resulting in a greater surface over which momentum can be exchanged so that the intensity of frictional dissipation is higher. In the presence of a curved channel, as in inlets B and C, the contact area is reduced in relation to the cross-sectional area of the water flow, thereby reducing the frictional dissipation. The more the draining water is concentrated into fewer large channels, the lower the frictional losses as long as the channels have collected sufficient water to fill the larger channel. By arranging the draining flow of an area into a tree-like channel network structure, the water flow can thus alter where and how much the flow is exposed to losses by frictional dissipation.

At the same time, it is important to note that river networks are formed by sediment transport. Sediment transport involves work to be performed to detach, lift, and keep the sediment in suspension. This work is derived from the water flow and is part of the frictional dissipation of the water flow. Thus, potentially more work can be performed to detach and lift sediment in areas of higher friction, while the channel network effectively transports the sediment out of the area. The evolution of a plain surface into an area dominated by a tree-like channel network can be understood by the same feedbacks as shown in Fig. 5.1 (Kleidon et al. 2013). The channel network reduces frictional dissipation within the network and pushes the areas of higher dissipation towards the edges of the network where the water flows are not yet channelized. This is similar to the case of the convection cell, where frictional dissipation is reduced within the convection cell, but enlarged at the boundary of the system. Channel networks can thus be seen as structures that maximize the rate of sediment transport such that they deplete their topographic driving gradient at the maximum possible rate. The growth of such river networks can be understood in terms of the same feedbacks that shape the dynamics to the maximum power limit of convection so that such networks can be expected to grow spontaneously and affect energy conversions within the system in a comparable way.

The vascular networks in plants have a similar effect as river networks in that they reduce frictional dissipation of water flow within the plant (Fig. 5.5b). Plants take up water from the soil, transport it into the canopy, where it is evaporated into the atmosphere. The driving gradient for the water flow is described by the difference in chemical potential between the water bound in the soil and that of the unsaturated air of the atmosphere, with the resulting evaporative flux depleting this gradient (the formulation of water fluxes is more fully described in Chapter 8). At both boundaries of the plant network, the transport of water is limited by diffusive transport. The water uptake by plants from the soil pores takes place by diffusion (inlet C in Fig. 5.5b), so that a high contact area enhances the diffusive uptake of water. Water is then transported by the root network, vessels of differing sizes in the xylem (inlet B), into the leaves. Within the leaf, water is distributed by a network of leaf veins (inlet A), diffuses into the mesophyll from where it evaporates into the interior air space of the leaf. The moistened air from the leaf's interior is then exchanged with the atmosphere through the stomata of the leaf. What this arrangement of a plant's vascular network achieves is that it maintains a transport network that reduces interior dissipation within the plant – just as in the case of the river network – and combines this network with large surface areas at the boundary of the network to enlarge the contact area for the comparatively slow, diffusive transport at the boundary. In contrast to river networks, however, these networks are not built out of the material transported by water, but rather by the products of photosynthesis. Observations show that this hydraulic network is intimately linked with a plant's ability to exchange gas and achieve a high photosynthetic capacity (Sperry 2000; Brodribb et al. 2007). An effective vascular network thus would appear to support greater rates of photosynthesis, which feeds back to the growth of the plant.

There are a lot more aspects involved in these dynamics that are only briefly mentioned here. Commonly, there are scaling laws associated with observed structures, for instance related to the characteristic length of streams to their drainage area (Rinaldo et al. 1998) or the branch length and plant biomass (West et al. 1999). A broad range of scaling laws were derived quantitatively from the assumption of minimum frictional dissipation in vascular networks in conjunction with a space-filling requirement (West et al. 1997), resulting in fractal networks as illustrated in Fig. 5.5. There are also many more scaling laws observed in natural processes, particularly in turbulence. Given that studies on river networks and vascular networks have successfully linked scaling laws to minimization of frictional dissipation within these networks, it would seem plausible that such scaling laws are linked with an optimized thermodynamic system that operates near its thermodynamic limit. This would, of course, need to evaluated further.

Also relevant, yet not mentioned here, are critical thresholds, e.g., the transition from laminar to turbulent flow or the threshold shear stress needed to lift sediment. These thresholds are characterized by a variety of dimensionless numbers, such as the Reynolds or Rayleigh number in fluid dynamics. They play an important role in that they can constrain the dynamics within a system so that it may not be able to evolve to its thermodynamic limit.

5.8 Structures and maximization in the Earth system

In summary, in this chapter, the potential feedbacks and mechanisms were described by which systems may reach their thermodynamic limit. Two dynamic feedbacks were described that directly relate to the expressions of power, which generally combine a flux with a gradient. While a fast, positive feedback enhances the generation rate within the system by a greater flux, a slower, negative feedback counteracts this increase by the successive depletion of the driving gradient. This combination of the two feedbacks appears to be rather general and is not restricted to the thermodynamic maximum power limit, but to the general shape of energy conversion rates that combine a flux of an extensive variable with a driving gradient of an intensive variable.

The maximum power limit then reflects a state in which the flux strongly interacts with its driving gradient. The means by which a system can achieve the maximum power limit appears to be through structure formation within the system. Structure formation results in a redistribution of where dissipation occurs within the system, resulting in a non-uniform arrangement of dissipation that is far from the homogeneous state that would be expected in a near-equilibrium state. This structure formation appears to generally push dissipation towards its system boundary where mixing maintains the flux across the boundary (an aspect not considered here), so that little dissipation occurs within the system. By forming such dissipative structures, it would seem that systems can enhance fluxes, dissipate the driving gradients to a greater extent, up to the point at which they reach the maximum power limit with its associated fluxes and gradients. This would suggest that such dissipative structures play a critical role in allowing systems to reach their maximum power limit. As power equals dissipation in steady state, this state would then make the system as dissipative as possible, thus accelerating the dynamics in a system along the direction imposed by the second law to the maximum possible extent. We may then also hypothesize that when such structures are not maintained and decay, the thermodynamic state of the system is altered, making it less dissipative and bringing it closer to equilibrium. In a steady state in which such structures are maintained, this also suggests that the effect of these structures may not necessarily need to be explicitly treated when we want to describe a complex system because if it operates at its thermodynamic limit, we may simply infer its characteristics directly from the maximum power limit.

In the following chapters, several thermodynamic limits are derived for Earth system processes to infer the magnitudes of fluxes and rates of energy conversions. These estimates are performed without explicitly considering the conditions under which the limit can be reached or describing the mechanisms and the dynamics by which these limits are being achieved. In Chapter 12 we then place these estimates back into a planetary perspective. There we will evaluate the implications of such a thermodynamic evolution toward limits for the whole Earth system, and whether it makes the Earth as dissipative as possible.

6
Radiation

6.1 The main driver of the Earth system

Radiation – obviously – plays a major role in the Earth system. It is by far the most important driver for energy conversions on Earth, both in terms of the sheer magnitude of the energy flux of solar radiation, and also in terms of its quality, as solar radiation represents radiation with a very low entropy. In the thermodynamic view of the Earth system shown in Fig. 1.5, it is the start point and endpoint for most energy conversions taking place within the Earth system. Spatial and temporal variations in the absorption of solar radiation are the causes for various heat fluxes and associated dynamics that distribute imbalances in radiative heating and cooling rates. Variations in radiative forcing in combination with dynamics shape most of the observed climatic variations, from the seasons in mid-latitudes to the large-scale variation of surface temperature from the tropics to the poles. The focus of this chapter is to describe the thermodynamic nature of radiation which is then used to understand the dissipative nature of radiative transfer processes, to derive the limits of energy conversions from radiation to other forms, and to describe radiative transfer as a dominant process shaping the environmental conditions that affect other energy conversion processes.

In thermodynamic terms, the radiative exchange between the Earth and space generates the most important driving gradients, and it exports the entropy that is being produced by Earth system processes to space. The entropy of radiation is represented mostly by the spectral composition of the radiative flux. The spectral compositions of the incoming solar radiation and the outgoing radiation from the Earth system are shown in Fig. 6.1. Solar radiation is composed mostly of visible light which is characterized by relatively short wavelengths. This spectral composition essentially corresponds to the composition of the radiation when it was emitted from the Sun at a high temperature of about 5760 K. When solar radiation is absorbed by the Earth system, it is subsequently reemitted to space

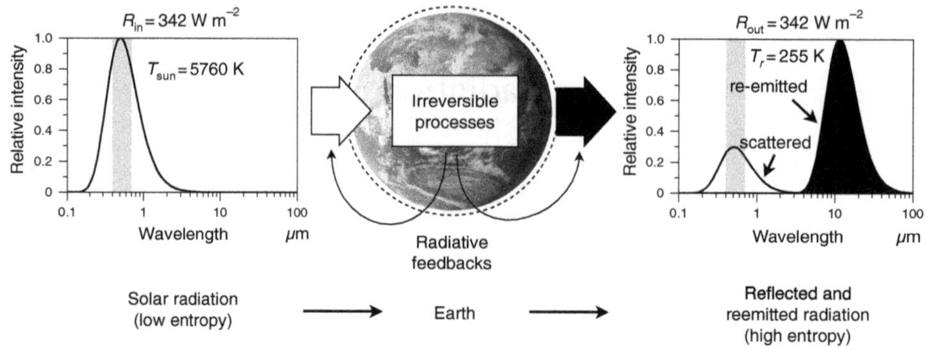

Figure 6.1 Radiative exchange of the Earth system by absorption and reflection of solar radiation and the reemission of terrestrial radiation into space in terms of the radiative flux, R_{in} and R_{out}, as well as the characteristic radiative temperatures of the Sun and Earth, T_{sun} and T_r. The entropy exchange associated with radiation is reflected mostly by the spectral composition of the radiation, with solar radiation consisting of shorter wavelengths in the visible range, while terrestrial radiation consists of wavelengths in the infrared. The shaded area in the graphs indicates the wavelengths of visible light. Source of Earth image: NASA.

at a much lower temperature of about 255 K, which is approximately the Earth's radiative temperature. The associated spectral composition of the emitted radiation is centered at much greater wavelengths in the infrared, so that this radiative flux has a markedly different spectral composition. Because the spectrum of the emitted radiation from the Earth is well separated from the solar spectrum, it is referred to as terrestrial radiation. The difference in spectral composition is associated with a difference in radiative entropy, with solar radiation being associated with low entropy at the Earth's orbit around the Sun, while the emitted, terrestrial radiation is associated with high entropy representative of thermodynamic equilibrium with the Earth's prevailing temperatures (see also Section 2.4.2). The radiative entropy is proportional to the inverse of the temperature at which radiation was emitted, as it is the case for thermal entropy. The entropy exchange of the Earth system to space is thus governed by the large difference of the inverses of the radiative temperatures of the Sun and the Earth system. This is the main aspect of the substantial entropy exchange that is associated by radiative exchange between the Earth and space.

As the planetary energy balance in steady state requires that the incoming radiative flux balances the outgoing flux, the actual consequences of the dynamics that occur in the Earth system is reflected primarily in the entropy exchange with space. It is this substantial entropy exchange by radiation that links to the "struggle for entropy that becomes available through the flow of energy from the hot Sun to the cold Earth," as was already expressed by Boltzmann (1886). This difference in entropy between the absorbed, solar and the emitted, terrestrial radiation allows for substantial entropy production, mostly by radiative processes. This entropy production is associated with the scattering and absorption of solar radiation, which are both irreversible radiative processes (note that reflection of radiation is a reversible process). The most dominant contribution to the entropy production is that of various absorption processes, from photodissociation and photoionization, electronic absorption, to the vibrational, rotational, and translational modes of absorption by molecules. These different absorption processes result in energy conversions at the microscopic scale of atoms, electrons, and molecules, in which radiative energy is converted into chemical energy, e.g., by ozone dissociation in the stratosphere or by photosynthesis at the surface, into electric energy, e.g., by photovoltaic panels, or into thermal energy, e.g., by the vibrational and rotational modes of absorption of water vapor and carbon dioxide in the atmosphere. These absorption processes start the conversion chains of radiative energy into other forms of energy that drive Earth system processes.

The second law of thermodynamics acts to limit the extent to which radiative energy can be converted into forms of free energy. The derivation of this limit requires the consideration of the entropy associated with radiative energy fluxes. This is equivalent to how the Carnot limit was derived for a heat engine, although here we deal with a "radiative engine", or radiative converter that converts radiative energy into other forms, instead of heat being converted into mechanical work. A Carnot-like limit sets the limit for energy conversion from radiation into other forms that do not involve heat. This is, for instance, the case for stratospheric ozone reactions, in which ultraviolet radiation breaks up the bonds of molecular oxygen and thereby generates chemical free energy. Photosynthesis and solar panels are also subject to this conversion limit as these two processes convert radiation directly into other forms of energy (chemical and electric) without an intermediate step of converting radiation into thermal energy. As we will see here, the limit of converting solar radiation is very high as the resulting maximum efficiency involves the difference between the radiative temperatures of the Sun and the Earth. This theoretical limit is above any natural or technological process, but sets the theoretical limit to photochemical conversions as well as solar-based renewable energy technologies.

The other mechanism by which solar radiation is converted into forms of free energy involves the intermediate step of radiative heating and cooling. When absorption of solar radiation and emission of terrestrial radiation take place at different locations and times, this difference in radiative heating and cooling can drive a heat engine that then generates mechanical power. A primary example of this type of conversion is the heating of the surface of the Earth by the absorption of solar radiation. The resulting radiative heating generates a temperature difference to the overlying atmosphere, which is then used by the atmospheric heat engine that generates convection. The limit for this form of energy conversion is much lower than the direct conversion explained earlier because the temperature difference between the surface and the atmosphere is much smaller than the difference in radiative temperatures between the Sun and the Earth. Yet it is this form of radiative heating and cooling that drives most of the dynamics within the climate system, such as vertical convection and the large-scale circulation of the atmosphere.

Radiative transfer plays an additional role as radiative exchange of terrestrial radiation affects surface temperature as well as its spatial and temporal variations. As temperature differences play a critical role in shaping the Carnot limit of heat engines, this role needs to be accounted for. Radiative transfer for both, solar and terrestrial radiation, is furthermore strongly affected by Earth system processes. The temperature of the Earth's surface, for instance, affects the extent of ice and snow cover and thereby how much of the incident solar radiation is reflected, which is described by the surface albedo. The radiative exchange in terrestrial radiation is strongly shaped by the absorption and subsequent reemission of radiation within the atmosphere by gases such as water vapor and carbon dioxide, which result in the atmospheric greenhouse effect and additional radiative heating of the surface. As the concentrations of water vapor and carbon dioxide result from the continuous cycling of water and carbon within the Earth system, the characteristics for the transfer of terrestrial radiation are not fixed, static properties of the atmosphere, but are actively shaped by the dynamics of the Earth system. These interactions between radiative transfer and geochemical cycles result in feedback processes that affect the radiative entropy exchange with space (as shown by the arrows in Fig. 6.1) and thus the overall, dissipative behavior as well as the emergent, thermodynamic state of the Earth system.

In this chapter, radiative entropy is first introduced, including the means to calculate radiative entropy. Radiative exchange processes are then described in terms of their irreversibility and entropy production, and in terms of how they are used later in the derivation of estimates. This is then followed by two thermodynamic conversion limits for radiation: (i) the limit relating to the direct conversion of radiative energy into forms of free energy, and (ii) the limit relating to the indirect conversion through the intermediate step of radiative heating and cooling. The sensitivity of these limits to the radiative forcing is then briefly described. We also see how the radiative fluxes change in the presence of work being performed. The chapter closes with a global synthesis of the radiative forcing and the associated entropy budget of the Earth system.

6.2 Thermodynamics of radiation

The derivation of radiation laws, including the formulation of radiative entropy, is based on the concept of a photon gas. A photon gas is similar to an ideal gas, except that it does not deal with molecules and thermal energy, but with discrete quanta, called photons, that carry electromagnetic energy of different wavelengths or frequencies. This assumption allowed for a statistical treatment of radiation and resulted in a breakthrough that was pioneered by Planck (1906) at the onset of the twentieth century, a time in which formulations that described radiation as an electromagnetic wave could not fully describe the range of observations. His statistical treatment of radiation together with the maximum entropy assumption led to a range of predictions of semi-empirical radiation laws, most notably the Stefan–Boltzmann law that describes the emission of radiation from a blackbody, and also the formulation of radiation entropy. In the following, we first describe the photon gas, how energy is distributed across frequencies, and how it relates to its radiative entropy and then make the connection to radiation laws. What is described here is a brief summary. Most of the expressions used are not needed for the further description of radiative transfer in the following chapters, but the description provides the relevant background for the derivation of the limit for the direct conversion of radiative energy into free energy in Section 6.5. More details on how radiation laws are derived from statistical physics can be found in general physics textbooks (e.g., Feynman et al. 1966) or in specific applications to radiative transfer in atmospheres (Wildt 1956, 1972; Essex 1984; Callies and Herbert 1988; Stephens and O'Brien 1993; Goody and Abdou 1996; Wu and Liu 2010). There are also accessible descriptions of the thermodynamics of radiation found in the engineering literature (e.g., Landsberg and Tonge 1980; Bejan 1997; Kabelac 1994).

The basis for describing the photon gas is the photon, which is characterized by a discrete amount of radiative energy $U_{\text{photon}} = h\nu$, which depends on the frequency ν of the radiation (with $h = 6.63 \cdot 10^{-34}$ J s being Planck's constant). The frequency ν of the photon is connected to the wavelength λ by the speed of light, $c = \lambda\nu$, so that frequencies or wavelengths can be used interchangeably. We consider a large number of photons that form the photon gas inside a cavity (Fig. 6.2) that is in radiative equilibrium, a form of thermodynamic equilibrium, with the surrounding wall at a certain temperature T. Equilibrium implies that the rates of emission and absorption of the wall are equal, so that the cavity or the photon gas neither gain nor lose energy. Furthermore, it is assumed that the wall is a perfect absorber and emitter, that is, it is a blackbody with an emissivity and absorptivity of one. The temperature of the blackbody is then characteristic of the spectral composition of the photon gas, and this characteristic is then referred to as the radiative temperature of the photon gas.

The statistical treatment is similar to the one of an ideal gas, as in Section 2.4, except that the particles are described as being indistinguishable. This treatment results in a somewhat different distribution, which is referred to as the Bose–Einstein statistics. The Bose–Einstein statistics describes the probability for a

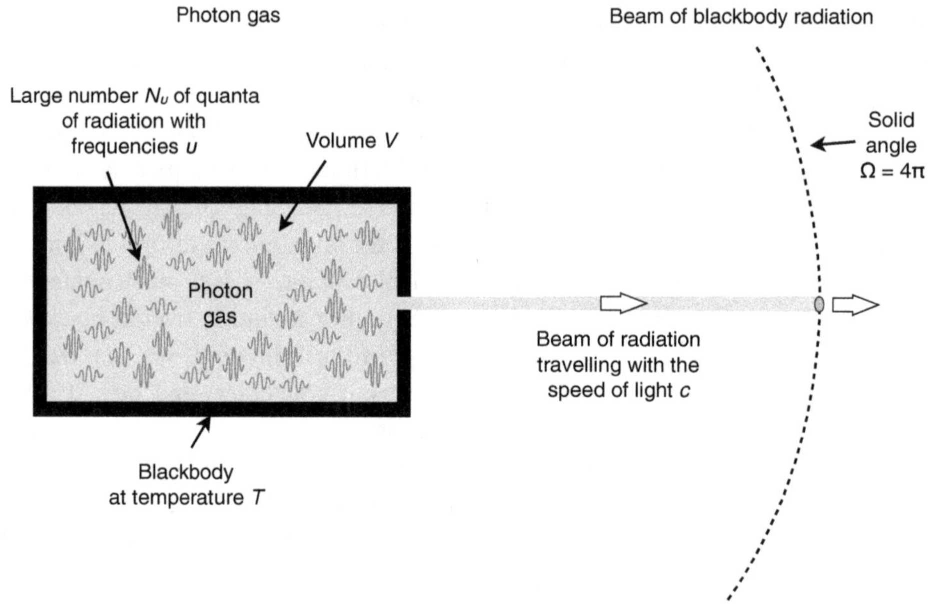

Figure 6.2 Illustration of a blackbody enclosing a photon gas that is in thermodynamic equilibrium with the temperature T of the blackbody. A small opening of the body allows the photon gas to escape and forms a beam of radiation that contains the signature of blackbody radiation.

photon to carry an energy of $h\nu$, and takes the form

$$N_\nu = \frac{1}{e^{h\nu/k_bT} - 1} \tag{6.1}$$

where k_b is the Boltzmann constant and T is the radiative temperature. When this probability is multiplied by the energy of the photon $h\nu$, and the density of states $8\pi\nu^2/c^3$, we arrive at the energy distribution u of the photon gas:

$$u(\nu) = \frac{8\pi h\nu^3}{c^3}\frac{1}{e^{h\nu/k_bT} - 1} \tag{6.2}$$

which describes the radiative energy per unit frequency and unit volume. Note that this expression characterizes unpolarized radiation, which is represented by the factor 8π instead of 4π.

The total energy U of the photon gas inside the volume V of the cavity is obtained by the integration of this energy density over the volume and all frequencies, and yields

$$U = V\int_0^\infty u(\nu)d\nu = aVT^4 \tag{6.3}$$

with a radiation constant $a = 8\pi^5 k_b^4/(15c^3h^3)$. This expression is already very close to the well-known Stefan–Boltzmann law for the total emission of a blackbody. Yet, the expression for U describes the internal energy of the photon gas, and we get to its relationship to radiative fluxes a little further on.

Just as an ideal gas, the photon gas also exerts a pressure p, against its environment. This radiation pressure is related to the momentum of photons, which can be inferred from the photon energy, $U_{\text{photon}} = h\nu = p_{\text{photon}}c$, where p_{photon} is the momentum of the photon. The total pressure by the photon gas can then be inferred from the momenta of the photons, yielding an expression for the radiation pressure p

$$p = \frac{1}{3}\frac{U}{V} \tag{6.4}$$

The entropy of the photon gas is then derived from the total energy of the photon gas and its radiative temperature:

$$S = \frac{U + pV}{T} = \frac{4}{3}\frac{U}{T} \tag{6.5}$$

This expression is almost the same as for an ideal gas, except that it includes an additional term $1/3 \cdot U/T$ that is due to radiation pressure (the pV term in eqn. 6.5).

The description of a photon gas can now be applied to a beam of radiation. To do so, we consider a tiny hole in the blackbody through which a thin beam of radiation escapes to the surroundings. The beam exits the blackbody at the speed of light and has a negligible spread in terms of the solid angle. The solid angle represents the three-dimensional equivalence of an angle and is defined as the fractional area of an enclosing sphere divided by its radius squared. The expressions for the density of energy states, that is, the Bose–Einstein probability distribution N_ν, multiplied by the density of states, then yields an expression for the spectral energy density of the beam K_ν:

$$K_\nu = \frac{2h\nu^3}{c^2} N_\nu \tag{6.6}$$

Note that this expression differs from the energy density of the photon gas by a factor of $4\pi/c$ because we now consider a radiation flux instead of the gas. The specific entropy density L_ν, is then given by (Planck 1906; Rosen 1954)

$$L_\nu = -\frac{k_b \nu^2}{c^2}\left(N_\nu \log(N_\nu) - (N_\nu + 1)\log(N_\nu + 1)\right) \tag{6.7}$$

To illustrate these expressions, Fig. 6.3 shows the functions K_ν and L_ν for the radiative temperatures of the Sun and Earth, respectively. We note that the shapes of the spectra are similar, but the entropy spectra are shifted towards higher wavelengths

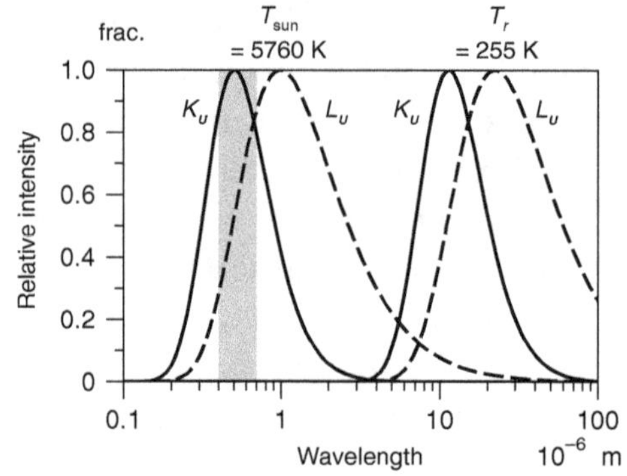

Figure 6.3 Spectral distribution of solar and terrestrial radiation in terms of energy (K_υ, solid lines) and entropy (L_υ, dashed lines). The area shaded in grey indicates the wavelengths of visible light. For comparison, the distribution of the spectra is scaled in respect to their respective maximum value.

and do not decline as fast. This shift in the entropy spectrum is attributable to the greater probability N_υ at greater wavelengths even though the density of states as well as the energy is lower at these greater wavelengths.

These expressions for the spectral energy and entropy density can be integrated to yield the characteristics of the overall radiative flux. When K_υ is integrated over the whole frequency range and over the whole solid angle Ω, this yields an expression for the total radiative energy flux R

$$R = \int_0^\infty \int_\Omega K_\upsilon \cos\theta d\Omega d\upsilon = \sigma T^4 \quad (6.8)$$

with the Stefan–Boltzmann constant $\sigma = 2\pi^5 k_b^4/(15h^3c^2) \approx 5.67 \cdot 10^{-8}$ W m^{-2} K^{-4}. This expression represents the well-known Stefan–Boltzmann law and describes the overall radiative energy emitted by a blackbody at temperature T per unit area of emitting surface in units of W m^{-2}. The same integration of L_υ yields the associated radiative entropy flux J_s:

$$J_s = \int_0^\infty \int_\Omega L_\upsilon \cos\theta d\Omega d\upsilon = \frac{4}{3}\sigma T^3 = \frac{4}{3}\frac{R}{T} \quad (6.9)$$

Note that the expression for the entropy flux associated with radiation is almost the same as for a heat flux, except for the factor 4/3. The additional contribution of 1/3 originates from the radiation pressure (see also eqn. 6.5). This factor has led to contrasting descriptions of the entropy flux of radiation, where some studies have omitted this extra contribution (e.g., Peixoto et al. 1991; Ozawa et al. 2003; Kleidon 2009c). The omission is justified in some cases depending on the focus of the application. This contribution does not play a role when radiation is used to describe heating and cooling sources, that is, processes that deal with the conversion of thermal energy and not directly with the conversion of radiative energy. When radiation is converted into thermal energy by absorption, the contribution of radiation pressure to the entropy of radiation does not add to the thermal entropy of a system. The contribution needs, however, to be considered when the direct conversion of radiation into free energy is considered, e.g., by photochemistry or photovoltaics, or when the entropy production of scattering and absorption is considered. The role of this contribution will become more apparent when we derive the two different limits of solar energy conversion.

6.3 Solar radiation

When we apply thermodynamics to the radiative processes of the Earth, the natural starting point is the emission of solar radiation at the surface of the Sun, after which we can trace it to the Earth until it is absorbed or reflected back into space. A sketch of the relevant properties for this description is shown in Fig. 6.4. We follow the thermodynamic processes involved and estimate how the entropy of solar radiation is affected until it is absorbed and heats the Earth.

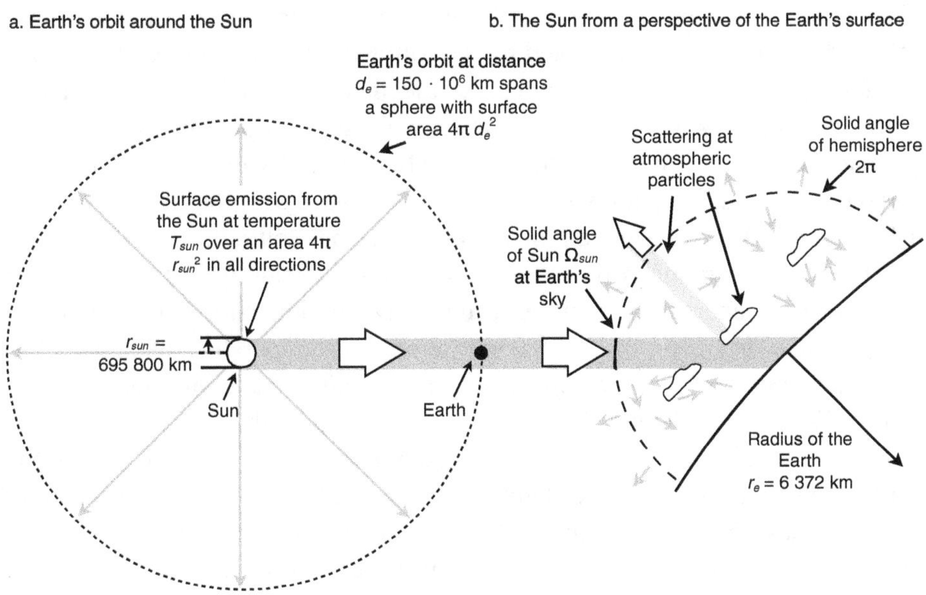

Figure 6.4 Schematic diagram of (a) the Sun–Earth system and (b) the Sun as it appears from the Earth's surface to describe the properties of solar radiation in terms of its thermodynamic disequilibrium as well as the effects of atmospheric scattering. Not drawn to scale.

6.3.1 From Sun to Earth

Solar radiation refers to radiation that is emitted from the Sun. Its spectral composition indicates that the emission process is near a radiative equilibrium with the surface temperature of the Sun with a temperature of about $T_{\text{sun}} = 5760$ K. Using the Stefan–Boltzmann law, this yields a radiative flux of $R_{\text{sun}} = 62.4 \cdot 10^6$ W m^{-2} at the Sun's surface. With the Sun having a radius of $r_{\text{sun}} = 695.8 \cdot 10^6$ m and a surface area of $A_{\text{sun}} = 4\pi r_{\text{sun}}^2 \approx 1.2 \cdot 10^{19}$ m^2, this results in a total emission of solar radiation of the Sun of $R_{\text{sun,tot}} = 7.6 \cdot 10^{26}$ W, which is referred to as the solar luminosity.

As the solar radiation passes space and reaches the Earth's orbit at about $d_{\text{earth}} = 150 \cdot 10^9$ m, little has changed in terms of the total emitted energy as well as its entropy. After all, there were no processes that altered the composition of the solar radiation as it passes through space, so its radiative flux as well as its spectral composition remains unchanged. The total radiative flux that passes through a sphere at the distance d_{earth} is the same as the solar luminosity, that is, $R_{\text{sun,tot}} = 4\pi d_{\text{earth}}^2 R_{s,\text{in}}$, where $R_{s,\text{in}}$ is the solar radiative flux per unit area at the Earth's orbit at the top of the atmosphere. Solving for $R_{s,\text{in}}$ yields a value of about 1340 W m^{-2}. This value is known as the solar constant even though it is not a constant as there are variations in solar luminosity as well as the Earth's orbit that affect its value. This estimate is close to observations, which yield a value of $R_{s,\text{in}} = 1370$ W m^{-2}.

The large thermodynamic disequilibrium of the incoming solar radiation at the Earth's orbital distance is illustrated by the following considerations. The radiative flux of $R_{s,\text{in}} = 1370$ W m^{-2} represents the flux emitted by a blackbody with an emission temperature of 394 K. When we also consider that solar radiation illuminates the cross section of the Earth (which is πr_e^2 with $r_e = 6.372 \cdot 10^6$ m being the Earth's radius), although emission takes place over the surface area of the Earth (which is $4\pi r_e^2$), the mean blackbody radiation would drop down to $R_{s,\text{in}}/4$. Note that this also requires that the Earth rotates sufficiently fast and has means to store thermal energy to maintain a uniform emission rate. This radiative flux corresponds to an even lower blackbody emission temperature of about 279 K. These blackbody temperatures are much lower than the emission temperature of the Sun. The solar radiative flux at the Earth's orbit is thus in a strong state of thermodynamic disequilibrium that is reflected in the difference between the Sun's emission temperature and the radiative temperature that would result when solar radiation is absorbed and reemitted at the distance of the Earth's orbit. Important effects, such as the reflectivity of the Earth, have been neglected so far in these estimates; they were only done here to illustrate that solar radiation is far from its radiative equilibrium at the orbit of the Earth.

The thermodynamic disequilibrium in solar radiation is reflected in two aspects: first, its spectral composition is substantially different, consisting of shorter wavelengths as would be expected from a blackbody that emits $R_{s,\text{in}}$ or less (as already shown in Fig. 6.1). The second aspect that reflects disequilibrium is less obvious and deals with the narrow solid angle into which solar radiation is concentrated in. This aspect is illustrated in Fig. 6.4b. From the perspective of the Earth's surface, the sky covers a solid angle of 2π (a hemisphere), yet the Sun covers only a minute fraction of the sky. The solid angle of the Sun, Ω_{sun}, reflects the size of the emitting surface of the Sun in relation to the size of the sphere that encompasses the Earth's orbital distance. It is thus given by $\Omega_{\text{sun}} = (r_{\text{sun}}^2/d_{\text{earth}}^2)\pi \approx 6.8 \cdot 10^{-5}$ sr. This means that solar radiation covers a tiny fractional area of the sky, of less than 0.001%. When we combine the mean radiative flux of $R_{s,\text{in}}$ with the tiny area over which this flux is concentrated, we recover the original intensity at which radiation was emitted at the Sun's surface (which can be seen by following the equations). In other words, when the concentration of solar radiation into the narrow solid angle is being considered, we note that neither the intensity nor the spectral composition of sunlight has changed when it reaches the Earth's orbit. This high concentration of solar radiation is another manifestation of solar radiation being in a strong thermodynamic disequilibrium at the Earth's orbit.

6.3.2 Scattering

Before solar radiation is absorbed, it passes through the Earth's atmosphere and is subject to scattering. Scattering is an irreversible process as it broadens the solid angle even if the spectral composition does not change (it does, as discussed later). To deal with scattering and its effect on the radiative entropy, the concept of diluted radiation was introduced (Press 1976; Landsberg and Tonge 1979). To do so, the radiative flux is written in the form

$$R_{s,\text{in}} = B\epsilon\, \sigma T_{\text{sun}}^4 \tag{6.10}$$

where B is a geometric factor and ϵ is a dilution factor. For incoming solar radiation at the top of the atmosphere, $B = \Omega_{\text{sun}}/4\pi$ and $\epsilon = 1$. If we consider a scattering process that does not affect the radiative flux itself, but only broadens its solid angle, this can be accomplished in the formulation by a greater value of B and a lower value of ϵ in such a way that the product of these two factors remains unchanged. In the context of the photon gas, this process leaves the relative distribution of photons unchanged, yet it spreads them over a greater space and results in a uniform reduction of N_ν by the factor ϵ. This, then, alters the expression for L_ν as N_ν affects several logarithmic terms (cf. eqn. 6.7) and thus the entropy of the radiative flux.

The effect of this dilution on the entropy of the radiative flux can be done analytically (Press 1976; Landsberg and Tonge 1979; Stephens and O'Brien 1993; Goody and Abdou 1996). The effect of dilution is represented by a function $X(\epsilon)$ in the

Table 6.1 *Estimates for the entropy increase by scattering of solar radiation due to dilution. The entropy increase is given in relative terms of the flux of scattered radiation,* $J_{s,\text{scat}}$*, in relation to the flux before scattering,* J_s*.*

Variable	$\Omega = \Omega_{\text{sun}}$	$\Omega = 2\Omega_{\text{sun}}$	$\Omega = \pi$	$\Omega = 2\pi$	$\Omega = 4\pi$
$B(-)$	$5.41 \cdot 10^{-6}$	$1.08 \cdot 10^{-5}$	0.25	0.5	1
$\epsilon(-)$	1	0.5	$2.17 \cdot 10^{-5}$	$1.08 \cdot 10^{-5}$	$5.41 \cdot 10^{-6}$
$X(-)$	1	0.59	$8.55 \cdot 10^{-5}$	$4.48 \cdot 10^{-5}$	$2.35 \cdot 10^{-5}$
$J_{s,\text{scat}}/J_s(-)$	1	1.18	3.95	4.14	4.33

expression of the entropy flux:

$$J_s = \frac{4}{3} B X(\epsilon)\, \sigma T_{\text{sun}}^3 \tag{6.11}$$

A numerical approximation of the function $X(\epsilon)$ is given by (Landsberg and Tonge 1979)

$$X(\epsilon) \approx \epsilon(0.9652 - 0.2777 \ln \epsilon + 0.0348\epsilon) \tag{6.12}$$

Example cases for different intensities of dilution are given in Table 6.1. What the examples show is that in the extreme case, the dilution due to scattering increases the entropy of solar radiation by a factor of up to 4.33. The entropy production by the scattering process is then given by

$$\sigma_{s,\text{scat}} = J_{s,\text{scat}} - J_{s,\text{in}} \tag{6.13}$$

where J_s and $J_{s,\text{scat}}$ represent the entropy fluxes of the radiation before and after the dilution takes place. With $T_{\text{sun}} = 5760$ K, eqn. 6.11 yields an influx of solar radiation entropy of about $J_{s,\text{in}} \approx 78$ mW m^{-2} K^{-1}. As this is undiluted radiation, it can equally be calculated by $J_{s,\text{in}} = 4/3 \cdot (R_{s,\text{in}}/4)/T_{\text{sun}}$. At most, solar radiation is scattered up to the whole solid angle. In this case, the entropy of the scattered, solar radiative flux is given by $J_{s,\text{scat}} = 339$ mW m^{-2} K^{-1}. This yields an upper bound of the entropy produced by the scattering of solar radiation of $\sigma_{s,\text{scat}} \approx 261$ mW m^{-2} K^{-1}. While the entropy of solar radiation is low even when scattered, this entropy increase due to dilution is an important factor that needs to be considered when we deal with solar conversion limits in Section 6.5.

In addition to the dilution of radiation, atmospheric scattering also changes the spectral composition as the process is wavelength dependent. Under clear-sky conditions, the intensity of Rayleigh scattering by which solar photons scatter at the electrons of atmospheric molecules depends on the inverse of wavelength to the fourth power (i.e., λ^{-4}). This scattering results in blue skies on clear days. The scattered solar radiation is then not just diluted, but also preferentially in the blue range of the visible spectrum. Since blue light corresponds to highly energetic photons of low entropy, the scattered radiation by Rayleigh scattering has a lower entropy than what would be expected from the dilution effect alone. On the other hand, the entropy of the remaining, undiluted radiation is enhanced due to this preferential scattering. The scattering of solar radiation that involves larger objects (Mie and geometric scattering), such as aerosols and cloud droplets, is also wavelength dependent, although the effect is not as pronounced. The actual magnitude of scattering depends on the path length and the concentration of scattering material as solar radiation passes through the atmosphere. It hence depends on the zenith angle as well as the concentration of aerosols and clouds. To distinguish between the components of solar radiation that experience scattering and those that do not, these components are referred to as diffuse radiation and direct radiation. Diffuse radiation is the diluted radiation that experienced scattering and is of higher entropy compared to the direct radiation that still represents most of the disequilibrium of the incident solar radiation at the top of the atmosphere.

6.3.3 Absorption

Solar radiation is absorbed by different processes in the atmosphere and at the surface. The type of process involved is generally related to the energy of the photon that is being absorbed, and thus on wavelength ranges.

The absorption of highly energetic photons with wavelengths in the visible range (400 nm $< \lambda <$ 700 nm) and shorter ($\lambda <$ 400 nm) result in chemical alterations and ionization of molecules, thus generating chemical and electrical energy. In this range, the energy of a photon is in the order of the binding energy of chemical compounds or of electrons. Thus, it is able to alter the chemical configuration of molecules. Examples include photoionization in the Earth's ionosphere, the breakup of the chemical bonds of molecular oxygen by photodissociation in the stratosphere which results in ozone production, and the photolysis of water by which photosynthesis produces chemical energy. In the visible range, absorption takes place mostly in the form of electronic absorption. Photons have an energy that is comparable to the energy difference between the electronic states; thus it is able to result in electronic transitions in atoms and molecules. These processes are revisited later on when we deal with chemical energy generation by photochemical processes in Chapter 9.

Longer wavelengths in the infrared and beyond (with $\lambda >$ 700 nm) are mostly absorbed by vibrational, rotational, and translational modes of molecules. Molecules that have an uneven distribution of electrons and thus a permanent dipole moment, such as CO_2 and H_2O, interact with radiation by absorbing and emitting photons. This absorption generally results in the direct conversion into thermal energy, but not into chemical alterations. This lack of ability to convert into non-thermal forms of energy is related to the long wavelength and the associated low photon energy.

The actual amount of absorption depends on the ability of a substance to absorb or reflect radiation. This ability is characterized by the albedo α, which ranges in values from zero to one. It describes the reflectivity of a substance, with greater values corresponding to greater reflectivity. Albedo values can vary greatly among different surface covers and clouds, but also depend on, e.g., the angle of the incident solar radiation or on wavelength. A summary of typical values of albedo for the reflection of solar radiation is given in Table 6.2. As can be seen by the

Table 6.2 *A selection of different albedo values to describe the reflectivity to solar radiation for different atmospheric and surface properties. After Hartmann (1994).*

	Range (%)	Typical value (%)
Planet		
Planetary albedo		30
Atmosphere		
Cirrus clouds		21
Cumulus clouds		48
Stratus clouds		69
Ocean		
Deep ocean, small zenith angle	3–10	7
Deep ocean, large zenith angle	10–100	
Ice and snow		
Sea ice	30–45	30
Fresh snow	70–95	80
Old snow	35–65	50
Old snow, forested	11–35	25
Land (bare)		
Desert	20–45	30
Wet sand	20–30	25
Dry sand	30–45	35
Wet clay	10–20	15
Dry clay	20–40	30
Concrete	15–35	20
Asphalt pavement	5–10	7
Land (vegetated)		
Tundra	18–25	15
Grassland	16–26	19
Coniferous forest	5–15	12
Deciduous forest	10–20	17
Evergreen forest	12–25	13
Cropland		18

examples in the table, there are many aspects that affect how much solar radiation is absorbed by the Earth system. These include atmospheric characteristics, specifically the amount and type of clouds, and surface characteristics such as snow, ice, and vegetative cover on land. These differences in the albedo are important because they alter the magnitude of the radiative forcing and thereby the thermodynamic forcing. When integrated over the whole planet for present-day conditions, the Earth's planetary albedo is about 30%. It is important to keep in mind that this value is not a fixed system property, but emerges from the internal dynamics of the Earth system that shape, for instance, the formation of clouds, the abundance of snow and ice, or the presence of vegetative cover on land.

6.3.4 Estimates

The naming convention for solar radiative fluxes that is used here is summarized in Fig. 6.5. We consider an incoming, solar radiative flux at the top of the atmosphere $R_{s,\mathrm{in}}$, which corresponds to the flux we derived earlier. A fraction of this influx is absorbed within the atmosphere, $R_{s,a}$, e.g., by stratospheric ozone, water vapor, aerosols, or clouds. The downward flux of solar radiation reaching the surface is described by $R_{s,\mathrm{down}}$ and includes both the direct and diffuse components.

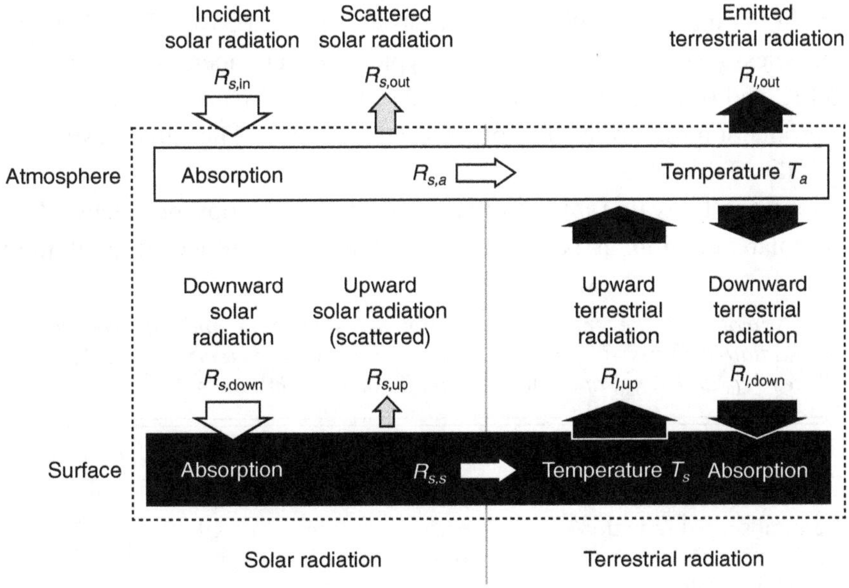

Figure 6.5 Schematic diagram of the fluxes of solar and terrestrial radiation between the Earth's surface and atmosphere. The role of convective fluxes is explored separately.

The fraction that is reflected at the surface is expressed by $R_{s,\mathrm{up}}$, so that the net absorption of solar radiation at the surface is expressed by

$$R_{s,s} = R_{s,\mathrm{down}} - R_{s,\mathrm{up}} \tag{6.14}$$

As surface reflection is described by the surface albedo α_s, with $R_{s,\mathrm{up}} = \alpha_s R_{s,\mathrm{down}}$, the net heating of the surface by absorption of solar radiation is given by

$$R_{s,s} = (1 - \alpha_s) R_{s,\mathrm{down}} \tag{6.15}$$

The total absorption of solar radiation is given by the sum of the surface and atmospheric absorption. This total absorption corresponds to the difference in solar radiative fluxes at the top of the atmosphere, for which we use the symbol $R_{s,\mathrm{toa}}$:

$$R_{s,\mathrm{toa}} = R_{s,s} + R_{s,a} \tag{6.16}$$

This notation will be used in the remainder of the book to describe fluxes and absorption of solar radiation.

Estimates for the present-day climate place these absorption rates at $R_{s,s} = 165$ W m^{-2} and $R_{s,a} = 75$ W m^{-2} in the global, climatic mean (Stephens et al. 2012). These numbers yield a total, planetary absorption of $R_{s,\mathrm{toa}} = 240$ W m^{-2}, which corresponds to a planetary albedo of about 30% combined with the mean, incident solar influx of $R_{s,\mathrm{in}}/4$. These rates clearly vary greatly spatially and temporally at a given location due to the geographic as well as diurnal and seasonal variations of the incident sunlight and variations in cloud cover. The global means can nevertheless be used together with the typical mean temperatures of the surface, $T_s = 288$ K, and of the atmosphere, $T_a = 255$ K, to estimate approximate rates of entropy production due to absorption of solar radiation. The global estimates are summarized in Table 6.3.

The first three lines in Table 6.3 list estimates for entropy production by scattering of solar radiation. It is assumed that the overall reflected solar radiation

Table 6.3 *Estimates of the entropy production by scattering and absorption of solar radiation. Radiative fluxes R_s, are given in units of W m^{-2}, temperature T, in K, and entropy fluxes $J_{s,\mathrm{in}}$ and $J_{s,\mathrm{out}}$, and entropy production σ, in mW m^{-2} K^{-1}.*

Process	R_s	T	$J_{s,\mathrm{in}}$	$J_{s,\mathrm{out}}$	σ
Scattering of reflected radiation	102.5		24	98	74
Scattering of absorbed radiation	82.5		19	79	60
Scattering (total)	185		43	177	134
Atmospheric absorption	75	255	17	392	375
Surface absorption	165	288	98	764	666
Absorption (total)	240				1041
Absorption at radiative temperature	240	255	115	1255	1140

associated with the Earth's planetary albedo ("scattering of reflected radiation") as well as half of the solar radiation absorbed at the surface (a rough approxiation, "scattering of absorbed radiation") is scattered into the solid angle of a hemisphere (see Table 6.1). This yields an estimated radiative flux of about 185 W m^{-2} with an entropy production of $\sigma_{s,\text{scat}} = 134$ mW m^{-2} K^{-1}.

For the entropy production by atmospheric absorption of solar radiation, we assume that direct solar radiation is absorbed and subsequently emitted at a temperature T_a into the whole solid angle. The entropy influx is then given by $4/3 \cdot R_{s,a}/T_{\text{sun}} \approx 17$ mW m^{-2} K^{-1}, which is the value $J_{s,\text{in}}$ estimated earlier, reduced to the value of $R_{s,a}$. When emitted, the entropy export is given by $4/3 \cdot R_{s,a}/T_a \approx$ 392 mW m^{-2} K^{-1}. Hence, this results in an estimate for the entropy production of solar absorption in the atmosphere of about $\sigma_{s,abs,a} = 375$ mW m^{-2} K^{-1}. For the entropy production by surface absorption of solar radiation, we follow the same procedure, except that we assume that half of the absorbed solar radiation has been scattered (a reasonable approximation; we use observations in Section 11.4 for more detailed estimates related to solar renewable energy). Hence, the entropy influx is $(1/2) \cdot (4/3) \cdot R_{s,s}/T_{\text{sun}} + 79$ mW m^{-2} K^{-1} $\approx$ 98 mW m^{-2} K^{-1}. The entropy export at emission is given by $4/3 \cdot R_{s,s}/T_s \approx 764$ mW m^{-2} K^{-1}, so that the entropy produced at the surface is $\sigma_{s,abs,s} = 666$ mW m^{-2} K^{-1}. When these two estimates are combined, the total entropy production due to the absorption of solar radiation is $\sigma_{s,abs} = 1041$ mW m^{-2} K^{-1}.

Also given in Table 6.3 is the estimate for the entropy production if all absorption and reemission of solar radiation takes place at a temperature 255 K ("absorption at radiative temperature"). Then, the total entropy production by absorption would be about 1140 mW m^{-2} K^{-1} and thus quite a bit greater, by about 99 mW m^{-2} K^{-1}, than the estimate using the numbers from earlier. As we will see, this additional potential to produce entropy relates to the radiative transfer of terrestrial radiation and convective heat fluxes within the atmosphere, which downgrades the solar radiation absorbed at the surface to the lower temperature of the atmosphere.

6.4 Terrestrial radiation

At the planetary level, the total absorption of solar radiation balances the emission of radiation to space in a climatological mean state. Yet, what sounds so simple at the planetary scale gets more complicated by the radiative transfer of terrestrial radiation within the atmosphere, which involves substantial amounts of re-absorption of the already emitted radiation. The magnitude of this effect manifests itself in the observed downward flux of terrestrial radiation at the surface of $R_{l,\text{down}} = 346$ W m^{-2} (Stephens et al. 2012). Given that the absorption of solar radiation at the surface is about $R_{s,s} = 165$ W m^{-2}, this downward flux of terrestrial radiation contributes more than twice to the radiative heating of the surface. This effect is well established and represents the atmospheric greenhouse effect.

From a thermodynamic perspective, the relevance of this radiative transfer of terrestrial radiation is twofold: first, it strongly shapes the vertical temperature difference between the surface and the atmosphere, a difference which is used to derive work for atmospheric convection; and second, it strongly shapes surface temperature, with implications for energy conversions, e.g., associated with hydrologic cycling or biotic processes. To obtain a minimum representation of radiative transfer, we use three radiative temperatures: the radiative temperature of the planet T_r, that characterizes the emission of radiation to space, the temperature of the surface T_s, and an effective radiative temperature of the atmosphere T_a. In most cases, we assume $T_a \approx T_r$ for simplicity, which assumes that there is a sufficient mass of absorber in the atmosphere.

The radiative temperature of the Earth T_r, is a useful and common characterization of the planetary energy balance. Its value is simply derived from the mean planetary energy balance by

$$R_{s,\text{toa}} - \sigma T_r^4 = 0 \tag{6.17}$$

With a value of $R_{s,\text{toa}} = 240$ W m^{-2}, this yields a value of the Earth's radiative temperature of approximately $T_r = 255$ K. As the radiative temperature characterizes the planetary energy balance, it provides an important reference temperature for the formulation of temperature differences, and also broadly sets the magnitude of the planetary radiative entropy balance (see e.g., Aoki 1983). We revisit this planetary entropy balance in Section 12.3 when we deal with the insights that thermodynamics can provide for habitable environments.

We derive an expression for the surface temperature T_s, from the surface energy balance. At this stage, we neglect convective heat fluxes, as we only deal with the extent to which radiative exchange shapes the temperature difference between the surface and the atmosphere and consider the effect of convection when we deal with its thermodynamic limit (Section 6.6). The surface energy balance is then described by the absorption of solar radiation at the surface $R_{s,s}$, the emission of terrestrial radiation into the atmosphere $R_{l,\text{up}}$, and the absorption of terrestrial radiation that was emitted downward by the atmosphere $R_{l,\text{down}}$:

$$R_{s,s} - R_{l,\text{up}} + R_{l,\text{down}} = 0 \tag{6.18}$$

We assume that the surface emits like a blackbody, so that $R_{l,\text{up}} = \sigma T_s^4$. For $R_{l,\text{down}}$, we use the grey atmosphere approximation in which the absorption of terrestrial radiation is assumed to be independent of wavelength. This approximation describes the cumulative effect of absorption and reemission of terrestrial radiation within the atmosphere (see e.g., Ozawa and Ohmura 1997; Lorenz et al. 2001, for the use of a grey atmosphere and thermodynamics). Using this approximation, the downward flux is expressed as:

$$R_{l,\text{down}} = \frac{3}{4}\tau R_{s,\text{toa}} \tag{6.19}$$

where τ is the optical depth, or optical thickness of the atmosphere for terrestrial radiation. It is an integrated measure of the absorbing mass, so that a greater mass of absorbers translates into a greater optical depth, which in turn results in a greater extent of absorption and reemission. The optical depth depends on the concentration of atmospheric greenhouse gases, such as water vapor and carbon dioxide, but also cloud cover. A greater optical depth then results in a greater greenhouse effect. Note that eqn. 6.19 essentially expresses the greenhouse effect as a function of the total absorbed solar radiation, with the multiplier being $3/4 \cdot \tau$. We can use eqn. 6.19 to infer the present-day value of the optical depth by using $R_{l,\text{down}} = 346$ W m^{-2} and $R_{s,\text{toa}} = 240$ W m^{-2}. These values yield an optical depth of about $\tau = 1.9$ for present-day conditions.

When these formulations are combined, the surface temperature can be inferred from the surface emission, $R_{l,\text{up}}$, and the surface energy balance given by eqn. 6.18:

$$\sigma T_s^4 = R_{s,s} + \frac{3}{4}\tau \cdot R_{s,\text{toa}} \tag{6.20}$$

To get a better impression of the factors that shape the surface temperature, surface emission is linearized by a Taylor expansion around the radiative temperature, so that

$$R_{l,\text{up}} \approx R_{s,\text{toa}} + k_r(T_s - T_r) \tag{6.21}$$

with $k_r = 4\sigma T_r^3 = 4R_{s,\text{toa}}/T_r$ and which has a value of about $k_r = 3.76$ W m^{-2} K^{-1} for present-day conditions. We can then express the net exchange of terrestrial radiation, $R_{l,\text{net}}$, in linearized form as well:

$$R_{l,\text{net}} = R_{l,\text{up}} - R_{l,\text{down}} = R_{l,0} + k_r(T_s - T_r) \tag{6.22}$$

where $R_{l,0} = (1 - 3\tau/4) \cdot R_{s,\text{toa}}$. This linearized expression is used later in the derivation of thermodynamic limits. One should note that this linearized form is a gross simplification and yields biases, which we accept here for the sake of simplicity.

We can then obtain an approximation of the surface temperature from eqn. 6.20, which we express by T_s' to note the approximation and obtain

$$T_s' \approx T_r \left(\frac{3}{4} + \frac{R_{s,s}}{4R_{s,\text{toa}}} + \frac{3\tau}{16}\right) \tag{6.23}$$

This expression shows that atmospheric absorption of solar radiation ($R_{s,a}$) acts to reduce surface temperature. If all absorbed solar radiation were absorbed at the surface, the second term on the right hand side of eqn. 6.23 would be greater, thereby enhancing the surface temperature. We also note the role of the greenhouse effect on surface temperature, which increases T_s' because the third term depends linearly on τ. This expression thus illustrates that surface temperature is shaped by the magnitude of solar radiation, where it is being absorbed, and by the atmospheric greenhouse effect.

To illustrate the role of the greenhouse effect on the resulting temperature difference, $T_s - T_a$, which we use later for the derivation of a maximum power limit, the sensitivity of T_s and T_a to τ is shown in Table 6.4 for the exact expression of T_s as well as the approximated form, T_s'. In the case of no absorption of terrestrial radiation in the atmosphere ($\tau = 0$), the surface temperature corresponds to the radiative temperature, with $R_{l,\text{up}} = R_{s,\text{toa}}$. In this case, the planetary energy balance is represented by the surface energy balance. With an increased optical thickness of the atmosphere and a stronger greenhouse effect, the surface temperature and the temperature difference increases to greater values while the radiative temperature T_a is fixed by the planetary energy balance.

When the downward flux of terrestrial radiation matches the observed value at $\tau = 1.9$, the surface temperature is much greater by 38 K than the observed surface temperature of 288 K. As we will see further on, this bias results from the absence of convective cooling in these estimates that is associated with mechanical work being derived from the temperature difference $T_s - T_a$. By affecting the surface temperature, it furthermore affects the rates of entropy production associated with

Table 6.4 *Estimates of global radiative fluxes between the surface and the atmosphere and associated temperatures using the grey atmosphere approximation for different values of the optical depth, τ. Note that these estimates do not include the important effects of convective heat transport in the atmosphere.*

τ	0.0	0.5	1.0	1.5	1.9	2.0	2.5
$R_{s,\text{toa}}$ (W m^{-2})	240	240	240	240	240	240	240
$R_{s,s}$ (W m^{-2})	240	165	165	165	165	165	165
$R_{s,a}$ (W m^{-2})	0	75	75	75	75	75	75
$R_{l,\text{down}}$ (W m^{-2})	0	90	180	270	342	360	450
$R_{l,\text{up}}$ (W m^{-2})	240	255	345	435	507	525	615
T_s (K)	255	259	279	296	308	310	323
T_s' (K)	255	259	283	307	326	331	355
T_a (K)	–	255	255	255	255	255	255
$T_s - T_a$ (K)	0	4	24	41	52	55	68

radiative transfer, even though the planetary rate of entropy production is strongly constrained by the planetary energy balance and the radiative temperature. We thus postpone the estimation of entropy production by absorption of terrestrial radiation to the end of this chapter and deal with conversion limits first. At the end, we then include the effect of convective cooling by using the maximum power limit of convection.

6.5 Limits to converting radiative energy

We first derive a thermodynamic limit to radiative energy conversion that concerns the direct conversion of solar radiation into a form of free energy, as in the case of e.g., photochemistry or photovoltaics. The derivation of this limit follows basically the derivation of the Carnot limit of a heat engine, but we consider a "radiative converter" instead (Fig. 6.6), so the conversion is not driven by heat fluxes and thermal entropy, but by radiative fluxes and radiation entropy. The derivation of this limit follows the work of Petela (1964, 2003), Press (1976), and Landsberg and Tonge (1980), but is kept at a simpler level and in a way that is consistent with how energy balances were formulated before.

The converter receives a solar radiative influx $R_{s,\mathrm{in}}$, that is balanced by a net radiative flux $R_{l,\mathrm{out}}$, a net convective (or conductive) heat flux J_{out}, and generated power G. The first law states that in steady state, these terms balance:

$$R_{s,\mathrm{in}} = R_{l,\mathrm{out}} + J_{\mathrm{out}} + G \tag{6.24}$$

The second law requires that during this energy conversion, at least as much entropy is exported as the converter receives. The entropy balance considers the addition of entropy by the influx of radiation with an entropy $J_{s,\mathrm{in}}$, the removal of entropy $J_{s,\mathrm{out}}$, by the net radiative flux $R_{l,\mathrm{out}}$, and by the heat flux J_{out}/T, where T is the temperature of the converter. In the ideal case in which there is no internal entropy production

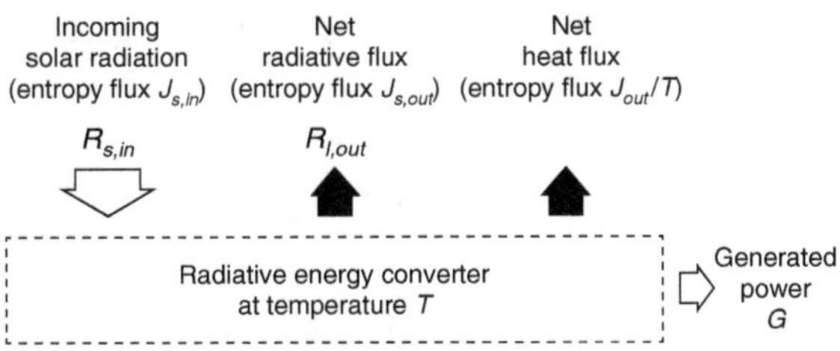

Figure 6.6 Schematic diagram of a radiative energy converter. Incoming solar radiation $R_{s,\mathrm{in}}$ with an associated entropy flux $J_{s,\mathrm{in}}$ is converted into power G, a net radiative flux $R_{l,\mathrm{out}}$ with entropy flux $J_{s,\mathrm{out}}$, and a heat flux J_{out}, which exports thermal entropy at a rate J_{out}/T, with T being the temperature of the converter.

within the converter, the entropy balance in steady state is described by

$$J_{s,\text{in}} = J_{s,\text{out}} + \frac{J_{\text{out}}}{T} \tag{6.25}$$

The combination of these two balances yields an expression of the Carnot limit for the power, $G_{\text{Carnot,rad}}$, that can be derived from the conversion process

$$G_{\text{Carnot,rad}} = R_{s,\text{in}} - R_{l,\text{out}} - T(J_{s,\text{in}} - J_{s,\text{out}}) \tag{6.26}$$

This derivation so far has been very general. To get a specific expression that involves radiative temperatures, we need to make certain assumptions on the form of the radiative entropy fluxes. Based on eqn. 6.11, the entropy influx is expressed as $J_{s,\text{in}} = 4/3 \cdot X/\epsilon \cdot R_{s,\text{in}}/T_{\text{sun}}$. In case of direct solar radiation in which it is constrained to the solid angle Ω_{sun}, we have $X = 1$ and $\epsilon = 1$. In case scattering occurred and the entropy of radiation has increased or the narrow solid angle cannot be used by the conversion process (as is the case for photochemistry and photovoltaics), X and ϵ have smaller values and the ratio X/ϵ is enhanced, as in Table 6.1. For the outgoing radiative flux, we use $J_{s,\text{out}} = 4/3 \cdot R_{l,\text{out}}/T$, but do not specify $R_{l,\text{out}}$ as in Petela (1964) because the net radiative cooling rate is not simply σT^4 but is affected by the atmospheric greenhouse effect.

With these assumptions of the functional form of the radiative entropy fluxes, we can use eqn. 6.25 to express J_{out} as a function of $R_{s,\text{in}}$ and $R_{l,\text{out}}$:

$$J_{\text{out}} = \frac{4}{3}\frac{X}{\epsilon}\frac{T}{T_{\text{sun}}} R_{s,\text{in}} - \frac{4}{3} R_{l,\text{out}} \tag{6.27}$$

We now consider two cases: In the first case, all entropy export is in radiative form with $J_{\text{out}} = 0$. This yields a limit of

$$G_{\text{Carnot,rad}} = R_{s,\text{in}} \left(1 - \frac{X}{\epsilon}\frac{T}{T_{\text{sun}}}\right) \tag{6.28}$$

The other case exports entropy entirely by the heat flux J_{out}, so that $R_{l,\text{out}} = 0$. In this case, the limit is given by:

$$G_{\text{Carnot,rad}} = R_{s,\text{in}} \left(1 - \frac{4}{3}\frac{X}{\epsilon}\frac{T}{T_{\text{sun}}}\right) \tag{6.29}$$

Note that the only difference is in the factor of 4/3, which reduces the limit when the entropy export is accomplished by a heat flux instead of a radiative flux.

These expressions of $G_{\text{Carnot,rad}}$ are very similar to the Carnot limit, except for the factors X/ϵ and $4/3 \cdot X/\epsilon$. As this factor is not less than 1, the efficiency can be smaller than what would be expected for a Carnot engine operating with the same temperatures except in the first case for direct radiation. Yet, because T/T_{sun} is very small on Earth, the maximum efficiency is nevertheless very high.

The highest efficiency can be achieved when the narrow solid angle of solar radiation is used and the entropy export is accomplished by radiation. In this case, $X(\epsilon)/\epsilon = 1$ and using typical temperatures of $T = 288$ K and $T_{sun} = 5760$ K, this yields a maximum efficiency for the conversion of direct solar radiation of $\eta_{direct} = 95\%$. For the second case in which the entropy export is accomplished by the heat flux, this efficiency is reduced to $\eta_{direct} = 93\%$. Nevertheles, the efficiency is near one, indicating that almost all of the direct solar radiation could in theory be converted into another form. This would, however, require a mechanism that makes explicit use of direct solar radiation. One of such a mechanism is a form of solar energy technology called direct concentrated solar (DCS), which concentrates direct radiation by using mirrors, a topic we will get back to when dealing with the limits of renewable energy in Section 11.4. Yet, natural processes typically cannot exploit this form of energy conversion.

The maximum conversion efficiency is less when the small solid angle is not explicitly used or if the solar radiation is already scattered and has a higher entropy. In this case, the starting point of energy conversion is diluted solar radiation, and the dilution enhances the factor $X(\epsilon)/\epsilon > 1$. Using the numerical approximation for $X(\epsilon)$ for a scattering to a solid angle of 2π from earlier (eqn. 6.12 or Table 6.1) yields a maximum efficiency for the conversion of diffuse solar radiation of $\eta_{diffuse} = 79\%$ and $\eta_{diffuse} = 72\%$ for the two cases. This is still a very high efficiency, although it is notably less than the maximum efficiency of using direct solar radiation.

Using a mean absorption of 240 W m^{-2}, these efficiencies yield limits on generation rates into other forms of energy of 173 - 228 W m^{-2}, which is substantially higher than generation rates of natural processes and what can currently be achieved by technology, as we will see later on. Examples of processes that are subjected to this maximum efficiency are photochemistry and photovoltaics.

6.6 Limits to converting radiative heating

A much more relevant limit for natural processes in the Earth system is the thermodynamic limit related to how much mechanical power can be generated out of radiative heating and cooling differences. In this case, we do not deal with radiation entropy, but rather the heating and cooling rates after absorption or before emission. As this limit deals with conversion of thermal energy into other forms of free energy, it is subjected to the same maximum power limit that was described in Section 4.5. The only modification in the following is that we combine this limit with the treatment of radiative exchange more explicitly.

What we ask in the following is how much power can be extracted from the vertical temperature difference that is maintained between the surface and the atmosphere by radiative exchange. For this, we consider a convective heat engine that operates with a certain heat flux J, and is driven by the temperature difference $T_s - T_a$. To derive the limit on the power that this engine can generate, we use the Carnot limit. We also need to consider the effect of the heat flux on the driving temperature difference. This is accomplished by adapting the surface energy balance from eqn. 6.18 to account for the convective heat flux J that passes through the engine and that represents an additional cooling term:

$$R_{s,s} - R_{l,\mathrm{up}} + R_{l,\mathrm{down}} - J = 0 \tag{6.30}$$

The corresponding energy balance for the atmosphere is given by

$$R_{s,a} + R_{l,\mathrm{up}} - R_{l,\mathrm{down}} - \sigma T_a^4 + J = 0 \tag{6.31}$$

Note that the combination of the two energy balances yields the planetary energy balance, $R_{s,\mathrm{toa}} = R_{s,s} + R_{s,a} = \sigma T_a^4$. For $R_{l,\mathrm{down}}$, we use the grey atmosphere approximation as before.

To derive the maximum power limit of the heat engine, we start with the Carnot limit of a dissipative heat engine:

$$G = J \cdot \frac{T_s - T_a}{T_a} \tag{6.32}$$

As in Section 4.5, we use the surface energy balance to express the temperature difference. To do so, we use the linearized approximation from eqns. 6.22 and 6.23 and obtain

$$T_s - T_a = \frac{R_{s,s} - R_{l,0} - J}{k_r} \approx \left(\frac{3}{16}\tau - \frac{1}{4}\frac{J + R_{s,a}}{R_{s,\mathrm{toa}}} \right) \cdot T_a \tag{6.33}$$

Note how the inclusion of convective heat fluxes J reduces the temperature difference as it enters as a negative term on the right hand side, resulting in convective cooling. This cooling effect represents the trade-off by which a greater convective heat flux diminishes the temperature difference that results in the maximum power limit.

When this expression for $T_s - T_a$ is used in the Carnot limit and maximized by $\partial G/\partial J = 0$, it results in a maximum power limit of

$$G_{\mathrm{max}} = \left(\frac{3}{16}\tau - \frac{1}{4}\frac{R_{s,a}}{R_{s,\mathrm{toa}}} \right)^2 \cdot R_{s,\mathrm{toa}} = \eta_{\mathrm{heat}} \cdot R_{s,\mathrm{toa}} \tag{6.34}$$

The associated optimum efficiency, η_{heat}, with this conversion into mechanical work is directly given by the term in the brackets:

$$\eta_{heat} = \left(\frac{3}{16}\tau - \frac{1}{4}\frac{R_{s,a}}{R_{s,toa}}\right)^2 \quad (6.35)$$

Using the values given before, with $\tau = 1.9$, $R_{s,toa} = 240$ W m^{-2}, and $R_{s,a} = 75$ W m^{-2} yields a value of $\eta_{heat} \approx 8$ % and a maximum power of about $G_{max} = 19$ W m^{-2}. Note that this limit is substantially less than for the case in which solar radiation is converted before it turns into thermal energy, where the maximum efficiencies were 72% and 93%.

Note that the optimum efficiency is less than a naive application of the Carnot efficiency, which would predict an efficiency of $\eta = (T_s - T_u)/T_u \approx 14\%$. This is because η_{heat} also includes the restriction that not all of the total solar radiative heating, $R_{s,toa}$, can act as a heat flux to drive the heat engine. If this were the case, the surface temperature would be substantially reduced as $R_{s,toa}$ would not heat the surface, but rather be converted into mechanical energy. This would reduce the temperature gradient, possibly even inverting it so that no work can be generated. This conclusion does not rely on the simplifications made here, but is a general consequence of the combined consideration of the Carnot limit and the associated energy balance. It emphasizes that thermodynamic limits and their efficiencies need to be viewed in its complete thermodynamic setting of the Earth system.

Another aspect we can see in the expression of the efficiency is that more power can be generated with a stronger greenhouse effect, as τ enters as a positive contribution in the expression. Also, the fraction of solar radiation that is absorbed within the atmosphere reduces the maximum power limit. Hence, it is not just radiative heating and cooling that shapes the maximum power limit, but also where solar radiation is being absorbed and how strong the atmospheric greenhouse effect is.

As we will see in the following chapters, this mechanical power drives most of the climate system, particularly convective mass exchange and hydrologic cycling, although the specific applications to dry and moist convection include some more details that are described in Chapters 8 and 10. There are also other radiative sources in the form of spatial and temporal variations from which mechanical power can be derived, although of lesser magnitude. These include the large-scale differences in insolation due to the different orientation of the Earth's surface to the incident solar radiation, and temporal effects, such as the diurnal and seasonal cycles of insolation. These aspects yield additional power, with somewhat different expressions, although the magnitude is considerably less. As this form of power, its limit, and the consequences of performing this work play critical roles in the climate system, we next look closer at the associated sensitivities and feedbacks involved in affecting this maximum power limit.

6.7 Sensitivity to radiative forcing

The maximum power limit of converting radiative heating into mechanical work is rather strongly dependent on the radiative forcing, yet the performance of mechanical work and the associated convective heat flux also alter the radiative conditions. We will explore these two aspects next as these describe the basic components for feedbacks on the radiative forcing in the following chapters.

We first evaluate the effects of the different aspects that constitute the radiative forcing on the partitioning of the surface energy balance and on surface temperature with special emphasis on how these affect the maximum power limit. The radiative forcing in our simplified description is characterized by three aspects: the total amount of absorbed solar radiation, $R_{s,\text{toa}}$, the fraction of which is absorbed in the atmosphere, $R_{s,a}/R_{s,\text{toa}}$, and the optical depth, τ. These sensitivities are shown in Fig. 6.7. They vary one of the three aspects in isolation, while keeping the others at the respective value of the present day. We should also note that the formulations used here are linearized and extremely simplified, so that we focus on the qualitative aspects of these sensitivities.

To make the following interpretation easier, it is instructive to look at the expressions associated with the maximum power limit for the optimum heat flux, J_{opt}:

$$J_{\text{opt}} = \frac{R_{s,s} - R_{l,0}}{2} = \frac{3}{8}\tau R_{s,\text{toa}} - \frac{1}{2}R_{s,a} \tag{6.36}$$

For present-day forcing, this expression yields a flux of $J_{\text{opt}} = 135$ W m^{-2}, which is somewhat too high, yet of the right order of magnitude when compared to the estimate of $J = 112$ W m^{-2} based on observations (Stephens et al. 2012). This flux forms a substantial component of the surface energy balance in terms of its magnitude. The associated expression for surface temperature, T_s, at maximum power is given by

$$T_s = \left(1 + \frac{3}{32}\tau - \frac{1}{8}\frac{R_{s,a}}{R_{s,\text{toa}}}\right) \cdot T_a \tag{6.37}$$

Using present-day conditions, this yields a surface temperature of $T_s = 291$ K, which is close to the mean surface temperature of 288 K. Convective cooling considerably shapes this magnitude of surface temperature, which can be seen by deriving the difference in surface temperature between the maximum power state and the state without convection, $\Delta T_s = T_s(J_{\text{opt}}) - T_s(0)$. This expression is given by

$$\Delta T_s = \left(-\frac{3}{32}\tau + \frac{1}{8}\frac{R_{s,a}}{R_{s,\text{toa}}}\right) \cdot T_a \tag{6.38}$$

For present-day conditions, this cooling effect is about $\Delta T_s = -36$ K, which is substantial.

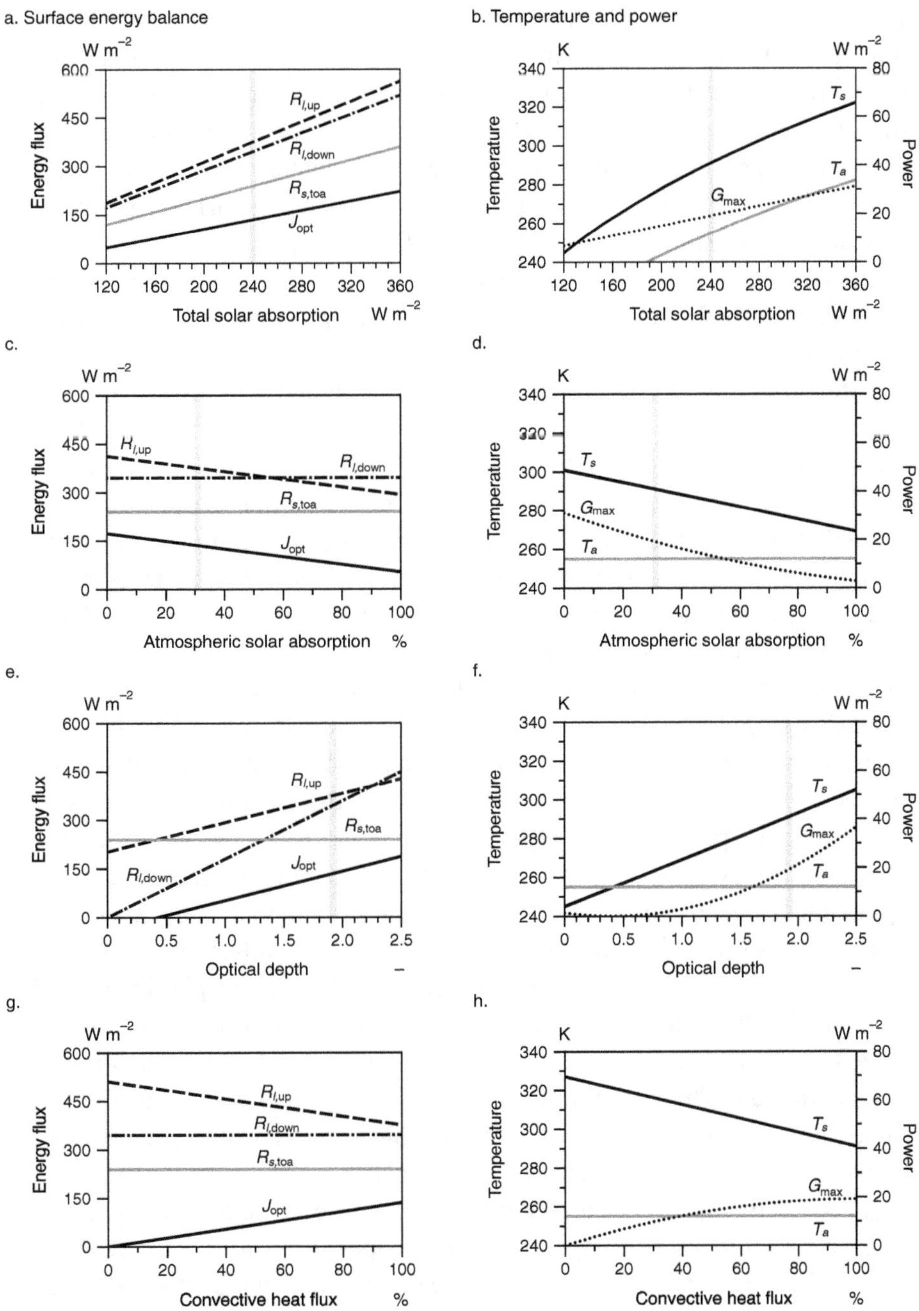

Figure 6.7 Sensitivity of (left) the surface energy balance and (right) surface and atmospheric (radiative) temperatures, T_s and T_a, and the maximum power limit, $G_{\max}$, to (a) and (b): total absorption of solar radiation, $R_{s,\text{toa}}$, (c) and (d): atmospheric absorption, expressed by the ratio $R_{s,a}/R_{s,\text{toa}}$, (e) and (f): optical depth τ, and (g) and (h): the convective heat flux J, expressed in relative terms to its optimum value, J/J_{opt}. The areas marked grey represent values of the present day.

The sensitivity to the total amount of solar radiation $R_{s,\mathrm{toa}}$, is shown in Fig. 6.7a,b. The partitioning of $R_{s,\mathrm{toa}}$ into surface and atmospheric absorption is held at a constant fraction in this sensitivity. A change in $R_{s,\mathrm{toa}}$ can be caused by external factors, particularly the distance to and the luminosity of the Sun, and internal factors that shape the planetary albedo, such as the extent of ice and cloud cover. With greater values of $R_{s,\mathrm{toa}}$, we note that the fluxes of the surface energy balance scale with $R_{s,\mathrm{toa}}$ in similar proportion, including the optimum heat flux, J_{opt}. The atmospheric and surface temperatures are enhanced as well. After all, T_a directly depends on absorbed solar radiation as given by the planetary energy balance (eqn. 6.17). As surface temperature T_s, directly relates to T_a, it varies in similar proportion so that the difference, $T_s - T_a$ does not change by much. The increase in power is thus mostly driven by the value of $R_{s,\mathrm{toa}}$ and varies linearly with $R_{s,\mathrm{toa}}$. The range of sensitivity represents a variation of $R_{s,\mathrm{toa}}$ from 50% to 150% of its present-day value, the surface temperature varies by almost 70 K, while the power varies by a factor of 3 over this range. These sensitivities are not surprising. After all, a stronger solar radiative forcing should increase the fluxes of the surface energy balance, it should be associated with warmer surface conditions, and it should yield a greater value of the maximum power limit.

Fig. 6.7c and d show how it matters where the solar radiation is being absorbed. The sensitivity is expressed in terms of the fraction $R_{s,a}/R_{s,\mathrm{toa}}$, with a value of 0% describing the case in which all solar radiation is absorbed at the surface, while a value of 100% implies that all of $R_{s,\mathrm{toa}}$ is absorbed in the atmosphere. As atmospheric absorption of solar radiation relates mostly to the presence of ozone, water vapor, aerosols, and clouds, this fraction is set by the internal composition and dynamics of the atmosphere that affect the absorbers in the atmosphere. Across the range of this sensitivity, the surface conditions differ rather substantially. While $R_{s,\mathrm{toa}}$, $R_{l,\mathrm{down}}$, and T_a remain constant, a greater share of atmospheric absorption is accompanied with reduced emission of terrestrial radiation from the surface $R_{l,\mathrm{up}}$, and a reduced flux J_{opt}. We note how the surface temperature T_s, declines linearly with this fraction. The maximum power limit G_{max} decreases with this fraction quadratically, as both the surface solar heating $R_{s,s}$, as well as the temperature difference $T_s - T_a$, are being reduced. This can also be seen directly in the expression for G_{max} given by eqn. 6.34, in which $R_{s,a}$ reduces power and enters quadratically. Over the range of this sensitivity, the surface temperature varies by more than 30 K. This sensitivity is not quite so straightforward to understand. It results from the characteristics of radiative transfer in which absorption of solar radiation in the atmosphere makes a difference as the resulting emission is directed both into space and to the surface. The surface heating is then substantially reduced, resulting in the reduced surface temperature and the reduced ability to generate power.

Figure 6.7e and f show the sensitivity to the atmospheric greenhouse effect in terms of the optical depth. This sensitivity relates to different concentrations of greenhouse gases in the atmosphere, which result from geochemical cycling of the involved substances. In this sensitivity to optical depth, $R_{s,\mathrm{toa}}$ as well as its partitioning among surface and atmospheric absorption remains unchanged, but both components of terrestrial radiation are increased with greater values of τ. While the increase of $R_{l,\mathrm{down}}$ with τ is directly evident from eqn. 6.19, the increase of $R_{l,\mathrm{up}}$ results from the warmer surface temperature, as is expected of a stronger greenhouse effect. Yet, the radiative temperature remains unchanged, since $R_{s,\mathrm{toa}}$ remains unchanged. The optimum convective heat flux is increased with greater values of τ as the surface is heated more strongly by the downward flux of terrestrial radiation. The combination of the greater heat flux and the greater temperature difference results in an increase of the maximum power limit that is approximately quadratically with optical depth. This can, again, be seen directly in the expression for G_{max} given by eqn. 6.34, in which τ enters quadratically. Surface temperature here spans a range of more than 50 K. This sensitivity is relatively straightforward to understand as it directly relates to an increased absorption of the radiation emitted from the surface, so that the atmosphere becomes less conductive for the exchange of terrestrial radiation between the surface and the atmosphere.

These three sensitivities show that the radiative forcing and the radiative properties of the atmosphere strongly shape the environmental conditions from which mechanical power can be derived. Yet, the performance of mechanical work and the associated convective heat flux affects surface energy partitioning and surface temperature rather strongly as well. This is shown in the sensitivity shown in Fig. 6.7g and h, in which the convective heat flux is varied from $J = 0$ to its optimum value, $J = J_{\mathrm{opt}}$. As the convective heat flux increases, the surface energy balance demands a decrease in the net cooling by terrestrial radiation. This reduction is associated with a decrease in surface temperature, which results in the decrease in temperature difference between the surface and the atmosphere. This effect of a larger convective heat flux, as discussed before, is the critical component that shapes the maximum power limit. At the maximum power limit, this temperature difference is reduced to half its value. In the sensitivity shown here, the associated difference in surface temperature is 36 K, as estimated earlier, which is quite substantial when compared to the other sensitivities shown in Fig. 6.7. This strong reduction of surface temperatures by convective cooling explains why T_s is so much greater in the purely radiative case shown in Table 6.4. It demonstrates that even though radiative processes provide the main driver, the conditions at the surface are strongly modulated by the consequences associated with the mechanical work being derived from the radiative heating difference in the surface–atmosphere system.

The mechanical work and the resulting convective exchange sets the basis for dynamical feedbacks within the Earth system that alter its radiative forcing. The convective heat flux involves the transport and exchange of mass, and this exchange sustains the cycling of water and geochemical components between the surface and the atmosphere. There are strong and important feedbacks related to this mass exchange that alter the radiative forcing. In terms of water, there are well-established feedbacks involving the presence of snow and ice at the surface that are linked to surface temperature, water vapor as the dominant atmospheric greenhouse gas, and clouds that alter the absorption of solar radiation and affect the strength of the greenhouse effect. These internal effects that alter the radiative forcing were not explicitly considered here. The cycling of carbon dioxide also alters the atmospheric greenhouse effect, yet its concentration is intimately linked to mass exchange, and biotic carbon uptake and release. These factors involve the conditions at the Earth's surface, particularly surface temperature, and depend on mass exchange, yet they alter the radiative forcing and thus the conditions in which mechanical work is being performed. Even though the model sensitivities shown here are highly simplified, they are nevertheless plausible and set the frame for these important feedbacks in terms of their thermodynamic implications. They will be explored further to some extent in the following chapters and at the end of this book when the different compartments are brought together to an integrated view of the thermodynamic state of the whole Earth system.

6.8 Radiation and the Earth system

This chapter dealt with the radiative forcing of the Earth system as the most dominant source from which energy can be converted. The formulations of radiative entropy were introduced to understand and quantify the very low entropy of solar radiation and the cause for this radiative disequilibrium. This low entropy is represented by two characteristics: First, solar radiation is confined to a very small area of the sky, and second, it is composed of very short wavelengths given the value of the radiative flux. Both characteristics relate to the distance of the solar radiative flux at the Earth's orbit to the emitting surface of the Sun. Radiative processes on Earth increase the entropy of the radiative flux by scattering into a broader solid angle and by absorption with subsequent reemission at a much lower temperature. The resulting radiative flux from Earth to space is scattered to a much wider solid angle, and most of the radiative flux is composed of longer wavelengths that represent a much lower emission temperature of the Earth system.

Using radiative fluxes from observations and the estimates from earlier, a planetary entropy budget is derived and shown in Fig. 6.8. In addition to the estimates performed, this budget includes the entropy production by atmospheric absorption

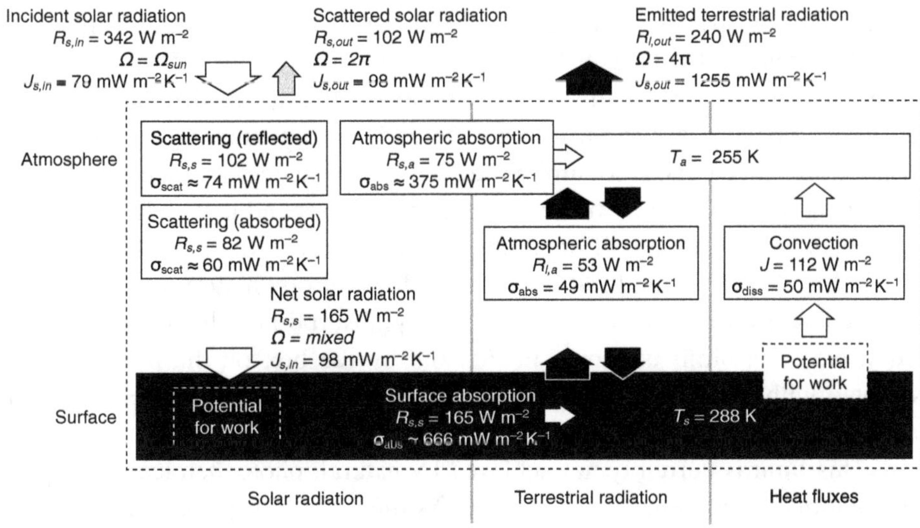

Figure 6.8 Estimate of the global radiative entropy budget for present-day conditions, using the energy balance components reported by Stephens et al. (2012) and Table 6.3.

of terrestrial radiation and by convection. The sum of these two processes is constrained by the solar radiative heating of the surface and the temperature difference between the surface and the atmosphere. The values in Fig. 6.8 are derived by using the convective heat flux from observations (Stephens et al. 2012), in which the sensible and latent heat flux are added together. The entropy production by this heat flux is given by its magnitude of 112 W m^{-2} and the respective temperature difference, while the remainder of the entropy exchange between the surface and the atmosphere is contributed by the radiative transfer of terrestrial radiation. The estimates broadly agree well with previous estimates (Aoki 1983; Peixoto et al. 1991; Stephens and O'Brien 1993; Goody 2000; Ozawa et al. 2003; Kleidon and Lorenz 2005; Kleidon 2009c; Wu and Liu 2010), noting that our estimates of radiative fluxes include the factor of 4/3 in the expressions for radiative entropy fluxes. Note that these estimates are rough approximations, particularly regarding scattering, the use of global-scale estimates of the fluxes, and the use of climatic means. The effects of the latter two aspects, as well as the contribution to entropy production by motion and other aspects of the dynamics of the Earth system are explored in latter parts of the book. As we will see later, these contributions slightly modify some of the numbers, particularly the contribution by convection and the radiative entropy export to space, but not the overall magnitude of the terms. This magnitude is, after all, set by the values of the radiative exchange of the Earth system to space. Hence, this radiative entropy budget describes the planetary thermodynamic constraint of the Earth system set by the exchange of radiative entropy with space.

The relevance of the radiative entropy budget is that it sets the starting point to quantify how much work can maximally be derived from solar radiation. Identical to the way in which the Carnot limit is derived for a heat engine, the combination of radiative energy and entropy fluxes and the first and second law yield the limits on work. There are two, rather different limits on work from radiation, as shown by the boxes in Fig. 6.8. The first limit is set by the low entropy nature of solar radiation before it is converted into thermal energy by absorption, which was derived in Section 6.5. This limit is shown in Fig. 6.8 by the box on the lower left. It is associated with a very high maximum efficiency of more than 70%, as the limit is set mostly by the difference in radiative temperatures between the Sun and the Earth. This limit is currently not achieved by natural, photochemical processes or by human-made solar energy technology. As this limit is far from currently being achieved, it is a hypothetical limit that we will get back to towards the end of the book when human activity and limits to renewable energy are described in Chapters 11 and 12.

The second limit is set by differential radiative heating and cooling and was derived in Section 6.6. This limit is shown by the box on the right in Fig. 6.8. It is much lower as it involves the much lower temperature difference between the surface and the atmosphere. This limit is highly relevant as this limit sets the bounds for the power that is associated with the dynamics of the climate system. Even though the entropy production is small when compared to radiative processes, it is this contribution that reflects the dynamics of the climate system in terms of its motion and other processes. The magnitude of the entropy production is thus not necessarily a relevant metric to characterize the importance of a process. The convective heat flux that is involved in this generation of work depletes the driving difference, resulting in a maximum power limit. By doing so, it substantially alters the conditions at the surface, as described in the section on sensitivities. The heat flux predicted by this limit is quite close to the observed magnitude of the turbulent heat fluxes of sensible and latent heat, indicating that this limit is highly relevant and that atmospheric convection is likely to operate near this limit.

The sensitivities described in Section 6.7 showed how important the basic characteristics of the radiative forcing are in setting the maximum power limit and the climatic conditions at the surface. These characteristics do not just refer to the total absorption of solar radiation, but also where it is being absorbed (at the surface or in the atmosphere), and how efficiently the surface can cool by emission of radiation, which is affected by the strength of the atmospheric greenhouse effect. The sensitivities showed that there are several constellations that can result in the present-day mean surface temperature of 288 K. These constellations involve the delicate interplay between the radiative forcing, the resulting dynamics within the Earth system, and emergent effects of Earth system processes that alter the radiative forcing and then shape the mean surface temperature. These interactions involve particularly the radiative effects of water in its different phases and geochemical cycling which alters the atmospheric composition and thus the atmospheric greenhouse effect. In the following chapters, we link the radiative forcing to motion, hydrologic and geochemical cycling, processes that affect radiative properties and thus feed back to the radiative forcing. At the end of the book, we then place the resulting interactions into the thermodynamic view of the Earth system.

7

Motion

7.1 Transporting mass on the planet

With the radiative forcing being described in thermodynamic terms, in this chapter, we link this forcing to motion as the next step in the cascade of energy conversions of the Earth system that was shown in Fig. 1.5. Motion transports energy, mass, and momentum between different places of the Earth system. It is through large-scale energy and mass transports associated with motion that the processes in one region affect processes elsewhere. Motion thus plays the role of the global connector, allowing regions to interact. It is this global connection that makes the Earth system a highly interactive, planetary system with large-scale material cycling of geochemical elements.

The purpose of this chapter is to describe motion, the limits that it is exposed to as well as its planetary consequences from a thermodynamic perspective. This description is quite different from the common approach in which the natural starting point is the momentum balance in the form of the Navier–Stokes equation of fluid dynamics. The view formulated here does not contradict this common approach, but rather supplements it by placing motion explicitly into the context of the thermodynamic, planetary setting. Furthermore, the consequences of motion are then evaluated to show that motion can be interpreted as the result of a system advancing to its state of thermodynamic equilibrium at an accelerated rate.

Our starting point for the description of motion is kinetic energy and the processes that generate and dissipate this form of energy. Kinetic energy is then directly related to the velocity associated with motion and the magnitudes by which thermal energy and mass is transported. The generation of kinetic energy originates from differential heating. This differential heating causes density differences in fluids, which affect the potential energy of the system. The tendency of the system to deplete its potential energy to a lower value is associated with buoyancy in the system, which then drives the onset of motion. This mechanism to generate motion is very general and applies to most forms of motion found in the atmosphere, oceans, and the solid Earth. There are, however, also other forms of motion that result from different conversions that are either indirectly related to this generation mechanism or to other drivers. An example of an indirect relationship is the wind-driven oceanic circulation, which relates to atmospheric motion and its generation. Another driver for motion are tidal forces that cause tidal movements, mostly in the oceans, although, as we will see later, this forcing is comparatively small compared to differential heating.

To illustrate how the generation of motion relates to the depletion of potential energy, we use the conceptual setup shown in Fig. 7.1. This figure shows two fluids of different densities, with a small parcel of a lighter fluid with a low density ρ_l being situated in a heavier fluid with a high density ρ_h. We use the indices "l" and "h" to refer to the properties of the two fluids. The two fluids are shown in three different states that correspond to different values of potential energy U_{pe}, of the system. The potential energy of the system is determined by the vertical position and the horizontal spread of the two fluids. These different states are described here qualitatively to emphasize the tight linkage between changes in potential and kinetic energy, and are detailed by calculations in the following section.

In state A, the lighter fluid is located at the bottom of the heavier fluid and is concentrated in the horizontal dimension to a width x_l. This occurs, for instance, when a fluid is heated from below, so that a small parcel of the fluid gains a density

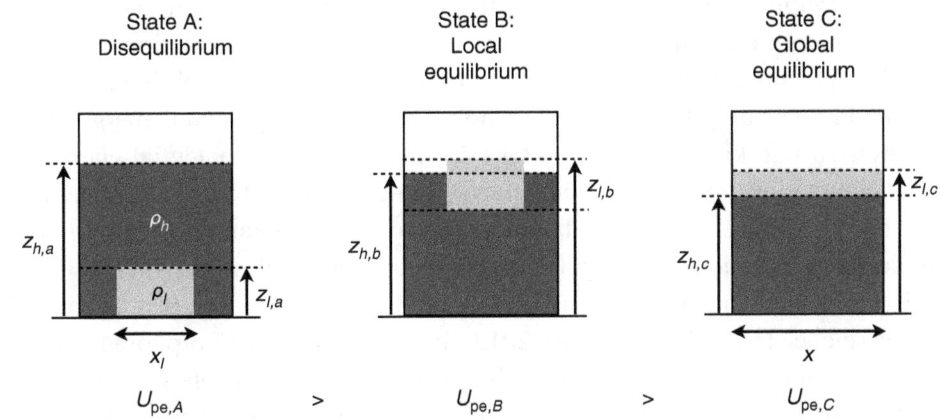

Figure 7.1 A conceptual setup to illustrate motion as a consequence of depleting the potential energy, U_{pe}, within the system. The system is composed of two fluids of different densities: a lighter fluid with a low density ρ_l, shown by the lighter grey, and a heavier fluid with a high density ρ_h, shown by the dark grey. The symbols x_l, x, z_l, and z_h refer to the width and upper position of the lighter and heavier fluids (indices "l" and "h") and are used in Section 7.2 to quantify the difference in potential energy of these states.

lower than the rest of the fluid. We treat the heated parcel as a separate fluid and consider the potential energy of the whole system in states A–C. Energetically, state A corresponds to a state of comparatively high potential energy, $U_{pe,A}$, because a parcel of low density is placed at a low geopotential height, while a parcel of equal volume of the heavier fluid is placed at a higher geopotential height. The second law applied to this system describes the tendency to deplete this high value of potential energy to a minimum value thereby increasing the thermal energy, as described in Section 3.6. State A is thus a state of thermodynamic disequilibrium. The tendency to deplete potential energy within the system is represented by the presence of a buoyant force, which accelerates the parcel of lighter fluid to a greater height. This tendency is identical to the well-known Archimedes principle that describes the buoyancy of an object submerged in a fluid. The parcel experiences this buoyancy until it floats on top of the heavier fluid, as shown in state B. The overall difference between states A and B is that in state B, the lighter parcel is located at a greater height, while its original volume is replaced by the heavier fluid. The potential energy of this state, $U_{pe,B}$, is thus reduced. In fact, the potential energy is even minimized with respect to the vertical position of the lighter parcel, so that this state represents a state of local equilibrium. This local equilibrium is also referred to as the hydrostatic equilibrium in atmospheric science. It is a local equilibrium because the potential energy can be further lowered to a value, $U_{pe,C}$, when the parcel of lighter fluid is spread vertically, as shown in state C.

This sequence of states, from an initial state A of disequilibrium to a state C of global equilibrium thus translates into a depletion of potential energy within the system, with motion being the means to accomplish this depletion. Upon this depletion, potential energy is converted into kinetic energy, which is then further converted into heat by frictional dissipation. Yet, all of the potential energy cannot be converted into kinetic energy. After all, even state C has potential energy, but it cannot be converted into another form. The amount that can be converted is described by the well-established concept of available potential energy in atmospheric science, which was originally described by Margules (1905) (see Tamura 1905, for an english summary) and established by Lorenz (1955). It is also used in oceanic science (e.g., Tailleux 2010, 2013), hydrology and geomorphology on land (e.g., Kleidon et al. 2013). This available potential energy depends on the rate at which the disequilibrium associated with state A is being generated, which in turn depends on the heating rate. The generation of available potential energy is thus limited by thermodynamics because this rate is limited to the heating rate, as we will see later.

While this sequence of states seems somewhat abstract, the different states can be directly linked to how motion is generated within the Earth system. State A represents the case in which air is heated near the surface, which then initiates vertical convection as a means of the atmospheric system to reach state B. The stronger heating of tropical surfaces compared to polar regions and the large-scale circulation that results can be seen as an example for a system to get from state B to state C. Atmospheric motion thus represents the means to deplete the potential energy within the system towards a minimum value. Equivalently, mantle convection in the Earth's interior or the uplift of continental crust represent the means of the solid Earth to get from state A to state B (Kleidon et al. 2013). Erosion of the continental crust and the lateral spread of sediments corresponds to the dynamics from state B to C, although it requires water as a medium for an effective transport. We can thus use the sequence of states shown in Fig. 7.1 as a conceptual image to understand how motion results as a means of a system to reach its state of thermodynamic equilibrium.

Yet, when describing large-scale motion in the Earth system, it may not always appear as if motion depletes the driving gradient. This is particularly the case for so-called geostrophic flow, a widely used concept in atmospheric and oceanic science. Geostrophic flow describes an idealized case in which flow is represented by a momentum balance composed of a pressure gradient force and a Coriolis force, which is due to the Earth's rotation. It is reflected in the westerly flow in the mid-latitudes, the atmospheric jet stream, and, in fact, most of the large-scale motion in the atmosphere and the ocean that is predominantly shaped by zonal flow. However, it is important to note that this flow results from the conservation of angular momentum that applies to the large-scale heat transport between the tropics and the poles, and that it is a comparatively small, but critical, aspect of the flow in the meridional, poleward direction that accomplishes the depletion of the differential radiative heating. This small contribution is critical because this part of the flow is associated with the generation of kinetic energy associated with the large-scale atmospheric general circulation. The dynamics of generation and dissipation of motion thus needs to be combined with the conservation of angular momentum to understand and interpret the relationship between the generation rate of large-scale motion, the kinetic energy that resides within the system, and the overall thermodynamic direction that underlies these dynamics.

The planetary consequences of motion in the Earth system go beyond the depletion of potential energy. As the generation of available potential energy is caused by heating, the depletion of the potential energy by motion is associated with heat fluxes; these affect the heating and cooling terms of the Earth system. We thus need to relate motion back to its original driver of differential radiative heating and cooling to understand its planetary consequences and their thermodynamic interpretation. Furthermore, large-scale motion accomplishes mass exchange and transport, which forms the main driver for geochemical cycles, such as the cycles of water and carbon. The dynamics associated with large-scale motion thus form a critical link between the radiative forcing of the Earth system described in the last chapter, and hydrologic and geochemical cycling of mass that is described in the following chapters.

In this chapter, we first quantify the example shown in Fig. 7.1 to demonstrate the depletion of potential energy and then relate this to the budget equations of kinetic energy. The application of these concepts to the large-scale circulation in the atmosphere–ocean climate system and the Earth's interior are then described in separate sections, where the limits to generating large-scale motion are estimated as well as the magnitude of the associated heat fluxes. Note that convective motion is dealt with separately in Chapter 8 as it deals intimately with hydrologic cycling. These estimates are then used to infer the rates of mass exchange, which then form the basis for evaluating limits to hydrologic and geochemical cycles in the following chapters. The planetary consequences of motion are then evaluated, including the feedbacks to the drivers. The chapter closes with a brief summary of the thermodynamic perspective on the causes and consequences of large-scale motion in the Earth system.

7.2 Depletion of potential energy

The ability to deplete potential energy within the system is the main driver for generating motion, as this lowers the potential energy within the system. This reduction in potential energy relates to an indirect manifestation of the second law, as discussed in Section 3.6 in terms of the reduction of kinetic energy in a system. If we focus not just on the potential and kinetic energy in the system, but also include the thermal energy within the system, then the conservation of energy implies that any depletion of the sum of potential and kinetic energy needs to be associated with an increase in thermal energy. The increase in thermal energy is associated with an increase in thermal entropy, so that the depletion of the two forms of energy are associated with an increase in entropy, as required by the second law.

In an isolated system, such as the one shown in Fig. 7.1, a state of thermodynamic equilibrium is then represented by a minimization of potential and kinetic energy, so that the greatest possible fraction of the system's energy is in thermal form. We next look at this depletion associated with the sequence shown in Fig. 7.1 in more detail by quantifying the potential energy of the states A–C using example values of a system.

The system is defined by two masses of different density, a light mass m_l with a low density ρ_l, and a heavier mass m_h with a high density ρ_h. For simplicity, the masses are assumed to be incompressible so that the densities do not change with height. The masses are spread out over a certain horizontal and vertical extent, with the horizontal extent of the system given by x and y. In the initial state A, the light mass is contained over a horizontal extent x_l and is positioned in the vertical between the lower boundary of the system and a height $z_{l,A}$. In the notation, we add the state to the index of the variable. The heavier mass is spread over x and up to a height $z_{h,A}$ except for where the lighter mass is situated. To illustrate this example, the following values are used in the calculations: $\rho_l = 500$ kg m^{-3}, $\rho_h = 1000$ kg m^{-3}, $m_l = 150$ kg, $m_h = 700$ kg, $x_l = 0.5$ m, $x = 1$ m, and $y = 1$ m. The densities taken here are similar to those of wood and water.

Before we start considering the differences in potential energy between the states, we first note that the mass of both masses is conserved across the three states. This requirement relates some of the vertical coordinates between the states with each other. For states A and C, we obtain the following expressions for the masses:

$$\begin{aligned} m_l &= x_l y \int_0^{z_{l,A}} \rho_l dz \\ &= \rho_l x_l y z_{l,A} = \rho_l x y (z_{l,C} - z_{h,C}) \end{aligned} \tag{7.1}$$

and

$$\begin{aligned} m_h &= (x - x_l) y \int_0^{z_{l,A}} \rho_h dz + xy \int_{z_{l,A}}^{z_{h,A}} \rho_h dz \\ &= \rho_h (x - x_l) y z_{l,A} + \rho_h x y (z_{h,A} - z_{l,A}) = \rho_h x y z_{h,C} \end{aligned} \tag{7.2}$$

Mass conservation is implied in the last expressions in eqns. 7.1 and 7.2, in which the two masses are formulated in terms of the positions associated with state C. Given the values of mass and horizontal dimensions in our example, these expressions yield the values $z_{l,A} = 0.6$ m, $z_{h,A} = 1$ m, as well as $z_{h,C} = 0.7$ m and $z_{l,C} = 1$ m.

The potential energy of a state is given by the vertical arrangement of the two masses and is determined by using eqn. 2.10:

$$U_{\text{pe,tot}} = U_{\text{pe},l} + U_{\text{pe},h} = \int \rho_l g z dV_l + \int \rho_h g z dV_h \tag{7.3}$$

Here, the total potential energy is separated into the potential energy of the light and heavy mass, and the integrations are performed over their respective volumes. In the initial state A, the potential energy of the two masses are then given by

$$U_{\text{pe},l,A} = x_l y \int_0^{z_{l,A}} \rho_l g z dz = \frac{\rho_l g x_l y}{2} z_{l,A}^2 \tag{7.4}$$

and

$$U_{\mathrm{pe},h,A} = (x - x_l)y \int_0^{z_{l,A}} \rho_h g z dz + xy \int_{z_{l,A}}^{z_{h,A}} \rho_h g z dz$$
$$= \frac{\rho_h g}{2} \left((x - x_l) y z_{l,A}^2 + xy (z_{h,A}^2 - z_{l,A}^2) \right) \quad (7.5)$$

For the given values, these yield values of potential energy of $U_{\mathrm{pe},l,A} = 450$ J and $U_{\mathrm{pe},h,A} = 4100$ J, so that the total potential energy of the system in state A is $U_{\mathrm{pe,tot}} = 4550$ J. This value of potential energy is given with respect to the lower boundary of the system.

In the transition from state A to state B, the potential energies of the two masses and of the total system change. We can see these changes when we consider what happens when the lighter mass is lifted by a small amount, Δz. Then, the potential energy of the lighter mass is, obviously, increased. The lifting corresponds to a change in which a slab of thickness Δz is taken from the bottom of the lighter mass and placed at the top of it. This increases the potential energy of the lighter mass by $(\rho_l xy\Delta z) g z_{l,A}$. The volume that the slab occupied at the bottom of the lighter mass is then replaced by the heavier mass. In other words, we swap a volume $xy\Delta z$ of the lighter mass with the heavier mass, placing the lighter mass at a height $z_{l,A}$ and the heavier mass at height 0. This then decreases the potential energy of the heavier mass by $-(\rho_h xy\Delta z) g z_{l,A}$. Thus, the gain in potential energy of the lighter mass is more than compensated for by the reduction of potential energy of the heavier mass. When we then consider the successive lifting of the lighter mass, we find that the potential energy of the whole system is lowered even though the potential energy of the lighter mass increases. For the values described earlier, this reduction in total potential energy during lifting is shown in Fig. 7.2, including the smaller increase in $U_{\mathrm{pe},l}$ compared to the greater decrease in $U_{\mathrm{pe},h}$.

The reduction in total potential energy continues up to the point when the lighter mass reaches the surface (i.e., when $z_l = z_h$, which are the variables without a reference to a state as this state is not shown in Fig. 7.1). At this point, the decrease in potential energy is reduced because the potential energy of the heavier mass is reduced not as much when the lighter mass is lifted outside the heavier fluid so that the total potential energy eventually reaches a minimum value. This minimum is described by state B. For the given values, the minimum is located at a height $z_l = 1.15$ m and corresponds to a total potential energy of $U_{\mathrm{pe,tot},B} = 3837.5$ J. Above this height, the potential energy of the light mass continues to increase linearly with the height, but the potential energy of the heavier mass does not decrease further at a height of $z_l = 1.3$ m. At this point, the light mass is lifted outside the heavier mass, so that the potential energy of the lighter mass increases further, but without a decrease in the potential energy of the heavier mass.

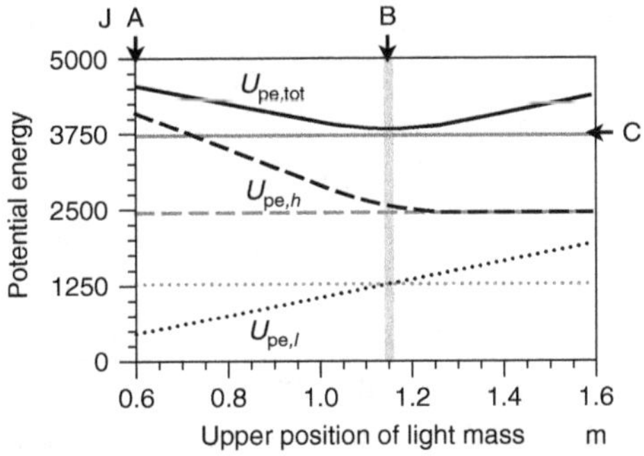

Figure 7.2 The potential energy of the light mass $U_{pe,l}$ (dotted line), the heavier mass $U_{pe,h}$ (dashed line), and the total potential energy $U_{pe,tot}$ (solid line) as a function of the upper position of the light mass z_l, of the simple system shown in Fig. 7.1. The values at states A and B are marked at the top. The grey shaded area represents state B at which the total potential energy gains a minimum value. The horizontal grey lines show the values of each component in state C at global equilibrium.

State B represents the state of local equilibrium at which the potential energy of the system cannot be reduced further by lifting the lighter mass. The difference in potential energy between the initial state A and state B, $U_{pe,A} - U_{pe,B}$, thus represents the available potential energy of the system. It represents the amount of energy that can be converted into kinetic energy associated with the upward motion of the lighter mass. As vertical motion is generally described as convection, the form of available potential energy represents the convective available potential energy, or CAPE, in the system, a concept widely used in atmospheric science to describe atmospheric convection (e.g., Brunt 1941; Renno and Ingersoll 1996).

If the shape of the lighter mass can be changed, the potential energy can be lowered further by uniformly spreading it over the whole width x of the system. In the atmosphere or ocean, this spreading is related to horizontal pressure differences that cause a horizontal pressure gradient force that initiates horizontal motion. The state of a uniform spread of the light mass is represented by state C, which is a state of global equilibrium and has the lowest total potential energy of the three states. It is shown in Fig. 7.2 by the grey horizontal lines and is accomplished by the horizontal redistribution of the masses; it lowers the potential energy of the heavier mass. The difference between states B and C thus corresponds to available potential energy, or APE, as it represents the amount of potential energy that can be converted into kinetic energy for lateral exchange. This lateral exchange results in dynamics towards state C. The state C then represents the state of thermodynamic equilibrium in which the potential energy cannot be depleted further by the redistribution of the two masses.

To summarize this example, the states A–C shown in Fig. 7.1 thus represent evolutionary steps towards thermodynamic equilibrium. This evolution is not as direct as for the case in which thermal energy is distributed within the system, as described in Section 3.4, because it first converts potential energy into kinetic energy before it is converted into heat and increases the entropy of the system. It demonstrates how this conversion is a manifestation of the second law and involves the generation and dissipation of motion as the means to reach thermodynamic equilibrium.

Let us now apply this example to motion in the Earth system in steady state. A continuous generation of available potential energy is required within the system to sustain steady-state circulations and the associated kinetic energy. This generation of available potential energy is mostly achieved by differential heating. The dominant example for this differential heating is the absorption of solar radiation at the Earth's surface, where air is heated at a lower geopotential and higher pressure, while it is cooled at a higher geopotential and lower pressure within the atmosphere by the emission of radiation to space. As warmer air has a lower density, the heating at a lower geopotential is associated with thermal expansion and lifting of the mass above, which generates potential energy. This generation of potential energy can directly be seen in the equations for potential energy (e.g., eqn. 2.11) in which the potential energy of an atmospheric column directly depends on its temperature. Cooling by emission of radiation results in thermal contraction, a reduction in the density, and thus a reduction of the potential energy. Differential heating and cooling, particularly by absorption and emission of radiation, thus represent a dominant way to alter potential energy and thus drive the disequilibrium reflected in available potential energy and kinetic energy within the Earth system.

Note that the link between heating and cooling with generation and reduction of potential energy corresponds to steps 1 and 3 of the Carnot cycle of Section 4.3, which determines the ability to perform mechanical work within the system. The Carnot limit thus sets an upper limit to the generation of potential energy and to the dynamics associated with the sequence shown in Fig. 7.1. We also need to take the consequences of motion into account, particularly regarding its effect to deplete the driving gradient by the transport of heat, similar to the case of the maximum power limit in Section 4.5. Applied to the sequence shown in Fig. 7.1, this implies that we can use heating rates and thermodynamic limits to infer and estimate limits to the generation and dissipation of kinetic energy as well as the mean intensity of the flow.

7.3 Maintaining motion

The next step is to link the maintenance of motion to the thermodynamic trade-offs that result in the maximum power limit. To do so, we first look at the well-established framework in atmospheric science of the Lorenz energy cycle

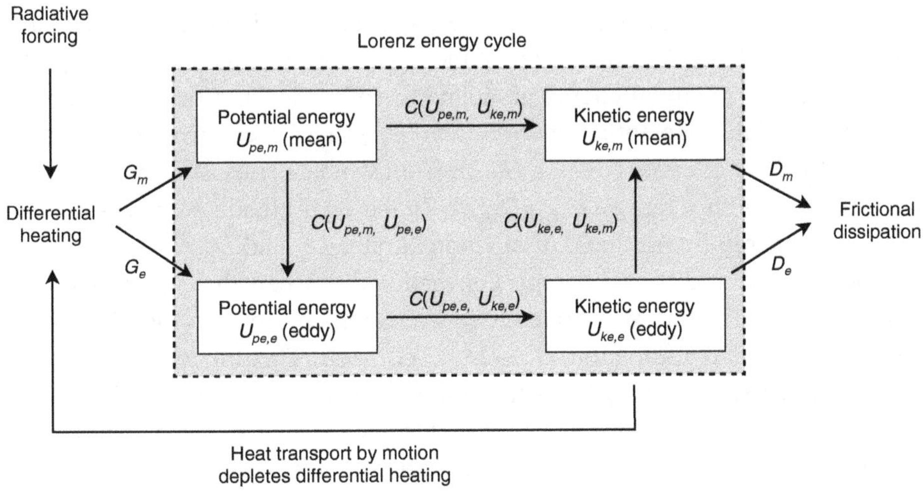

Figure 7.3 The Lorenz energy cycle (grey box) that describes the generation and dissipation of large-scale motion in terms of conversion rates C between potential and kinetic energy associated with mean flow and eddies. The diagram shows how these conversion rates are related to radiative forcing, differential heating, and the effects of motion on the differential heating. The link to thermodynamics is described by the generation and dissipation rates, $G = G_m + G_e$ and $D = D_m + D_e$, which relate to the maximum power that is derived from differential heating of the system.

(Lorenz 1955), which describes the generation and conversion rates of potential and kinetic energy. The linkage of this framework to the thermodynamic drivers and effects is shown in Fig. 7.3. The center of the figure shows the Lorenz energy cycle (Lorenz 1955). The framework is very general and well-established so in the following, it is only briefly explained and then linked to the thermodynamic setting of the system. Further details can be found in atmospheric science textbooks, e.g., Peixoto and Oort (1992). This framework has also been used to describe the generation of large-scale motion in oceans (Oort et al. 1989; Tailleux 2013).

The Lorenz energy cycle describes motion in terms of its mean flow as well as the circular motion associated with eddies that are associated with the large-scale high and low pressure systems of the mid-latitudes and that is characterized by the variations from the mean flow. Because of the distinction between mean flow and eddies, there are in total, four forms of energy involved in the description of the Lorenz energy cycle: the available potential energy associated with the mean flow and eddies, $U_{\text{pe},m}$ and $U_{\text{pe},e}$, as well as the kinetic energy of the two components, $U_{\text{ke},m}$ and $U_{\text{ke},e}$. The dynamics of these forms of energy are described by the respective conversion rates. These conversion rates are shown as fluxes C, in Fig. 7.3, with the direction of the arrows given as they take place in the Earth's atmosphere. The starting point of the conversions are the generation rates of potential energy of the mean and eddy components, G_m and G_e. The mean available potential energy is then converted into the kinetic energy of the mean flow at a rate $C(U_{\text{pe},m}, U_{\text{ke},m})$ and into the available potential energy associated with eddies at a rate $C(U_{\text{pe},m}, U_{\text{pe},e})$. The eddy available potential energy then generates the variations in motion described by $U_{\text{ke},e}$ at a rate $C(U_{\text{pe},e}, U_{\text{ke},e})$, from which a fraction is converted back into mean flow with a rate $C(U_{\text{ke},e}, U_{\text{ke},m})$. In the end, kinetic energy is converted into thermal energy by frictional dissipation at rates D_m and D_e.

In the climatic mean, the energy conversions between the different forms of energy balance the generation rates, $G_m + G_e$, which are then also balanced by the rates of frictional dissipation, $D_m + D_e = G_m + G_e$. The link to thermodynamics is then established by recognizing that the total generation rate, $G = G_m + G_e$, is thermodynamically constrained by the differential heating of the system. This generation rate is subjected to a maximum power limit similar to convection as described in Section 4.5, because the generated flow transports thermal energy, and thus depletes the differential heating of the system. Hence, the factors that constrain the Lorenz energy cycle are situated outside of its description as it involves the energy balances that describe differential radiative heating and include the effects of heat transport by motion. Such a limit on the generation rate within the Lorenz energy cycle was already alluded to by Lorenz (1960) for the same reasons as the factors that result in the maximum power limit. This limit is quantified in the next section after we deal with the budget equation of kinetic energy.

In the following, we focus merely on the totals of available potential energy and kinetic energy in the system, U_{pe} and U_{ke}, and its conversion rates that generate and dissipate motion. In the climatic mean, these rates are then linked by $G = C(U_{\text{pe}}, U_{\text{ke}}) = D$, which are then constrained by G_{max}. As the maximum power limit is associated with an optimum heat flux, J_{opt}, and a characteristic temperature difference, ΔT, we can in principle infer not just the limit to the rates within the Lorenz energy cycle, but also the magnitudes of the forms of energy involved as well as the characteristic velocity associated with the motion.

To start identifying these connections, we first consider the kinetic energy budget of a system in its simplest form,

$$\frac{dU_{\text{ke}}}{dt} = G - D = 0 \tag{7.6}$$

which is zero in the case of the climatic mean steady state. Here, G refers to the generation rate of kinetic energy, that is, we skip the conversion to available potential energy and directly focus on kinetic energy generation. Note that eqn. 7.6 does not include other terms that remove kinetic energy and convert it into another form other than thermal energy.

The frictional dissipation in the system, D in eqn. 7.6, is mostly turbulent. Turbulence involves the generation of turbulent kinetic energy that is associated with the small-scale circular motion, its successive conversions into eddies of smaller sizes, until it is dissipated by viscous friction into thermal energy of the fluid, resulting in dissipative heating. Without going into the details of these conversions of turbulent kinetic energy, the overall frictional dissipation is expressed as the product of a friction force per unit area (or shear stress), F_{fric}, and the velocity v at a reference height

$$D = F_{\text{fric}} \cdot u = \rho C_d v^3 \tag{7.7}$$

where C_d is the drag coefficient that describes surface properties, such as its aerodynamic roughness, as well as air stability and others. More generally, both, the force and the velocity are vectors, so that D results from the vector product of these quantities. This has been excluded here for simplicity. The friction force in eqn. 7.7 is described as a drag force, or bulk formula, of the form

$$F_{\text{fric}} = \rho C_d v^2 \tag{7.8}$$

The expression for the drag coefficient is typically based on Monin–Obukhov similarity theory (e.g., Arya 1998), yet it has semi-empirical components. Another way to express surface friction that is often used in micrometeorology is to express the exchange in the form of an aerodynamic conductance, k_a, which describes how efficiently turbulence depletes the momentum difference between the moving fluid and the surface at rest. This formulation is equivalent to the aforementioned form when $k_a = C_d v$, so that the momentum flux, or force, due to turbulent friction is written as

$$F_{\text{fric}} = k_a(\rho v) \tag{7.9}$$

What is nice about formulating friction in this way is that we can directly see that friction represents a momentum flux that is driven by the momentum difference between the fluid with a momentum ρv and the surface at rest, with the conductivity given by k_a. It is thus directed to deplete kinetic energy and follows the second law, as was already described in Section 3.6.

With frictional dissipation being constrained by the maximum power by which kinetic energy can be generated, this constraint is insufficient to inform us about the particular set of variables (k_a, v) that characterizes the state of the system. There are, in principle, various combinations of this set of variables that can yield a given rate of frictional dissipation. It thus requires an additional constraint to infer the aerodynamic conductance and the characteristic velocity of motion, and thus U_{ke} of the system. There are different additional constraints that can be used, which partly depend on the characteristics of the system that is being evaluated. The most straightforward constraint is the momentum balance of the system, which in steady state requires that the friction force is balanced by the accelerating force in the system. If the accelerating force is expressed in terms of the thermodynamic driving gradient, the momentum balance can provide an additional constraint which may be sufficient to then infer the values of the variables (k_a, v) from the maximum power limit.

An additional constraint is provided by fluxes of heat, mass, or angular momentum that are associated with motion in a system. Since the heat flux that is associated with the maximum power limit involves motion and thus the transport of thermal energy at a certain rate, the heat flux is related to v, for instance in the form $J = c_p \rho v (T_h - T_c)$ (as in Section 5.2). In this case, the expression of the optimized heat flux together with the resulting temperature difference can be used to infer the mass exchange, ρv, in the system, as well as its kinetic energy (which is related to v) and its available potential energy (which is related to $T_h - T_c$). It is through the indirect consequences of motion that additional constraints should provide sufficient means to fully constrain the properties of a system that is maintained at the thermodynamic limit of maximum power. This, however, requires that we consider the processes that are located outside the grey box shown in Fig. 7.3, specifically the energy and entropy balances that describe the differential heating of the system as well as the effects of motion in depleting the differential heating. This is what we do in the following sections.

7.4 Limits to atmospheric and oceanic motion

We start with the planetary setting of the atmosphere and ocean system that is forced by a differential heating and cooling by absorption and emission of radiation. This differential radiative forcing is reflected in solar radiation being absorbed predominantly at the surface, and more absorption taking place in the tropics than in the polar regions due to the inclination of the surface while emission to space takes place predominantly from the atmosphere. We first link this difference in the forcing with the sequence shown in Fig. 7.1 before the thermodynamic limit to this sequence is estimated.

The state A of disequilibrium in Fig. 7.1 is generated by differential heating from below and cooling from above. Air in the lower atmosphere is heated either by condensation of moisture at the base of clouds in the atmosphere or, more directly, by heating the surface by absorption of solar radiation. In both cases, the heating causes a decrease in air density, which increases the potential energy of the air column by raising the center of mass. The resulting buoyancy then results in an updraft that is fueled by the decrease in potential energy. This updraft results in the transition towards state B in Fig. 7.1. The transition from state A to B is characterized by vertical exchange of mass of different densities and represents atmospheric convection. The transition to state C relates to the large-scale circulation of the atmosphere–ocean system. As this circulation transports heat from the tropical to polar regions, it is directed towards depleting the large-scale differences in potential energy that are caused by differences in heating that results from the differential absorption of solar radiation across the Earth system.

We can thus associate the transition from states A to B and from B to C in Fig. 7.1 with two types of motion in the climate system: the vertical exchange is represented by atmospheric convection, and the horizontal exchange by the large-scale circulation of the atmosphere and ocean. The two types occur at very different scales: convection takes place at the local to regional scale, while horizontal exchange necessarily involves larger scales. This separation is very common in atmospheric science. It can be justified on thermodynamic grounds because the heating gradients involved are quite different: atmospheric convection is caused by the heating of the atmosphere from below, while large-scale circulation results from differential horizontal heating.

Both types of motion relate to absorption of solar radiation, and the limits to their strengths can be estimated by thermodynamics. To do so, we need to consider the radiative fluxes that heat and cool different regions of the Earth system. When the Carnot limit is applied to the redistribution of heat in this setting, it yields directly the mechanical power that can maximally be derived from the heating and cooling and that generates kinetic energy. In fact, we already obtained a first-order estimate of the maximum power and associated heat fluxes for convection in Section 6.6, where the limit to the power from a vertical difference in radiative heating and cooling was obtained. This estimate for the limit on atmospheric convection is further refined in the following chapters on hydrologic cycling and on land (Chapters 8 and 10), as this refinement requires a specific consideration of the hydrologic processes for formulating moist convection (i.e., convection driven by condensational heating) and it requires the characteristics on land to describe dry convection (i.e., convection driven by surface heating). In the following, we will thus focus first on the limits to the large-scale circulation and its associated poleward transport of heat.

To estimate the limit on generating the large-scale circulation of the atmosphere–ocean system, we consider motion as being the result of a heat engine that is driven by large-scale differences in the solar radiative forcing. The schematic setup is illustrated in Fig. 7.4a. The figure shows a hemisphere that is separated into two regions of equal surface area, with the 30° latitude being the boundary between the regions that follows from geometric considerations. The region with latitudes below 30° latitude is referred to as the tropical region with index t in the variables, while the region poleward of 30° is referred to as the polar region with index p.

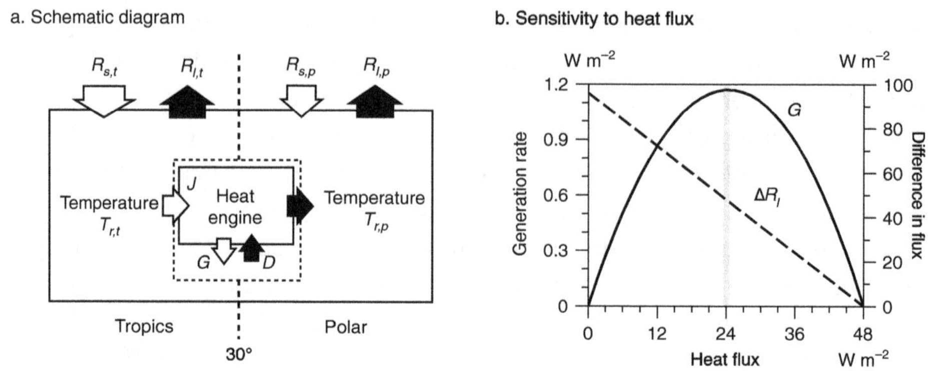

Figure 7.4 Schematic diagram (a) of poleward heat transport J, in the Earth's atmosphere–ocean system driven by the difference in solar radiative heating between tropical and polar regions, $R_{s,t}$ and $R_{s,p}$. We consider a separation of the planetary surface into two regions of equal area, with the separation represented by the 30° latitude indicated by the dashed line. The emission of terrestrial radiation to space by the two regions is described by $R_{l,t}$ and $R_{l,p}$ with respective radiative temperatures $T_{r,t}$ and $T_{r,p}$. The sensitivity of the power G, from this radiative forcing is shown in (b) as well as the difference in the emission of terrestrial radiation, $\Delta R_l = R_{l,t} - R_{l,p}$. The area shaded in grey in (b) indicates the state of maximum power.

The regions are heated by solar radiation, but at different intensities due to the different inclination of the surfaces to the incoming solar radiation. The poleward transport of heat J, fuels the heat engine, and acts to level out the differences in solar radiative heating. This heat flux is accomplished mostly by the large-scale atmospheric circulation, although oceanic currents contribute to this heat transport as well. The following formulation is similar to previous approaches in which either entropy production or power associated with J was maximized (Paltridge 1975, 1979; Lorenz et al. 2001; Kleidon and Renner 2013b). As we will see, the application of the maximum power limit results in an almost identical outcome as the maximization of entropy production (as was also already shown in Section 4.7).

The limit of the large-scale heat engine is given by the Carnot limit, eqn. 4.4, using the heat flux J and the temperature difference between the two regions:

$$G = J \cdot \frac{T_{r,t} - T_{r,p}}{T_{r,t}} \tag{7.10}$$

For simplicity, we use the radiative temperatures of the tropical and polar regions, $T_{r,t}$ and $T_{r,p}$ respectively, as these describe the respective temperatures of the entropy export to space. Note that the temperatures that drive the real atmosphere are likely to be higher than the radiative temperatures, which could affect the power associated with the maximum power limit, although the associated optimum heat flux is likely to be similar.

To obtain these temperatures and see how they relate to each other, we use the two energy balances of the regions. The energy balances of the two regions in steady state are written as

$$R_{s,t} - R_{l,t} - J = 0 \tag{7.11}$$

$$R_{s,p} - R_{l,p} + J = 0 \tag{7.12}$$

where $R_{s,t}$ and $R_{s,p}$ are the absorption of solar radiation and $R_{l,t}$ and $R_{l,p}$ the longwave emission to space of the two regions. We use a linearized formulation for radiative emission of the form $R_l(T) = R_s + k_r(T - T_r)$, where R_s is the mean solar absorption; $R_s = (R_{s,t} + R_{s,p})/2$, $k_r = 4\sigma T_r^3 = 4R_s/T_r$ is a linearization constant, and T_r is the mean radiative temperature. Using this linearization, we obtain expressions for the radiative temperatures of the two regions

$$T_{r,t} = T_r + \frac{R_{s,t} - R_s - J}{k_r} \tag{7.13}$$

$$T_{r,p} = T_r + \frac{R_{s,p} - R_s + J}{k_r} \tag{7.14}$$

The temperature difference between the regions, $\Delta T = T_{r,t} - T_{r,p}$, is thus given by

$$\Delta T = T_{r,t} - T_{r,p} = \frac{\Delta R_s - 2J}{k_r} \tag{7.15}$$

with $\Delta R_s = R_{s,t} - R_{s,p}$ representing the difference in absorption of solar radiation between tropical and polar regions.

We follow the same procedure as before (Sections 4.5 and 6.6) and maximize the power of the Carnot limit. Using the expression for ΔT given by eqn. 7.15 in the Carnot limit, eqn. 7.10, we obtain

$$G = J \cdot \frac{\Delta R_s - 2J}{k_r T_{r,t}} \tag{7.16}$$

The maximization of power by $\partial G/\partial J = 0$ then results in a value of the maximum power, $G_{\max}$, an optimum heat flux J_{opt} that attains this maximum, and an associated optimum temperature difference, ΔT_{opt}:

$$G_{\max} = \frac{\Delta R_s^2}{8 k_r T_{r,t}} = \left(\frac{\Delta R_s}{32\, R_s} \frac{T_r}{T_{r,t}} \right) \Delta R_s \tag{7.17}$$

$$J_{\text{opt}} = \frac{\Delta R_s}{4} \tag{7.18}$$

$$\Delta T_{\text{opt}} = \frac{\Delta R_s}{2k_r} = \left(\frac{\Delta R_s}{8R_s} \right) T_r \tag{7.19}$$

As in the previous examples, this maximum power limit results from the trade-off between the heat flux J and the driving temperature difference ΔT. This trade-off is shown in Fig. 7.4b. Instead of using ΔT, the trade-off is shown in terms of the difference in the emission of terrestrial radiation, ΔR_l, given by

$$\Delta R_l = R_{l,t} - R_{l,p} = \Delta R_s - 2J = k_r \Delta T \tag{7.20}$$

Since we have used a linearized expression for the emission to space, this difference in radiation is directly proportional to the difference in radiative temperatures. We note that for no heat transport, $\Delta R_l = \Delta R_s$, which reflects a local radiative equilibrium between absorbed solar and emitted terrestrial radiation. With increasing values of J, this difference declines, and vanishes when half of ΔR_s is transported, which would yield a state of global radiative equilibrium with uniform radiative temperatures. The maximum power state reflects the state where this difference is depleted to half the value attained at local radiative equilibrium. The difference in radiative fluxes at the top of the atmosphere is well observed by satellites, and commonly used to infer the magnitude of poleward heat transport (see e.g., Masuda 1988), so that we test the maximum power limit next by comparing not just the magnitude of G, but also by the value of ΔR_l to observations.

We evaluate the maximum power limit using observations for present-day conditions. The difference in the incident solar radiation between the two regions is calculated from geometric factors. Due to the more perpendicular orientation of surfaces in the tropics, this region receives about 20% more solar radiation than the planetary average, while the polar regions receive about 20% less. We thus have $R_{s,t} = 1.2R_s$, $R_{s,p} = 0.8R_s$, and $\Delta R_s = 0.4R_s$. With a value of $R_s = 240$ W m^{-2}, the difference in solar radiative forcing between the regions is thus $\Delta R_s = 96$ W m^{-2}. With this forcing, we obtain a maximum power of about $G_{\max} = 1.2$ W m^{-2}, an optimum heat flux of $J_{\rm opt} = 24$ W m^{-2}, and a difference in radiative temperatures of $\Delta T_{\rm opt} = 13$ K that is associated with a difference of $\Delta R_l = 48$ W m^{-2}. The associated, optimum efficiency of the heat engine is given by the expression in the brackets in eqn. 7.17, and yields a value

$$\eta_{\rm opt} = \frac{G_{\max}}{\Delta R_s} = \frac{\Delta R_s}{32 R_s} \frac{T_r}{T_{r,t}} \tag{7.21}$$

which yields an estimate of $\eta_{\rm opt} \approx 1$ %. These estimates compare quite well to estimates based on observations. The efficiency of the atmospheric heat engine is estimated from observations to be about 2 % (Peixoto and Oort 1992) with a mean generation rate of about 2 W m^{-2} (Peixoto and Oort 1992; Li et al. 2007). This low efficiency of the large-scale atmospheric circulation has long been recognized (Ertel and Köhler 1948). To evaluate the magnitude of poleward heat transport, the top-of-atmosphere radiative fluxes were used in previous studies (Masuda 1988;

Trenberth and Caron 2001). Observations place the magnitude of the heat transport at about $6 \cdot 10^{15}$ W at a latitude of 30° per hemisphere. When dividing this by the surface area of the hemisphere, this yields a value of 48 W m^{-2}, which corresponds very well with our value of ΔR_l. Note that in the way we set up the model, the value of ΔR_l at a state of maximum power is related to J_{opt} by $\Delta R_l = 2J_{\text{opt}}$ (see eqn. 7.20), (see also Fig. 7.4b). Trenberth and Caron (2001) further estimated that $5 \cdot 10^{15}$ W of the poleward heat transport is accomplished by the atmospheric circulation, with oceans transporting heat of up to $2 \cdot 10^{15}$ W by the oceanic circulation.

Note that most of the power associated with this forcing is used to generate motion within the atmosphere, with only a small component being used to force the ocean. This has several reasons. Oceanic motion is generated by different means, predominantly by surface winds that exert a surface stress on the ocean surface, as well as buoyancy and tidal forces. The wind-driven part of the forcing relates to the formation of surface waves, similar to the example described in Section 4.6, and would in principle be subjected to a conversion limit as well, as described by the limits to converting kinetic energy in Section 4.6. Estimates place this transfer at a removal rate of about $65 \cdot 10^{12}$ W from the kinetic energy of the atmosphere, which is mostly used to generate wind waves, so that the wind forcing results in only about $2 - 7 \cdot 10^{12}$ W associated with the generation rates for oceanic motion due to wind forcing (Munk and Wunsch 1998; Ferrari and Wunsch 2009; von Storch et al. 2012). The tidal forcing contributes about $3 \cdot 10^{12}$ W (Ferrari and Wunsch 2009), and about $1 \cdot 10^{12}$ W is estimated to be due to buoyancy (Oort et al. 1994), although some estimates are considerably lower (Gnanadesikan et al. 2005) so that there is considerable uncertainty in these numbers. Converted to rates per unit area, these rates correspond to an order of magnitude of about 1 mW m^{-2}, and are thus much smaller than the estimated generation rates associated with atmospheric motion.

An important aspect to note is that seasonal variations in solar radiation affect the maximum power limit quite substantially. Seasonal variations in $R_{s,p}$ alter ΔR_s by about a factor of 2, with the difference being much reduced in summer, and about doubled in winter. While such variations do not affect the mean heat flux, J_{opt}, as it is linear in ΔR_s, the use of annual means underestimates the power, as ΔR_s enters the expression for $G_{\max}$ as $(\Delta R_s)^2$. When the doubled value of ΔR_s is used for half the year and $\Delta R_s = 0$ for the remaining part of the year, the estimate of $G_{\max}$ doubles in value, yielding a limit of about $G_{\max} = 2.4$ W m^{-2} and an efficiency of about 2%. The seasonality in ΔR_s then yields a seasonal variation in $G_{\max}$, with a greater generation rate in winter than in summer. This greater generation rate can directly be related to greater synoptic activity and greater wind speeds that are observed in winter compared to summer (e.g., Peixoto and Oort 1992). Considering these seasonal variations in the maximum power limit also brings the estimate of power closer to observations.

What is interesting about the limit to generating large-scale motion is that the expression given by eqn. 7.17 depends mostly on the difference in solar absorption, ΔR_s. This difference is to a large extent determined by purely geometric factors related to the angle of the absorbing surface to the incident solar radiation although the reflectivity of polar regions, e.g., as affected by ice and cloud cover, also play a role. This result is in line with the finding by Stone (1978) that the strength of poleward heat transport is largely determined by how solar radiation is being absorbed by the planet. The agreement of the predicted magnitudes with observations suggests that the large-scale poleward heat transport operates at this limit and is thus constrained by thermodynamics and by the tight interaction between heat transport and the driving temperature difference. Its interpretation is consistent with several studies that applied maximum entropy production (MEP, Section 4.7) to poleward heat transport (Paltridge 1975, 1978; Lorenz et al. 2001). The connection between the maximum power derived here and entropy production is made by noting that power equals dissipation in steady state, and entropy is produced by the frictional dissipation of kinetic energy at the prevailing temperature. Hence, we have $\sigma_{\text{fric}} \approx D/T_r = G/T_r$. Since T_r is fixed by the planetary energy balance, a maximum in power approximately corresponds to a maximum in the associated entropy production.

7.5 Inferring mass exchange from maximum power

There is more that we can infer from the maximum power limit about the dynamics of the system. As the maximum power limit predicts the value of the associated heat flux and temperature difference, and this heat flux is accomplished by the associated motion within the system, we can also infer the speed by which this motion takes place and the rate by which mass is exchanged between the two reservoirs. This is conceptually illustrated in Fig. 7.5. The heat flux is accomplished by motion that effectively exchanges thermal energy and thereby affects the temperatures of the two regions. When we view this exchange as the result of swapping certain amounts of mass between the two regions at a certain rate, as indicated by the boxes in the figure, then the heat flux is accomplished because the two masses hold different amounts of thermal energy. Technically, this relates closely to the well-established concept of eddy transport in atmospheric and oceanic sciences. What we do here can thus be seen as a highly simplified description of eddy transport. This transport of thermal energy can then be formulated as

$$J = (c_p \Delta T) J_m \tag{7.22}$$

where $c_p \Delta T$ is the difference in the amount of heat per unit mass and J_m is the rate by which mass is being exchanged between the regions. As the maximum power

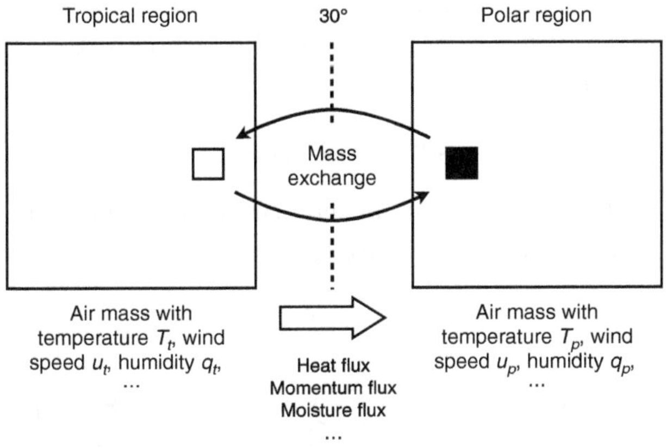

Figure 7.5 The heat flux inferred from maximum power is accomplished by a certain rate of mass exchange. By swapping a certain amount of mass, as represented by the small squares, between the two regions at a certain rate, not only thermal energy is transported, but also other properties, such as momentum and moisture of the regions. Hence, the maximum power limit also sets the magnitude of other exchange fluxes beyond thermal energy.

limit not only predicts the value of J, but also the associated value of ΔT, we can use eqn. 7.22 to infer the rate of mass exchange by:

$$J_m = \frac{J}{c_p \Delta T} \tag{7.23}$$

This mass exchange rate can then be used to infer the exchange of other properties between the reservoirs, for instance, momentum, angular momentum, moisture, or the concentrations of other substances of the system. One complicating aspect in eqn. 7.23 is that it is not just the heat capacity c_p, that determines the rate of heat exchange, but also, for instance, the moisture content as this represents latent heat. The mass exchange rate thus also depends on the magnitude of moisture transport, an aspect that is dealt with in Chapter 8. Nevertheless, what this shows us is that we can infer more than just the heat flux and the temperature gradient from the maximum power limit, as the associated mass exchange also exchanges other properties of the system.

This approach is illustrated using the example of poleward heat transport of the previous section. The optimized heat flux J_{opt} determined from maximum power passes through the cross-sectional area at the latitude $\phi = 30°$. The cumulative heat flux that is transported through this cross section is given by $J_{opt}A_t$, where A_t is the surface area of the tropical regions, which is half of the total surface area of the hemisphere, so that $A_t = \pi r_e^2$. The cross sectional area is given by $A_{30} = 2\pi r_e \cos\phi \cdot \Delta z_a$, where Δz_a is the height of the atmosphere of about $\Delta z_a = 10^4$ m. This heat flux is accomplished by a certain rate of mass exchange, which we refer to as $J_{m,\text{opt}}$. We thus write

$$J_{\text{opt}}A_t = c_p \Delta T_{\text{opt}} J_{m,\text{opt}} \tag{7.24}$$

where c_p describes the specific heat capacity of the mass that is being transported. We then obtain for the optimum mass exchange

$$J_{m,\text{opt}} = \frac{J_{\text{opt}}A_t}{c_p \Delta T_{\text{opt}}} \tag{7.25}$$

This mass exchange rate is also given by the rate by which mass is exchanged through the cross-sectional area A_{30}, at a certain rate v_{opt}, by meridional flow. As the mean density is given by the atmospheric mass per unit area $m_a = p_s/g$, divided by the height Δz_a, the mass exchange rate is also described by $J_{m,\text{opt}} = 2\pi r_e \cos\phi \cdot p_s/g \cdot v_{\text{opt}}$. We can thus derive an expression for the rate of v_{opt}:

$$v_{\text{opt}} = \frac{g A_t}{2\pi r_e \cos\phi \cdot c_p p_s \Delta T_{\text{opt}}} \cdot J_{\text{opt}} \tag{7.26}$$

Using the numbers from above, this equation yields an estimate of $v_{\text{opt}} = 0.7$ m s^{-1}, which compares very well to the observed magnitude of mean meridional velocities of about 0.5–1 m s^{-1} (Peixoto and Oort 1992).

This approach will also be used in later chapters to infer the exchange rates of substances, such as moisture or carbon, particularly regarding vertical, convective exchange.

7.6 Maximum power and geostrophic motion

While the magnitude of meridional motion can directly be inferred from the maximum power limit, meridional flow only forms a minor component of the large-scale circulation in the atmosphere. Most of large-scale motion is in the zonal direction, with its easterly trade winds in the tropics and westerly winds in the mid-latitudes. The thermodynamic perspective of the previous section, however, tells us that the driver for large-scale motion is the difference in solar radiative forcing between the tropics and the polar regions. The resulting heat transport depletes this difference in forcing, but takes place in the meridional, not in the zonal direction. It thus seems that most of the atmospheric flow does not deplete the driving gradient, contrary to what would be expected from thermodynamics.

This seeming contradiction is resolved when considering the conservation of angular momentum of the Earth system. When applied to the two box representation of the atmosphere, angular momentum consists of the combination of atmospheric mass rotating at the speed of the Earth's rotation rate as well as the relative motion that we denote by the zonal wind speeds of the two regions. Angular momentum conservation requires a full treatment of the planetary angular momentum balance, including oceans and the solid Earth as well as relative variations of the Earth's rotation rate (Oort 1989). When we use a simplified view and just apply angular momentum conservation to large-scale atmospheric motion, it can, however, tell us how zonal flow develops and how it is linked to maximum power. Let us assume that the zonal flow is represented by two zonal velocities, u_t and u_p, of the two regions. Conservation of angular momentum then already tells us that when motion in the atmosphere is generated, u_t needs to oppose u_p, so they should have opposite signs. Since the angular momentum also relates to the distance to the axis of rotation, u_p needs to have a greater value than u_t because the motion in the polar region is closer to the axis of rotation. The two boxes would thus have different values of angular momentum relative to the Earth's rotation. The mass exchange between the regions at a latitude of 30° then exchanges not just heat and mass, but also the relative angular momentum of the two boxes. This exchange constitutes a torque that generates and maintains the zonal flow and that is in steady state opposed by the torque exerted by surface friction. The zonal flow is thus caused by the meridional mass exchange, even though the resulting magnitude of the zonal flow is much larger than the meridional flow.

The effects of angular momentum conservation are typically accounted for by the Coriolis force, an apparent force that deflects horizontal motion towards the right in the northern hemisphere. This force relates to geostrophic flow, a concept that is widely used in atmospheric and oceanic science to describe the mostly zonal motion on a rotating sphere. Geostrophic flow results from a momentum balance consisting mostly of the meridional, accelerating force caused by a large-scale pressure gradient, F_{acc}, associated with available potential energy and the Coriolis force, F_c. The geostrophic balance of forces is illustrated in Fig. 7.6.

We next consider a simple representation of geostrophic flow, which is typically used in climatology textbooks (e.g., Peixoto and Oort 1992) to illustrate its relation to power and meridional heat transport. We use the formulation of Miller et al. (2011b) in the following, who used this formulation and evaluated energetic aspects of geostrophic flow. To do so, we consider a flow in the horizontal direction, with the two components of the velocity given by u for the zonal component and v for the meridional component. When we express the momentum balance at a given latitude in simple terms, it can be written as

$$m\frac{du}{dt} = fv - ku \qquad m\frac{dv}{dt} = F_{\mathrm{acc}} - fu - kv \tag{7.27}$$

where F_{acc} is the accelerating force due to a pressure gradient in the meridional direction, f is the Coriolis parameter with $f = 2\Omega \sin\phi$, $\Omega = 7.27 \cdot 10^{-5}\ \mathrm{s}^{-1}$

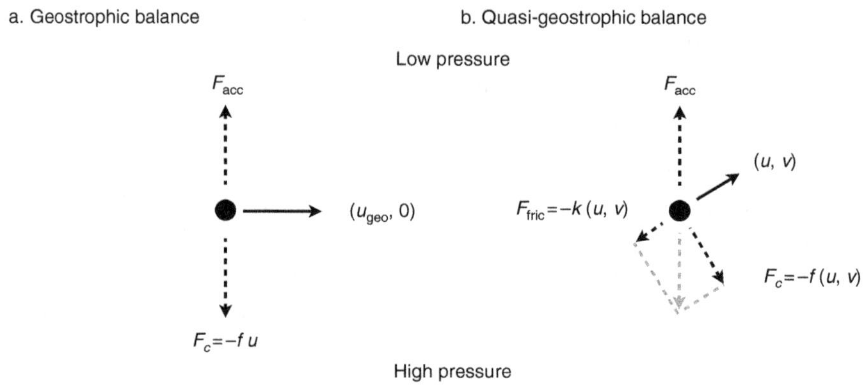

Figure 7.6 Illustration of (a) the geostrophic balance of forces consisting of an accelerating pressure gradient force, F_{acc}, and the Coriolis force F_c, acting on a parcel of air indicated by the black circle and (b) the effect of friction, F_{fric}, on the geostrophic balance. The diagram illustrates the formulation in terms of eqn. 7.27.

is the Earth's rotation rate, ϕ is the latitude, and where the friction force, F_{fric}, is expressed in a simplified way by $-ku$ and $-kv$ in the velocity components, with k being a friction coefficient.

In steady state, the time derivatives vanish, and geostrophic flow is described by the limiting case in which friction is neglected, so that $k = 0$ in eqn. 7.27. This yields a solution for geostrophic flow described by

$$u_{geo} = \frac{F_{acc}}{f} \qquad v_{geo} = 0 \tag{7.28}$$

This solution shows that the resulting flow is perpendicular to the accelerating force (Fig. 7.6a). This is well known, and is often observed in the free atmosphere where the wind direction follows the isobars and, for instance, in the jet streams in the upper atmosphere. Note, however, that geostrophic flow has no meridional component ($v_{geo} = 0$), so that no thermal energy is transported in the meridional direction.

When friction is not neglected, the steady-state solution for the equations of motion given by eqn. 7.27 is

$$u = \frac{f}{f^2 + k^2} \cdot F_{acc} \qquad v = \frac{k}{f^2 + k^2} \cdot F_{acc} \tag{7.29}$$

We note that in the presence of friction, there is a meridional flow component with $v > 0$ that comes at the expense of u being reduced. This component v, is the so-called ageostrophic component of the flow. It plays a critical role, because it is due to this comparatively small component that the driving pressure gradient and the distribution of potential energy is depleted towards a global minimum, as reflected in the transition of states B–C in Fig. 7.1.

We can also use this simple formulation of geostrophic flow and look at its energetics. The generation of kinetic energy is given by the product of the accelerating force, F_{acc}, and the meridional velocity, v, while the frictional dissipation is given by the friction force and the wind speed:

$$G = F_{acc} \cdot v = \frac{k}{f^2 + k^2} \cdot F_{acc} \qquad D = k \cdot (u^2 + v^2) = \frac{k}{f^2 + k^2} \cdot F_{acc} \qquad (7.30)$$

Note that both expressions are equal, so that the steady-state condition, $G = D$, is fulfilled. We also note that in a purely geostrophic flow with $k = 0$, $G = D = 0$. This is, of course, as expected, as friction is specifically excluded from the geostrophic balance. Yet this is a rather abstract situation. It means that geostrophic flow in steady state does not need power to be sustained because there are no frictional losses that would dissipate the kinetic energy. As such, geostrophic flow is thus not the consequence of a flow that depletes its driving gradient, as it is not associated with dissipation. It is merely a consequence of the conservation of angular momentum.

When we link these effects of angular momentum conservation back to maximum power, we would expect that the state of maximum power is associated with a zonal flow that is comparatively weak and the least geostrophic. A state of maximum power is associated with a comparable large value of k in eqn. 7.30 that is associated with a considerable meridional flow component to transport the thermal energy from which the kinetic energy for the large-scale circulation is generated and which depletes the driving pressure gradient associated with F_{acc}. This connection between maximum power and the intensity of the geostrophic flow component relates to an observation by Suomi (as cited in Catling 2005, and referred to as the "jet stream paradox") that the Earth has a comparatively slow jet stream, while other planets in the solar system have much higher characteristic jet stream wind speeds even though the thermal forcing of the other planets is comparatively weaker. When considering that jet streams are well described as geostrophic flow, they become weaker the greater the rates of generation and dissipation are. At a state of maximum power, this would suggest the weakest jets for a given radiative forcing. We will come back to the radiative forcing of different planets in Section 12.3 and see there that the Earth has the strongest solar radiative forcing of a planet with an atmosphere. With the strongest forcing, the most power can be generated, and the geostrophic flow has a comparatively large ageostrophic component and thus a comparatively weak jet stream. It would thus seem that the observation by Suomi is no paradox, but can be explained by the fact that the intensity of geostrophic flow does not relate to the strength of the radiative forcing as it requires no power to sustain it.

The linkage between maximum power and the angular momentum balance also suggests different roles of the two flow components: while the generation of kinetic energy is associated with the meridional flow, dissipation of the kinetic energy is associated mostly with the zonal flow, which has wind velocities that are typically at least an order of magnitude greater for present-day Earth than the meridional wind velocities (Peixoto and Oort 1992). This means that the magnitude of wind speed does not necessarily tell us how much power is associated with the flow. As we will see later in the book, that has important implications for the potential of wind power as a renewable energy resource. A quantitative analysis of this aspect would, of course, require an explicit link between the meridional flow associated with maximum power and the angular momentum balances of the two regions.

At a more general level, this linkage between the angular momentum balance and maximum power describes the role that the planetary rotation rate has in shaping planetary motion and which can potentially act as a constraint. It is through the angular momentum balance that planetary rotation rate affects the torques that are associated with meridional heat transport and surface friction. Hence, the resulting association of kinetic energy to its meridional and zonal flow components is affected. This dependence is, for instance, shown in recent climate model simulations (Pascale et al. 2013) that show different ways in which the atmospheric circulation is arranged, from a single, large meridional circulation cell at low rotation rate, to strong zonal flow at a high rotation rate. These simulations show that at high rotation rates, the angular momentum balance constrains the meridional heat transport to such an extent that it is no longer determined entirely by thermodynamic maximum power limit, but reduced to lower values. This aspect has previously been raised as a criticism to applications of MEP to poleward heat transport in that these applications do not account for the constraining effects of planetary rotation rate (Rodgers 1976; Goody 2007). The description provided here potentially allows us to address this criticism (see also Jupp and Cox 2010).

7.7 Limits to motion in the Earth's interior

We next consider motion in the Earth's interior, particularly mantle convection. The slow, creeping motion of solid material in the Earth's mantle is driven by different forms of heating related to radioactive decay, secular cooling, and the crystallization of the core. These heat sources result in a total heat production of less than 0.1 W m^{-2} at the Earth's surface (Davies and Davies 2010), although this rate was estimated to be substantially higher in the Earth's deep past (Stevenson et al. 1983). Mantle convection then results from the differential heating of the interior and cooling at the surface. It is thus subjected to thermodynamic limits. These limits have long been evaluated (Stacey 1967; Metchnik et al. 1974; Hewitt

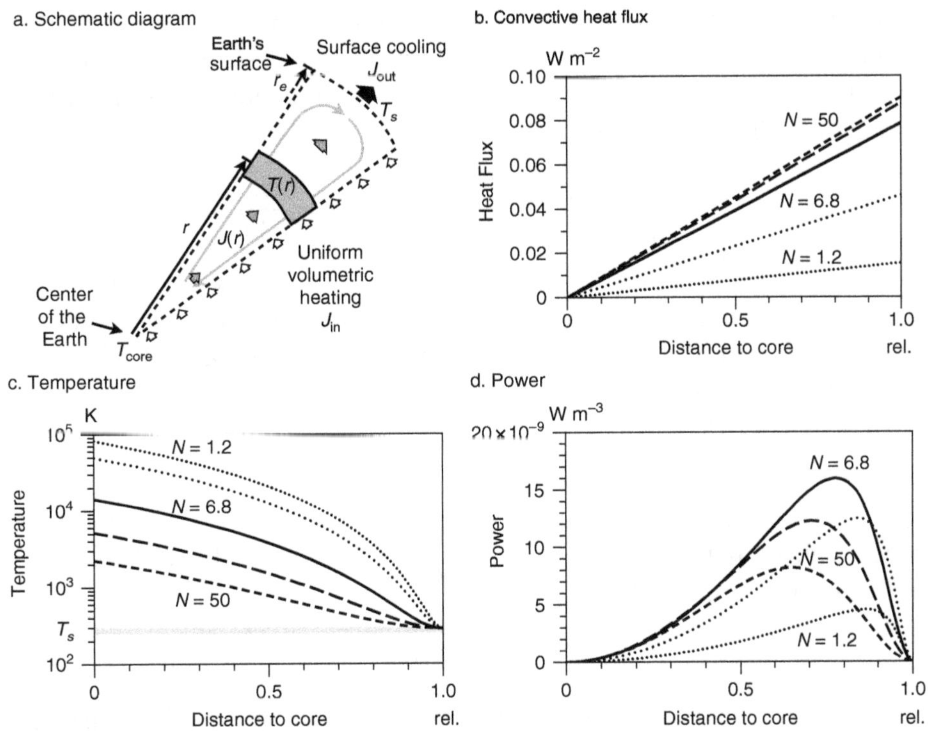

Figure 7.7 Illustration of (a) the schematic setup to derive the maximum power limit for interior motion. The plots show the variation of (b) the convective heat flux J_c, (c) temperature T, and (d) the contribution to the total power G_{max}, as a function of the distance to the core, r/r_e. The plots are shown for five values of the Nusselt number, N, that is, the ratio of the convective to diffusive heat flux, of 1.2 (weakest convective heat flux, narrow dotted line), 2, 6.8 (maximum power, solid line), 20, and 50 (greatest convective heat flux, narrow dashed line).

et al. 1975; Backus 1975; Verhoogen 1980) although without considering the effect on the driving temperature gradient; they have also been related to MEP by Vanyo and Paltridge (1981) and Lorenz (2002). Here, we follow the work of Lorenz (2002) and Dyke et al. (2011) in a slightly reformulated way and focus on the limit of maximum power that can be derived from the interior heating rate.

To derive the maximum power limit for mantle convection as well as its consequences, we consider a segment of the Earth as shown in Fig. 7.7a. We consider a setup of a uniform and fixed interior volumetric heating rate J_{in}, which is a simplification for the interior of the Earth, but which makes the following considerations relatively straightforward. The complicating factor in deriving the maximum power limit compared to the two-box representations of earlier derivations is that heat flux J, from the inside outwards to the surface of the sphere accumulates the heating rate

J_{in}, and successively passes through greater volumes with an increasing distance r, from the center of the sphere. We consider again a steady state, in which the total interior heating by J_{in} is in steady state with the cooling rate J_{out}, at the surface of the solid Earth. The connection is given by

$$J_{\text{out}} = \int_V J_{\text{in}} dV = \int_0^{r_e} J_{\text{in}}(4\pi r^2) dr = \frac{4\pi}{3} r_e^3 J_{\text{in}} \tag{7.31}$$

The total heat flux J, at any given radial distance r, to the center of the Earth is entirely determined in this setup, yet it is composed of a purely radiative–diffusive component J_d, and a convective component J_c, that is associated with interior motion. The total heat flux is then written as a multiple of the diffusive heat flux

$$J = J_c + J_d = -Nk_d \nabla T \tag{7.32}$$

where k_d is the heat conductivity of the mantle material, and N is the Nusselt number, defined by the ratio $N = J/J_d$, and T is the temperature. The convective heat flux is then described by

$$J_c = (N-1)/N \cdot J \tag{7.33}$$

The derivation of the maximum power limit is derived here not with respect to J_c, but rather with respect to N.

Before we derive the maximum power limit, we derive the distribution of temperature T as a function of r and N. The temperature distribution is derived from the energy balance of a given shell at a distance r to the center, as illustrated in Fig. 7.7a. It is given by the heating of the shell and the divergence of heat fluxes across the shell, yielding a diffusion equation for J:

$$J_{\text{in}} = \nabla \cdot J \tag{7.34}$$

Considering only the variation in the radial direction r, from the core outward to the surface (with $r = r_e$) and using the expression of the heat flux given by eqn. 7.32, the energy balance constraint is then expressed as

$$J_{\text{in}} = -\frac{2k_d N}{r}\frac{\partial T}{\partial r} - k_d N \frac{\partial^2 T}{\partial r^2} \tag{7.35}$$

It can be shown that the analytic solution of this diffusion equation is given by

$$T(r) = T_{\text{core}} - \frac{J_{\text{in}}}{6k_d N} \cdot r^2 \tag{7.36}$$

where T_{core} is the temperature of the core ($r = 0$). As we know the temperature at the surface very well, which is the temperature at a distance of $r = r_e$ and which is not affected by N because it is controlled by the radiative forcing, we can also express this temperature function as

$$T(r) = T_s + \frac{J_{\text{in}}}{6k_d N}\left(r_e^2 - r^2\right) \tag{7.37}$$

so that $T_{\text{core}} = T_s + J_{\text{in}} r_e^2/(6k_d N)$. The associated, convective heat flux is then given by

$$J_c = -k_d \frac{N-1}{N} \nabla T = \frac{(N-1)}{3N} J_{\text{in}} \cdot r \tag{7.38}$$

The maximum power limit is now derived from the entropy balance. The entropy balance of the system is given by the entropy export by the heat flux J_{out} at the temperature of the surface T_s, and the addition of heat J_{in}, at a prevailing temperature $T(r)$:

$$\begin{aligned} \sigma_{\text{tot}} &= \frac{J_{\text{out}}}{T_s} - \int_0^{r_e} \frac{J_{\text{in}}}{T(r)} (4\pi r^2) dr \\ &= \int_0^{r_e} J_{\text{in}} \left(\frac{1}{T_s} - \frac{1}{T(r)}\right) (4\pi r^2) dr \end{aligned} \tag{7.39}$$

where eqn. 7.31 was used for J_{out}. This total entropy production is caused by diffusion and convection, with the respective contributions given by the ratio $1/N$ and $(N-1)/N$. Entropy production by convection is, in the ideal case, entirely due to frictional dissipation. As frictional dissipation balances power in the steady state, $G = D$, this entropy production is given by

$$\sigma_{\text{conv}} = \frac{N-1}{N} \sigma_{\text{tot}} = \frac{G}{T_s} \tag{7.40}$$

which assumes, for simplicity, that the frictional dissipation takes place at T_s. The power G, can then be expressed as an integration of the respective contributions of heat addition at distances r:

$$G = \frac{N-1}{N} \int_0^{r_e} J_{\text{in}} \cdot \frac{T(r) - T_s}{T(r)} (4\pi r^2) dr \tag{7.41}$$

This corresponds to the Carnot limit applied to the incremental heating, $J_{\text{in}}(4\pi r^2)$ dr, that is added at a distance r.

This expression for the power G, has a maximum for a certain magnitude of the convective heat flux, with the same trade-off as in the previous examples. This maximum and the trade-off is shown by the plots in Fig. 7.7, which show the functions $J_c(r)$, $T(r)$, and the term in the integral of G for five different values of N. The example uses values of $J_{\text{in}} = 3.7 \cdot 10^{-8}$ W m^{-3}, which corresponds to a total heat flux at the Earth's surface of about $47 \cdot 10^{12}$ W (Davies and Davies

2010), or 0.09 W m^{-2}, and a thermal conductivity of $k_d = 3$ W m^{-1} K^{-1}. A greater convective flux is associated with a greater value of N (Fig. 7.7b), which increases the factor $(N - 1)/N$ in eqn 7.41. The efficiency of the conversion, as described by the factor $(T(r) - T_s)/T(r)$ in the integral, also depends on N, which can be seen in eqn. 7.37 or in Fig. 7.7c. The greater the value of N, the more efficient is the convective cooling, resulting in lower temperatures in the interior. This greater cooling reduces the efficiency, and thus the power as in the maximum power cases described before. Consequently, there is an optimum value of about $N_{opt} = 6.8$ at which the power is maximized. This is shown by the contributions at different distances in Fig. 7.7. The total maximum power in this example is about $G_{max} = 25.3 \cdot 10^{12}$ W or 0.05 W m^{-2}, a number that is consistent with Verhoogen (1980), although without considering the maximum power limit. This value of G_{max} is associated with an overall efficiency of about 54%.

Note, however, that this formulation of the thermodynamic limit of interior motion is grossly simplified. Heating does not take place uniformly in the interior, but relates to heat release by crystallization at the outer core, and a good fraction of radiogenic heating takes place in the continental crust near the surface. Also, a considerable fraction of the heat flux of the Earth's surface relates to secular cooling of the Earth, which is not related to a heating source. Another factor that is likely to affect the thermodynamic setting is that continental crust acts as an insulator at the top. Lenardic et al. (2005) suggested that this insulation by continents affects interior temperatures, the viscosity (which is less for higher temperatures), and thus may in fact enhance heat loss from the interior. Another factor that is omitted here is that there are other forms of work that are also generated by heat. This applies particularly to the generation of the magnetic field by motion within the core (Stevenson 1983; Christensen and Tilgner 2004), which is, however, estimated at about 0.2–0.5 $\cdot$ 10^{12} W and thus notably less in magnitude. Other forms of work, such as work done by earthquakes, folding, or the uplift of mountains, are of similar, small magnitude (Verhoogen 1980). The work of mantle convection in this example would thus correspond to the major component of work done in the interior. The example used here should be seen as a rough estimate of how the maximum power limit would apply to interior motion, which nevertheless results in some reasonable magnitude of heat transport by mantle convection.

7.8 Thermodynamic consequences of motion

After describing these limits of generating motion within the Earth system, we focus next on the thermodynamic consequences that are associated with motion. In Chapter 3, we already dealt with this aspect in more general terms in that the dynamics that take place within the system are directed such that these follow the second law. Even more so, we already explored to some extent before that the dynamics within a system act to accelerate the rate towards establishing thermodynamic equilibrium. Hence, we should thus find system-wide, thermodynamic consequences of motion within the system. We consider these consequences now in more detail in the context of the examples given in Sections 7.4 and 7.7.

To interpret thermodynamic consequences of motion within a system, it is important to recognize two aspects: first, recall that the thermal entropy of a system is composed of two contributions, with the first contribution represented by the total thermal energy content, and a second contribution that is associated with an uneven distribution of thermal entropy within the system that is typically much smaller (cf. Section 3.5). Thermodynamic disequilibrium within the system is related to the second contribution, while the first contribution relates to a potential disequilibrium of the system with its surroundings. Second, systems have boundary conditions that are either fixed or that can be altered by the dynamics taking place within the system. The relevance of this distinction is that in the first case, fixed boundary conditions are associated with fixed rates of entropy exchange. In this case, motion is able to change the relative partitioning of entropy production within the system, whereas the total entropy production is fixed by the entropy exchange that is set by the boundary conditions. In the second case, the boundary conditions are not fixed but can be altered by the dynamics within the system which can then alter the entropy exchange with the surroundings. In such systems, motion can not only alter the partitioning of the internal entropy production, but also enhance the overall entropy export, thereby making the whole system more dissipative. The thermodynamic consequences of motion can thus be reflected in altered overall thermal energy content and related thermal entropy, the extent of disequilibrium, as well as in altered entropy exchange across the system boundary. As the entropy exchange across the boundary is driven by some gradient as well, e.g., a difference in radiation entropy, an accelerated rate of establishing equilibrium would ultimately be achieved by an enhanced entropy export of the system.

The two examples that we considered in great detail in this chapter are two examples with different boundary conditions. In the example of motion in the interior of the Earth, the entropy export of the system was fixed. This fixed entropy export resulted from a prescribed overall heating rate, and by a fixed surface temperature thus fixing the temperature at which entropy leaves the system. The latter aspect is justified by recognizing that the magnitude of solar heating of the surface is much larger than the heat flux from the interior, so that the surface temperature is predominantly determined by the external conditions imposed by the climate system, and not by the dynamics of the interior. Motion thus cannot affect the entropy export of the system. It nevertheless results in a clear effect that can be seen in the sensitivity of T to N shown in Fig. 7.7. With greater motion and a greater convective heat flux, corresponding to a greater value of N, the temperature gradient in the interior is substantially reduced. This reduction results in a lowering of the entropy of the interior, simply because less thermal energy is stored inside the system with a greater value of N. Even though the entropy of the system is reduced, this does not imply a greater thermodynamic disequilibrium because this low entropy relates to the lower content of thermal energy. The maximum power limit would then describe the state in which the thermal energy content in the interior is reduced to the lowest possible value that can be achieved by motion. As the entropy exchange at the surface is fixed, this state would then also describe the state by which the difference between the content of thermal energy of the interior and the system's surroundings is leveled out to the greatest extent possible that can be achieved by the internal dynamics given the fixed boundary condition of the system.

The example of large-scale heat transport of the atmosphere–ocean system has quite different boundary conditions. While the radiative heating in the two regions was prescribed and determined by geometric differences in insolation, the entropy export of the system is affected by the amount of heat transport. This effect is shown in Fig. 7.8. The figure shows the entropy export of the system, $J_{s,\text{out}} = (4/3) \cdot (1/2) \cdot (R_{l,t}/T_{r,t} + R_{l,p}/T_{r,p})$, with the factor of 4/3 added to account for the specifics of radiative entropy fluxes. As greater poleward heat transport levels out the differences in temperatures, both regions emit radiation to space at more uniform temperatures. In the limiting case of an isothermal planet, the entropy export of the system would simply be given by $(4/3) \cdot R_s/T_r$, which is indicated by the horizontal line in Fig. 7.8. The large-scale motion resulting from the difference

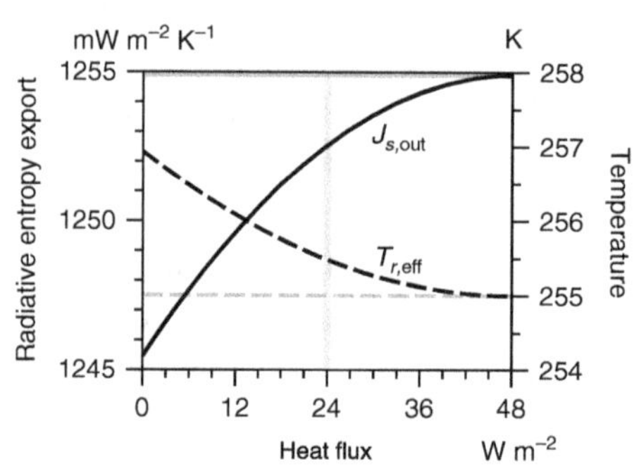

Figure 7.8 Thermodynamic consequences of poleward heat transport in terms of radiative entropy export, $J_{s,\text{out}}$ (solid line), and the effective radiative temperature of the planet, $T_{r,\text{eff}}$ (dashed line), using the same model as in Fig. 7.4. The vertical grey line indicates the state of maximum power. The horizontal grey lines show the limiting cases for radiative entropy export and the effective radiative temperature associated with an isothermal planet.

in solar insolation thus enhances the planetary entropy export to space. As this requires motion, and motion requires power to sustain it, the limiting case cannot be reached, but is constrained to the value associated with the maximum power limit.

Another way to express this is to determine an effective planetary radiative temperature. So far, we have determined the planetary radiative temperature by equating the absorbed solar radiation with blackbody emission at the planetary scale. In the example, the energy balances are separated among the two regions with different solar radiative forcing, which would each have different radiative temperatures. We can derive an equivalent planetary radiative temperature from the global entropy balance by $T_{r,\text{eff}} = R_s/(3/4 \cdot J_{s,\text{out}})$. This effective radiative temperature can also be seen in Fig. 7.8. It shows that poleward heat transport associated with motion results in a lowering of the effective radiative temperature. The limiting case of the radiative temperature that we have used so far, which is determined by $R_s = \sigma T_r^4$, can, however, not be reached by motion as it is associated with an isothermal temperature distribution from which no power could be derived to maintain motion and heat transport. This example thus shows that in the case of poleward heat transport, the maximum power limit is associated with an enhanced entropy export of the system, so that motion has consequences on the planetary rate of entropy exchange. Motion can thus be seen as a means to enhance the rate of establishing thermodynamic equilibrium at the planetary scale.

What we have not considered yet is that the internal dynamics of the atmosphere–ocean system can also affect the value of the Earth's albedo, and thereby the extent to which the planet is heated by absorption of solar radiation. The prime factors that shape the planetary albedo are the abundance of clouds and ice cover. These relate closely to the magnitude of hydrologic cycling and the strength of the greenhouse effect, and thus relate to material cycling. This aspect is dealt with in the following chapters, so that at the end of the book, we can deal with even larger effects on the entropy exchange of the Earth system.

7.9 Motion in the Earth system

At the end of this chapter, the thermodynamic setting in which large-scale motion is generated is briefly summarized as well as the resulting estimates on how much power is associated with sustaining planetary motion.

The chapter started with the introduction of buoyancy, the force that results in the acceleration of mass. We described how the mass exchange resulting from buoyancy acts to minimize the potential energy within the system by vertical exchange of mass, how horizontal motion depletes potential energy further, and how potential energy is generated and destroyed by heating and cooling. We focused here on horizontal motion, and deal with vertical convection in the following chapter. The differential heating and cooling rates thus provide the link to the thermodynamic constraints imposed on generating planetary-scale motion, and allow us to infer a maximum power limit for the associated poleward transport of heat. As this limit is associated not just with a heat flux, but also with a temperature difference, these can be used to infer the magnitude of mass exchange that is associated with motion. By considering the angular momentum balance, it was shown that not all of the kinetic energy reflects the direct consequence of depleting the thermodynamic driving gradient, but that a considerable fraction may result simply from the conservation of angular momentum that is then reflected in predominantly zonal flow. Even if the flow is mostly zonal in the Earth's climate system, it is nevertheless generated out of the differential solar radiative heating of the planet, which sets the maximum power limit to the generation of motion. A maximum power limit was also inferred for motion in the Earth's interior that results from interior heat sources. Motion has considerable impacts on the thermodynamic state of a system as it depletes its thermodynamic driving gradients. In the case of the poleward transport of heat, motion alters the radiative emission to space and thereby enhances the overall planetary export of entropy.

When the maximum power limit is evaluated using observed values for the forcing, it yields magnitudes of power and heat transport that match estimates based on observations rather well. The power associated with maintaining large-scale atmospheric motion of about $1000 \cdot 10^{12}$ W appears small compared to the solar radiative flux that is greater by a factor of 100, as summarized in Fig. 7.9. This strong reduction results from the fact that the large-scale circulation is driven by differences in solar radiative heating rather than by the mean heating and by the low efficiency of about 2% for converting the radiative forcing into kinetic energy. A comparably small fraction of large-scale atmospheric motion is then further converted into ocean waves and into motion in the ocean. These rates were previously estimated to be in the order of about $60 \cdot 10^{12}$ and $5 \cdot 10^{12}$W, respectively.

The low efficiency of the atmospheric circulation is well known and can be explained using the maximum power limit by the combination of a low temperature difference and the flux–gradient trade-off. That observed estimates can be obtained by the maximum power limit even with these rather simple models lends further support to the hypothesis in Chapter 5 that thermodynamic systems tend to evolve to their thermodynamic limit and are thus predictable by their limits. Complex general circulation models that simulate geophysical flow in much greater detail have evaluated this limit as well (Shutts 1981; Kleidon et al. 2003, 2006; Pascale et al. 2012) and support this hypothesis further.

In comparison, the power of interior motion of about $25 \cdot 10^{12}$ W is two orders of magnitude less than the power that drives the atmospheric circulation, although

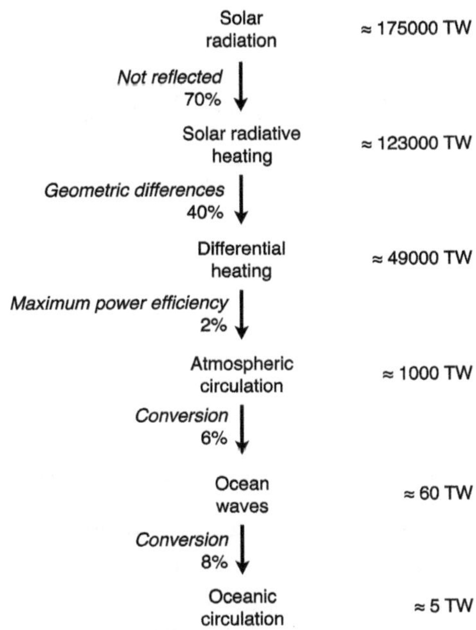

Figure 7.9 Summary of the conversions from radiative forcing to large-scale motion in the atmosphere–ocean system based on observed estimates and the maximum power limit, as described in Section 7.4.

it takes place mostly in the interior of the solid Earth. Over geologic time periods, this interior power is of importance in driving material exchange, for instance, in terms of mountain formation and sea-floor spreading at mid-oceanic ridges. Yet for understanding the drivers of the present-day climate, this energetic perspective clearly identifies atmospheric dynamics as the main player for material exchange in the Earth system.

The material exchange accomplished by motion sets the basis for geochemical cycling, specifically the cycling of water and carbon. It provides the essential transport mechanism to sustain biotic activity, and thus, indirectly, human activity. In the following chapters, we will evaluate this exchange in greater detail, as well as the feedbacks that hydrologic and geochemical cycling have on the generation of motion. As most of the material exchange is associated with frictional dissipation at the surface–atmosphere interface, we will particularly focus on the material exchange at this interface in the following chapters.

8

Hydrologic cycling

8.1 Thermodynamic roles of water

So far we have dealt with radiative forcing and how differences in radiative heating result in motion that depletes differential heating, thereby following the second law. Our next step in the sequence of energy conversions from radiation to Earth system processes shown in Fig. 1.5 is to relate the differential heating and the resulting motion to hydrologic cycling and associated thermodynamic directions and limits.

There are two aspects closely related to water that greatly shape the thermodynamic setting of the Earth system. First, hydrologic cycling involves energy conversions of substantial magnitude that are associated with the different phases of water, so that the latent heat involved in the different phases contributes a considerable share of the heat fluxes in the atmosphere. Second, the presence of water in its different phases is associated with the abundance of ice and clouds, two aspects that greatly affect the albedo, the magnitude of the solar radiative forcing, and thereby the radiative environment of the Earth system. Furthermore, water vapor in the atmosphere and clouds contributes substantially to the atmospheric greenhouse effect, so that hydrologic cycling affects both absorption and reflection of solar radiation as well as the radiative transfer of terrestrial radiation. These highly relevant aspects illustrate how water and the magnitude of its cycling affects the overall thermodynamic state of the Earth system.

The thermodynamic treatment of hydrologic cycling starts with the thermodynamics of phase transitions. The different phases of water – solid, liquid, and gas – correspond to different intensities by which water molecules are bound to each other. The water molecules are bound most strongly in their solid state, and are unbound when in the gaseous state. When we consider the entropy of a system in which water is present in two states, for instance, liquid water and vapor then the total energy of the system involves thermal energy, but also uncompensated heat related to the water vapor pressure as well as intermolecular binding energies.

Binding energies are described in thermodynamics by the mass and the respective chemical potential, as introduced in Section 2.3. The total energy of the system is thus spread over the thermal energy, the pressure of water vapor, and across the intermolecular bonds in the liquid state. This spreading of energy across different forms of entropy then result in an overall state of maximum entropy in which both phases coexist for a given temperature of the system. This state is reached when the air is saturated with water vapor, and it represents a state of thermodynamic equilibrium of the air in contact with liquid water. In this state of saturation, the total entropy of the system can no longer be increased by a molecule changing its state from liquid to gas or from gas to liquid. This state then represents a state in which there is no net loss or gain in water in either the liquid or the gaseous state. As the gain in the gaseous state relates to a net evaporation of water, while a loss in the gaseous state relates to condensation and precipitation, there is no net hydrologic cycling involved at the macroscopic scale when a system is in a state of thermodynamic equilibrium with respect to different phases of water.

The maintenance of hydrologic cycling requires a continuous removal of saturated air from the water surface and its replacement by unsaturated air. This mass exchange of air with different moisture contents maintains the thermodynamic disequilibrium with respect to the phases of water and thereby maintains the evaporative flux from the surface into the atmosphere and the condensation and subsequent precipitation of liquid water within the atmosphere. Hydrologic cycling is thus tightly linked to motion, and it is associated with a considerable transport of heat in latent form. The thermodynamic formulation of hydrologic cycling here essentially follows the example already used in Section 6.6, where we considered the maximum power that can be derived within a system which is heated by the absorption of solar radiation at the surface, while the atmosphere is cooled by the emission of longwave radiation to space. When applying this setup to hydrologic cycling, we need to adapt the formulation to account for the fact that the evaporated water transports a considerable amount of latent heat and that air is primarily heated once the latent heat is released upon condensation and not necessarily when radiation is being absorbed, which alters the operation of the heat engine.

To describe the thermodynamics of the different steps involved in hydrologic cycling and how these relate to the radiative forcing of the system, we use the schematic setup shown in Fig. 8.1. We focus here on the hydrologic cycling over the ocean surface, which is the most dominant source for the water cycle as oceans cover the greatest fraction of the Earth's surface. We start with the evaporation at the surface. As mentioned earlier, the evaporative process requires unsaturated air to sustain the flux. How this unsaturated air results from the cycle is an aspect that we get back to at the end of the description of this cycle. As the phase transition from liquid surface water to water vapor requires substantial amounts of energy,

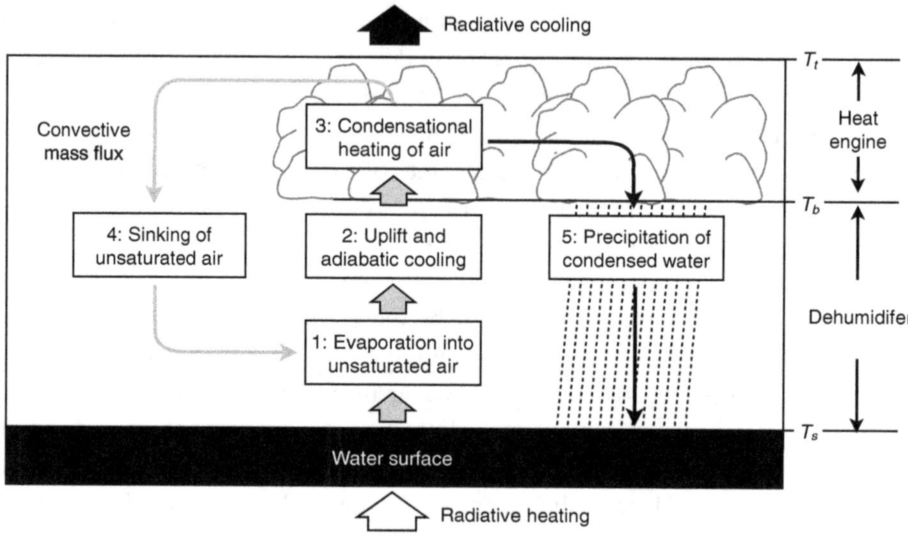

Figure 8.1 Illustration of the key steps involved in hydrologic cycling, which is driven by the radiative heating of the surface and the cooling of the atmosphere. Also indicated are the three temperatures, T_s, T_b, and T_t, of the surface, the cloud base, and the cloud top that are used in the text for the description of hydrologic cycling.

mostly required to break up the intermolecular bonds of the liquid, the evaporative flux consumes a sizable proportion of the radiative heating of the surface. When the evaporated vapor is then lifted by updrafts to greater heights and lower pressure, air cools due to the adiabatic ascent, as in step 2 of the Carnot cycle described in Section 4.3. This lifting proceeds until the air is cooled to the point at which water vapor reaches its saturation and condenses. As condensation is associated with the formation of water droplets, this point also sets the location for the cloud base, which is described in Fig. 8.1 by the temperature T_b. Condensation releases the heat that was taken up during evaporation, so that from now on the ascent of the air is no longer adiabatic. The heating of air by condensation results in buoyancy and this process generates the power to drive the convective motion. As this convective motion is dominantly generated by the release of latent heat rather than surface heating, it is referred to as moist convection.

Upon further ascent, moisture is successively removed from the air by condensation. The upward mass flux of moist air within the cloud is in steady state balanced by the downward mass flux of drier air and the mass flux associated with the precipitation of liquid water from the cloud. The descending air then warms adiabatically and since no moisture is added, it results in unsaturated air at the surface that sustains the evaporative flux.

We thus obtain a complete cycling of water, from surface evaporation to condensation and precipitation. It is sustained by the motion that is not generated by the direct heating of the surface by absorption of radiation, but rather by the heating caused by the condensation of water vapor in the atmosphere, which is indirectly connected to the radiative heating of the surface. This aspect of generating convective motion is different on land, where surface heating and the resulting sensible heat flux act as direct drivers for dry convection. The following description focuses on moist convection and neglects the role of the sensible heat flux (the sensible heat flux and dry convection is described later in Chapter 10). Note that this separation between moist convection, dry convection, and the large-scale circulation as three processes that generate motion in the atmosphere is somewhat artificial, but is done here to better focus on the individual generation mechanisms. We should keep in mind that these processes are linked and interact in the real atmosphere and that their contribution may vary across regions and time.

To apply the heat engine formalism as well as the maximum power limit to hydrologic cycling, the starting point is the condensational heating that takes place when water vapor condenses. This heating drives a heat engine that performs the work in generating motion; it also performs work in lifting the moisture to the height at which it condenses. When droplets of water form at a height as a result of condensation, these droplets have potential energy that dissipates once these droplets fall to the surface when precipitating. This form of dissipation has been estimated to be about 2 W m^{-2} in the tropical atmosphere (Pauluis et al. 2000; Lorenz and Renno 2002; Pauluis and Dias 2012) and thus forms a considerable fraction of the power involved in moist convection, so that this term needs to be taken into account. Several studies have approached moist convection as being the result of a heat engine (Emanuel 1986; Renno and Ingersoll 1996) and these approaches have been combined with work being performed in atmospheric dehumidification that is associated with the lifting of moisture (Pauluis and Held 2002a,b; Goody 2003). These studies use thermodynamic approaches to hydrologic cycling and moist convection and provide the basis for the description in this chapter. The description here is simplified and, more importantly, considers the flux–gradient trade-off by including the energy balances in the approach, which then yields the maximum power limit associated with hydrologic cycling (following Kleidon and Renner 2013b,a). This limit is similar to how the maximum power limit was derived for differences in radiative heating in Section 6.6, yet it includes a specific formulation of the latent heat flux associated with hydrologic cycling.

The maximum power limit of hydrologic cycling sets a basis to evaluate the interactions of the hydrologic cycle with other processes and the thermodynamic state of the Earth system. There are two immediate effects of hydrologic cycling: first, hydrologic cycling is inevitably associated with the presence of clouds, as these result from condensation. Clouds reflect substantial amounts of solar radiation by their brightness, and are associated with a strong greenhouse forcing so that they impact the radiative transfer of terrestrial radiation as well. Hence, clouds strongly feed back to the radiative forcing. Their fractional cover, however, is dynamic and may adjust to maximize power involved in hydrologic cycling as well by means that are not explored here. The second effect is related to the mass exchange in the system. As latent heat is highly efficient in transporting heat and the maximum power limit is associated with a certain optimum heat flux, hydrologic cycling strongly shapes the magnitude of mass exchange near the surface. This effect was already alluded to when describing mass exchange in the context of motion in Section 7.5. As we will see, this is an important effect that shapes the magnitude of mass exchange and its sensitivity to changes in the radiative forcing of the Earth system, which then provides a linkage to the following chapter on geochemical cycling.

This chapter starts with the thermodynamics of the phase transitions of water and describes how saturation of air with water vapor is a manifestation of a state of thermodynamic equilibrium. The hydrologic cycle is then described as a thermodynamic cycle and related to the forcing. We then derive the maximum power limit of hydrologic cycling by considering moist convection as a heat engine in combination with the energy balances of the system. The sensitivity of this limit of hydrologic cycling to changes in the radiative forcing is then evaluated. The broader consequences of hydrologic cycling to other Earth system processes are then described, particularly regarding the effects of clouds and greenhouse forcing on the radiative forcing of the system. The chapter closes with a brief summary and makes the connection to the processes described in the other chapters.

8.2 Thermodynamics of the phase transitions of water

The different phases of water are characterized by different intensities by which water molecules are bound together by intermolecular bonds. Water in its solid state has the strongest binding, while in water vapor, these bonds are absent. These bonds are associated with a negative binding energy, so that it requires the addition of energy to convert ice into liquid water, and liquid water into water vapor. The backward conversion is associated with the release of this binding energy. When we deal with phase transitions and the coexistence of different phases of water within a system, the total energy of the system includes, at a minimum, the thermal energy associated with temperature as well as this binding energy. The determination of the state of thermodynamic equilibrium of a system with different phases of water then requires us to evaluate how the entropy of the system is maximized by spreading the total energy among the thermal and binding energies of the different phases within the system.

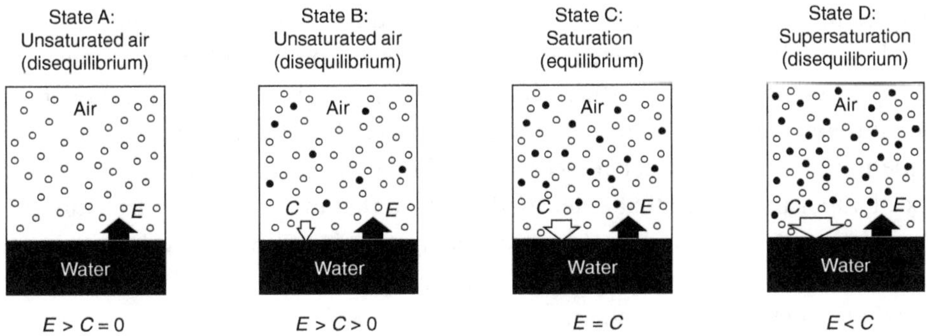

Figure 8.2 Four states of an isolated system consisting of water and air with different water vapor contents, indicated by the black circles. The states differ by their rates of evaporation E, and condensation C.

To illustrate this application of thermodynamics to phase transitions, we consider an isolated system in which liquid water and air are present. Fig. 8.2 shows this system in different states that differ by the amount of water vapor in the air. State A is an initial state with no water vapor in the air, state B has some water vapor in the air, state C represents a state in which the water vapor content is in thermodynamic equilibrium, while state D contains more water vapor than at saturation. The system has a fixed total amount of energy within the system which is characterized by a uniform temperature T, air pressure p, and by the binding energy involved in the different phases of water. The difference in the states is described by different partial pressures of water vapor e, with state C representing the state of saturation with vapor pressure e_{sat}. The rates by which water is converted among the two phases are described by the microscopic rates of evaporation E and condensation C, which differ across the four states. The goal of the thermodynamic application is to first determine the value of e_{sat} that characterizes the state of thermodynamic equilibrium in relation to the temperature and air pressure of the system. Before we get to this thermodynamic application, we first consider the microscopic dynamics that lead to the equilibrium state, following a conceptual model of saturation that is used in textbooks (e.g., Bohren and Albrecht 1998).

At the microscopic level, each water molecule has some kinetic energy associated with the random motion that characterizes the overall temperature of the system and, if it is in the liquid state, some binding energy. We start the considerations of how these two energies change by considering state A. This initial state has all water molecules bound in its liquid state. The thermal energy of the water molecules in their liquid state that is associated with temperature is not uniformly spread across all molecules, so there is some variation in the kinetic energy among the molecules. There is thus a probability that a few water molecules have kinetic energies that are greater than the molecular binding energy that holds them in the liquid state. These water molecules can thus detach from the others at the surface of the liquid, enter the gaseous state, and remove some thermal energy from the liquid. This flux of molecules represents the evaporative flux E shown in Fig. 8.2. Since in the initial state, no water molecules are in the vapor state, there is no opposite flux of vapor molecules that enter the liquid, so that the flux of condensation C, is zero. Hence, the temporal evolution of the system is towards state B with greater concentrations of water vapor, which is macroscopically described by a greater partial pressure of water vapor e.

Once the water vapor concentration starts to increase, there is an increasing chance that water vapor molecules move back into the liquid. This flux of condensation C, is thus related to the partial pressure of water vapor e. When a water vapor molecule enters the liquid, it can again bind to the other water molecules. The change in binding energy is compensated by an increase in the thermal energy, thereby increasing the temperature of the liquid. While E depends on the temperature as temperature describes the mean kinetic energy of the water molecules at the microscopic scale, the flux C primarily depends on the partial pressure e. The temporal evolution is thus directed towards increasing partial pressures of water vapor, which increases the flux C, and reduces the difference $E - C$.

A steady state C is reached when the fluxes balance with $E = C$. As we will see, this state is a state of thermodynamic equilibrium and has maximum entropy. In this state, the water vapor pressure reaches its saturation value, $e = e_{sat}$. This state represents the case in which it is equally likely for a water molecule to be unbound from the liquid state and add to the water vapor as it is for a water vapor molecule to enter the liquid state. If we consider the state D that has a higher water vapor concentration, $e > e_{sat}$, a state referred to as supersaturated, then it is more likely for a water vapor molecule to enter the liquid state as it is for a liquid water molecule to be unbound and enter the gaseous state.

From this simple model, we can also infer how the steady state of water vapor saturation changes when the temperature of the system is increased. The increase in temperature enhances the rate E, since the greater temperature is associated with a greater fraction of molecules having kinetic energies that exceed the binding energy of the liquid. As the steady state is characterized by $E = C$, and C increases with e, the new steady state is maintained with a greater value of e_{sat}. Hence, the saturation state of the system increases with temperature, as we will also see in greater detail.

We next relate this microscopic description to a thermodynamic interpretation. As described earlier, the fluxes E and C redistribute water between its liquid and vapor state, affect the value of e, and thus affect the distribution of thermal energy, uncompensated heat, and binding energy within the system. This redistribution of energy affects the total entropy of the system. When a certain amount of water is converted from liquid into gas, it lowers the thermal energy of the water and thus its entropy. Yet, as it adds this thermal energy to the vapor, and water vapor has a greater molar entropy than liquid water (cf. Table 2.3), this energy transfer increases the entropy of the system. On the other hand, an increase in e results in a greater amount of uncompensated heat stored in the vapor, and this increase in (eV) results in a decrease in the entropy of the system at a given temperature (see also eqn. 2.17), as it is opposite to a free expansion of a gas and constitutes the work required to compress the water vapor to a greater value of e. Due to these two opposing effects, there is a maximum in the entropy of the system that corresponds to the state of saturation at which e attains the value of e_{sat}. We will see this in equation form in the next section when we deal with this formulation quantitatively.

As we deal with net fluxes of moisture in the following, E and C will have a slightly different meaning. A net gain of water vapor is referred to as evaporation, while a net loss of water vapor due to the conversion into the liquid state is referred to as condensation or precipitation P. For simplicity, we assume that condensation and precipitation are equal in the following, thus dealing with precipitation and neglecting the processes involved from condensation to droplet growth to precipitation. With this definition of net hydrologic fluxes, evaporation and precipitation at saturation are zero, which is slightly different from the microscopic picture in which these fluxes are non-zero, but compensate each other.

8.3 Saturation of water vapor

The example of the last section showed qualitatively how the change of water from its liquid to vapor phase affected the energy and entropy of the system and how it relates to a state of maximum entropy. Such changes in energy that are caused by a change in mass are characterized in thermodynamics by the chemical potential μ. The chemical potential is defined as the change in Gibbs free energy (cf. Section 3.5) with a change in mass, that is

$$\mu = \frac{\partial G}{\partial m} \tag{8.1}$$

with the temperature and pressure of the system being fixed. As changes in the Gibbs free energy include both changes in TS as well as in eV, it captures the changes in thermal energy and in the uncompensated heat that result from the change of mass due to the phase transition. We also related the state of thermodynamic equilibrium to a minimum in free energy (Section 3.6), so that the equilibrium between the liquid and vapor state of the system is described by a minimum in the Gibbs free energy of the system.

The value of saturation vapor pressure is then derived by considering the minimum in the Gibbs free energy of the liquid–gas mixture. At this minimum, the change in chemical potentials of both phases are equal, $d\mu_l = d\mu_v$, a relationship referred to as the Gibbs–Duhem relationship. This represents the condition for thermodynamic equilibrium in that the Gibbs free energy has a minimum value and hence does not change when a small amount of mass is converted from liquid to vapor, or in the reverse direction. The changes of the chemical potentials are given by

$$d\mu_l = -S_{m,l}\, dT + V_{m,l}\, de_{\text{sat}} = -S_{m,v}\, dT + V_{m,v}\, de_{\text{sat}} = d\mu_v \tag{8.2}$$

where the values of entropy and volume are normalized by mass, so that these represent the molar entropies and volumes of liquid water and water vapor (with indices l and v referring to the liquid and vapor phase). Values for the molar entropy of water were already given in Table 2.3. When this equation is rearranged, it yields

$$\frac{de_{\text{sat}}}{dT} = \frac{S_{m,v} - S_{m,l}}{V_{m,v} - V_{m,l}} \tag{8.3}$$

As the difference in molar entropy is connected to the latent heat of vaporization λ, by $S_{m,v} - S_{m,l} = \lambda/T$, this yields the Clapeyron equation that describes how the saturation vapor pressure changes with temperature:

$$\frac{de_{\text{sat}}}{dT} = \frac{\lambda}{T \Delta V_m} \tag{8.4}$$

The difference in molar volume, $\Delta V_m = V_{m,v} - V_{m,l}$, can be approximated by the ideal gas law, because the molar volume of the gas is much greater than the molar volume of the liquid. With $e_{\text{sat}} V_{m,v} = R_v T$, where R_v is the gas constant for water vapor, we arrive at the Clausius–Clapeyron equation

$$\frac{de_{\text{sat}}}{dT} = \frac{\lambda e_{\text{sat}}}{R_v T^2} = s \tag{8.5}$$

The left hand side of this equation describes how the saturation vapor pressure increases with temperature, a property that we will see plays quite an important role in the description of hydrologic cycling. This derivative, or slope of the saturation vapor pressure curve, is abbreviated by the symbol s. As we will see, the relevance of s is that gradients in the saturation vapor pressure are to first order proportional to gradients in temperature, with the proportionality given by s.

Equation 8.5 can be solved to yield an analytic expression of the saturation vapor pressure, e_{sat}, as a function of temperature T, and yields

$$e_{\text{sat}}(T) = e_{\text{sat}}(T_0)\, \exp\left[\frac{\lambda}{R_v}\left(\frac{1}{T_0} - \frac{1}{T}\right)\right] \tag{8.6}$$

where $e_{sat}(T_0)$ is the saturation vapor pressure at a reference temperature T_0. In practice, however, an approximation based on observations is used for e_{sat}, for instance in the form derived by Magnus (1844). This approximation, referred to as the Magnus formula, has a similar form:

$$e_{sat}(T) = a \exp\left[\frac{bT}{c+T}\right] \tag{8.7}$$

with the empirical parameters $a = 611$ Pa, $b = 17.43$, and $c = 243.1$ K. The temperature is expressed in units of °C. The main reason for using an observations-based function is that the latent heat of vaporization has a temperature dependence, which would need to be taken into account in the integration. It is also important to note that the latent heat of vaporization depends on the pressure of the dry air, resulting in a greater value of λ with lower pressure. This effect manifests itself, for instance, in a lower boiling temperature at greater altitude. The boiling temperature is reached when the saturation vapor pressure matches the pressure of the dry air. The effect is important in the description of the hydrologic cycle because it applies to condensation in the atmosphere, which occurs at a lower pressure. The greater value of λ does not, however, imply that more thermal energy is being released, but rather that saturation is reached at a somewhat lower temperature. This effect can be accounted for by reducing the so-called dew point temperature, defined as the temperature at which condensation of unsaturated air would occur. This dew point temperature decreases with height at a rate of about $1.8 \cdot 10^{-3}$ K m^{-1} due to this pressure effect. More details on this aspect can be found in the more detailed treatment of phase transitions in the thermodynamic literature (e.g., Gyftopoulos and Beretta 1991). In this book, we neglect these pressure and temperature effects and assume a fixed value for the latent heat of vaporization of $\lambda = 2.5 \cdot 10^6$ J kg^{-1} K^{-1}.

To illustrate these relationships, the saturation vapor pressure as well as its slope as functions of temperature are shown in Fig. 8.3. If the value of the water vapor pressure at a given temperature is below or above e_{sat}, it implies that the vapor pressure is in disequilibrium with an open water surface. If $e > e_{sat}$, this would result in net condensation, which enhances water in the liquid phase and reduces water vapor to the saturation pressure. In the opposite case of unsaturated water vapor with $e < e_{sat}$, a net evaporative flux would proceed until the vapor pressure reaches the saturation value.

This derivation of the saturation vapor pressure curve from thermodynamics describes the thermodynamic equilibrium between moist air and an open water surface, but it can also be extended to other phase transitions. The specific reference here to the open water surface is of relevance, because there are relevant processes in the Earth system in which some modifications need to be included in the state of thermodynamic equilibrium. When water vapor condenses in the atmosphere,

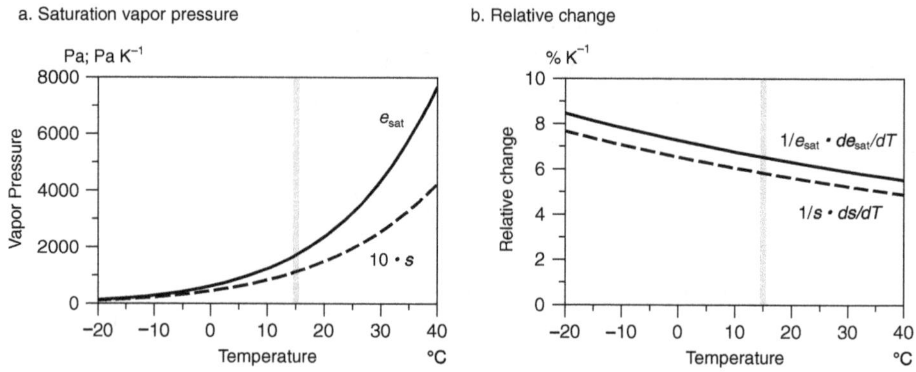

Figure 8.3 Graph (a) shows the saturation vapor pressure curve, e_{sat}, as a function of temperature as well as its slope s (multiplied by 10). Shown in (b) are the relative changes of both quantities with a change in temperature. The vertical grey line denotes the present-day global mean surface temperature.

it does not condense with an open water surface, but rather in the form of small droplets. These droplets have a certain surface tension associated with their size, and the associated binding energy of the curved water surface needs to be accounted for in the derivation of the state of thermodynamic equilibrium. When this aspect is included in the derivation (e.g., Wallace and Hobbs 1977), the water vapor pressure at which condensation occurs is shifted to higher values, so that the vapor pressure seems to be in a state of supersaturation. Condensation under these conditions is referred to as homogeneous nucleation. It is, however, modulated by the presence of cloud condensation nuclei, which present a surface at which water vapor can condense. In many areas, these condensation nuclei are present in the atmosphere, particularly over land, so that condensation in the atmosphere then occurs near the saturation vapor pressure associated with an open water surface. This case is referred to as heterogeneous nucleation.

There are also other examples of forms of energy that need to be considered in the saturation state. These examples include the chemical energy associated with dissolved ions in water, as in the case of seawater or dissolved sugar in plants, or the capillary binding energy of water in a soil matrix. These energies are typically small and the associated difference in the latent heat of vaporization is thus small as well. This aspect is dealt with in greater detail in Section 9.4 in relation to the generation of associated forms of chemical energy.

8.4 Maintaining hydrologic cycling

After establishing the reference state of thermodynamic equilibrium regarding the phase transitions of water, we next look into the question of how the water vapor concentration is maintained in a disequilibrium that allows for substantial cycling of water and how this is quantified. The maintenance of disequilibrium in hydrologic cycling involves several steps that are schematically shown in Fig. 8.1. These steps are: (1) the irreversible mixing of saturated water vapor with unsaturated air near the surface; (2) the work of expanding water vapor to the level at which it saturates as air rises; (3) the generation of motion by condensational heating at the cloud base that generates buoyancy. This kinetic energy is dissipated within the cloud and near the surface, where its turbulent dissipation maintains the mixing; (4) the descend of the dehumidified air back to the surface that maintains the unsaturated conditions of near-surface air and that maintains step (1); and (5) the frictional dissipation of the potential energy associated with the falling raindrops.

We start with step 1, which involves the evaporation of water from the ocean surface. Near the surface, air is essentially saturated, thus being in thermodynamic equilibrium with the liquid water phase. It is through the mixing of this surface air with the drier air of the atmosphere that this saturated air is exchanged with unsaturated air, thus resulting in a net flux of evaporation from the surface into the atmosphere. We refer to this flux as evaporation E, which is slightly different from the use in Section 8.2 because we now refer to the net rate of evaporation. Evaporation is associated with the irreversible mixing, and follows a gradient in chemical potential. It is associated with a latent heat flux λE, of

$$\lambda E = k \cdot \rho_v \cdot \Delta\mu \tag{8.8}$$

where k is a certain conductivity in units of m s^{-1} which is related to the rate of surface mass exchange as we will see later, ρ_v is the density of water vapor, and $\Delta\mu = \mu_s - \mu_a$ is the difference in chemical potentials of the ocean surface (with $\mu_s = 0$ because it is saturated), and of the unsaturated air of the atmosphere, μ_a. The chemical potential for air with a certain water vapor pressure e_a is derived from the work involved in expanding water vapor from its saturated value to the vapor pressure e_a. At a given temperature T, it is described by

$$\mu_a = R_v T \ln\left(\frac{e_a}{e_{sat}(T)}\right) \tag{8.9}$$

where μ_a is in units of J kg^{-1}, R_v is the gas constant for water vapor, and e_a/e_{sat} is the relative humidity of the air. This process is irreversible as this mixing of saturated air with unsaturated air can only be reversed by performing work to either compress water vapor to saturation or by cooling air to the point at which it condenses. The magnitude of this irreversibility is determined from the related entropy production, σ_{evap}, which is described by

$$\sigma_{evap} = \frac{E \cdot \Delta\mu}{T_s} \tag{8.10}$$

This entropy production relates to the increase of entropy of the atmosphere as the water vapor content of the atmosphere attains values closer to saturation and thus comes closer to equilibrium with the open water surface of the ocean.

We can estimate a global mean number of this entropy production using observations (Oki and Kanae 2006). Using a mean evaporative flux of about $E = 2.7$ mm d^{-1} (or kg m^{-2} d^{-1}), a mean atmospheric relative humidity of $e/e_{\text{sat}} \approx 60\%$, and a mean surface temperature of $T_s = 288$ K yields a value of $\sigma_{\text{evap}} = 7.4$ mW m^{-2} K^{-1}. Numerical estimates from a present-day climate model simulation (Kleidon 2008) yield a similar magnitude of a global mean value of about 8.3 mW m^{-2} K^{-1}. This entropy production results in the reduced ability of the system to perform work out of the latent heat flux. The magnitude of this reduced potential can be estimated directly from the value of σ_{evap}. If this entropy production was associated with frictional dissipation, it would correspond to a rate of dissipation of $D_{\text{loss}} = \sigma_{\text{evap}} \cdot T$, so that the reduced potential corresponds to D_{loss}. With the values from earlier, this potential converts to a value of about 2–2.5 W m^{-2}. Yet, at the same time, this evaporative flux is associated with a flux of latent heat of about $\lambda E = 88$ W m^{-2} (Stephens et al. 2012), which connects this process to the heat engine and energy balances, as we will see later.

Evaporation, however, is commonly not expressed in terms of a difference in chemical potential, but in terms of an aerodynamic conductance (or drag coefficient) and a difference in specific humidity, which describes the mass of water vapor. It then takes a similar form as in eqn. 7.9 for friction, with evaporation being described by

$$E = k_a \cdot \rho \cdot (q_s - q_a) \tag{8.11}$$

where k_a is the aerodynamic conductance for water vapor (which can also be expressed in terms of a drag coefficient by noting that $k_a = C_d\ v$, see Section 7.3), ρ is the air density, $q_s = q_{\text{sat}}(T_s)$ is the saturated specific humidity of the surface air, and $q_a = q_{\text{sat}}(T_b)$ is the specific humidity of the atmosphere, which can be expressed in terms of a temperature T_b at which air would reach saturation. Specific humidity is related to vapor pressure by $q = (R_a/R_v) \cdot e/p$, with R_a and R_v being the gas constants of air and water vapor, and p being air pressure. With this, the difference in specific humidity can be expressed as a difference in water vapor pressure, $e_{\text{sat}}(T_s) - e_{\text{sat}}(T_b)$, and this difference in water vapor pressure can be related to the temperature difference by using a linear approximation and the slope of the saturation vapor pressure curve described by the Clausius–Clapeyron equation, eqn. 8.5, as $e_{\text{sat}}(T_s) - e_{\text{sat}}(T_b) \approx s(T_s - T_b)$. With these modifications, the evaporation rate can then be expressed in terms of the temperature difference, $T_s - T_b$, by

$$\lambda E = k_a \cdot c_p \frac{s}{\gamma} \cdot (T_s - T_b) \tag{8.12}$$

where s is the slope of the saturation vapor pressure curve (cf. eqn. 8.5), and $\gamma = c_p\, p/\lambda \cdot (R_v/R_a) \approx 65$ Pa K^{-1} is the psychometric constant, evaluated at mean sea level pressure of $p = 101325$ Pa. This formulation of the evaporation rate in terms of the temperature difference is quite convenient, as we will see in the following derivations.

In step 2, water vapor is lifted by motion to a height z. We assume that this lifting occurs adiabatically without change in energy or entropy, as in step 2 of the Carnot cycle. Then, the increase in potential energy of the lifted, moist air and its expansion is compensated for by a decrease in temperature. This results in a close linkage between the gain in potential energy and the reduction in thermal energy, which is described by the dry adiabatic lapse rate Γ_d, with

$$\Gamma_d = \frac{g}{c_p} \tag{8.13}$$

where g is the gravitational acceleration and c_p is the heat capacity of air, which yield a value for the lapse rate of $\Gamma_d = 9.8 \cdot 10^{-3}$ K m^{-1}. The temperature, T, at height z is therefore cooler than the surface. Through this dry adiabatic lapse rate, the temperature difference $T_s - T$ is directly related to the height, z, given by

$$T_s - T = \Gamma_d \cdot z \tag{8.14}$$

We can use this relationship to determine the height z_b at which the lifted air condenses, so that $e_a = e_{sat}(T_{dew})$, with the temperature T_{dew} corresponding to the dew point temperature. Here, we need to consider the change of the latent heat of vaporization with height, which is commonly done by considering the lapse rate of the dew point, with $\Gamma_{dew} = 1.8 \cdot 10^{-3}$ K m^{-1} as mentioned before. The temperature T_b at which vapor condenses at height z_b then needs to be adjusted by this lapse rate of the dew point, so that $T_b = T_{dew} - \Gamma_d \cdot z_b$. We thus obtain the relationship between the temperature difference at which water evaporates and condenses with the height at which the condensation takes place:

$$T_s - T_{dew} = (\Gamma_d - \Gamma_{dew}) \cdot z_b \tag{8.15}$$

Using the global mean conditions from before yields a dew point temperature under surface conditions of $T_{dew} = 280.4$ K, which yield a height of about $z_b = 955$ m and a temperature $T_b = 278.6$ K.

Associated with this lifting are three changes that affect the energy and entropy of water vapor: First, as air is expanded when it is lifted, so is the water vapor expanded up to the saturation vapor pressure. This increases the chemical potential of the vapor, up to the point when it vanishes at the height z_b when condensation occurs.

Second, once the water vapor condenses at height z_b, the liquid water droplets are associated with a potential energy as the condensation occurs at a geopotential $g\, z_b$. Third, the condensation of water vapor releases latent heat, thereby warming the temperature of the ascending air mass and generating buoyancy. When we assume that the evaporated water from the surface condenses in a steady state, without a lateral net transport of water vapor, then this condensational heating is described by λE. At the same time, the condensation at height z_b is associated with the generation of potential energy of the liquid water droplets, G_{lift}. This rate is described by the work involved in lifting the mass flux E to the geopotential $g\, z_b$. As the height z_b is related to the temperature difference by eqn. 8.15, the generation rate G_{lift} is given by

$$G_{\text{lift}} = E \cdot (g\, z_b) = E \cdot \frac{g}{\Gamma_d - \Gamma_{\text{dew}}} \cdot (T_s - T_b) \tag{8.16}$$

When we again use global mean values from earlier, this work corresponds to a global mean of about $G_{\text{lift}} = 0.3$ W m^{-2}. Estimates from observations (Pauluis and Dias 2012) place this work to be about 1.8 W m^{-2} in the tropical atmosphere, so it is somewhat greater than our estimate. The discrepancy can be attributed to some of our simplifying assumptions, e.g., that condensation occurs at the height z_b rather than throughout the cloud and using global means, which should result in an underestimation of G_{lift} compared to the tropical atmosphere in which convection reaches to greater heights.

In step 3, we consider the power that can be derived from the condensational heating that occurs at the base of the cloud. We abbreviate this rate by λP, where P is the precipitation rate. In steady state, the condensational heating matches surface evaporation, so that $\lambda P = \lambda E$, noting that in most cases, this condition is not met at all places and times due to temporal variations and net horizontal transport of water vapor. We also assume for the sake of simplicity that all condensation takes place at the base. In this case, we deal with a heat engine inside the cloud that is driven by the release of thermal energy by λE and the temperature difference $T_b - T_t$ between the cloud base and the cloud top. We refer to this heat engine as a moist heat engine, because it is not driven directly by radiative heating of the surface, but rather by the release of latent heat at the cloud base. This heating causes the air at the cloud base to be warmer than in the air in the surrounding of the cloud, resulting in buoyancy, which results in an updraft within the cloud and generates motion. This engine is similar to the heat engine that we described in Section 6.6, as it also involves a heating rate at the bottom, and radiative cooling from the top of the cloud. This heat engine is subjected to the Carnot limit, as before, so that the power G_m, generated by this engine is constrained by

$$G_m = \lambda E \cdot \frac{T_b - T_t}{T_t} \tag{8.17}$$

We used again the limit of a dissipative heat engine, assuming that most of the dissipation of the engine in fact occurs inside and near the cloud, although some of this motion sustains the downdraft as well as the turbulent mixing at the surface. Note that our simplifying assumption that all condensational heating takes place at the cloud base is likely to overestimate the power. Since some of the water vapor condenses at a greater height and at a lower temperature, the temperature difference from which power is generated is smaller, so that condensation at greater height results in a smaller contribution to the overall power than what is expressed by eqn. 8.17. Using the global mean values from earlier, and assuming that the cloud top temperature is approximately the radiative temperature, $T_t = 255$ K, yields an estimate for the power of the moist heat engine of about $G_m = 7.3$ W m^{-2}. Note that the use of the radiative temperature here may overestimate the cloud-top temperature, but it is used here because it reflects the constraint imposed by the planetary energy balance. This power performs the work of lifting water vapor, G_{lift}, while the remaining part is dissipated by friction. This frictional dissipation results in the irreversible entropy production of $\sigma_{\text{fric}} = (G_m - G_{\text{lift}})/T_b = 25$ mW m^{-2} K^{-1}. A substantial part of this frictional dissipation likely takes place within the cloud between the updraft and the cloud droplets (Xu et al. 1992; Romps and Charn 2015).

Step 4 deals with the downdraft that brings dry air to the surface and maintains the evaporative flux of step 1. This downdraft is associated with a certain mass flux of air that becomes increasingly unsaturated as this air warms adiabatically upon descent. This mass flux connects the downdraft to the transport of latent heat from the surface to the base of the cloud as well as the heat flux within the moist heat engine. In steady state, these mass fluxes need to balance each other. This consideration of the mass flux through the different steps provides a connection between the heat fluxes in sensible and latent form, and thus a constraint for the temperatures T_s, T_b, and T_t.

The mass flux of the downdraft is referred to as J_m and consists of comparatively dry air. It is balanced by the mass flux of the updraft, which proceeds in two steps: in the first step, this mass flux transports water vapor from the surface to the cloud base. The overall magnitude of this mass flux is thus enhanced by E. The flux of latent heat λE, can be expressed in terms of the mass flux by

$$\lambda E = J_m \cdot \lambda \left(q_{\text{sat}}(T_s) - q_a \right) \approx J_m \cdot c_p \frac{s}{\gamma} (T_s - T_b) \tag{8.18}$$

where the relationship of the specific humidity to the water vapor was used as before. In step 2, this mass flux transports released latent heat upon condensation from the cloud base to the cloud top, which we can express as

$$H_c = J_m \cdot c_p(T_b - T_t) \tag{8.19}$$

Assuming that both heat fluxes are in steady state, so that $\lambda E = H_c$, we then obtain a condition of the form

$$\frac{s}{\gamma}(T_s - T_b) = (T_b - T_t) \tag{8.20}$$

which yields an expression for T_b as a function of T_s and T_t:

$$T_b = \frac{s}{s+\gamma}T_s + \frac{\gamma}{s+\gamma}T_t \tag{8.21}$$

For global mean values, $s = 111$ Pa K^{-1} and $\gamma = 65$ Pa K^{-1}, so that the equation would yield a value of $T_b = 276$ K. This value is quite close to the value of T_b derived earlier which was determined from the mean observed relative humidity.

The relevance of the constraint associated with the convective mass flux J_m is that it allows us to partition the overall temperature difference, $T_s - T_t$, into a component that is associated with the irreversible entropy production of the boundary layer and a contribution that generates motion by moist convection, as illustrated in Fig. 8.1:

$$T_s - T_b = \frac{\gamma}{s+\gamma}(T_s - T_t) \tag{8.22}$$

and

$$T_b - T_t = \frac{s}{s+\gamma}(T_s - T_t) \tag{8.23}$$

We can thus roughly estimate that a fraction $\gamma/(s+\gamma) \approx 37\%$ of the potential power $\lambda E(T_s - T_t)/T_t$ is lost by irreversible evaporation near the surface, while a fraction $s/(s+\gamma) \approx 63\%$ of this potential results in the generation of moist convective motion.

Note also that with a given value for the evaporation, eqn. 8.18 can be used to derive an expression for the convective mass flux J_m, as a function of the latent heat flux and the temperature difference, $T_s - T_t$:

$$J_m = \frac{s+\gamma}{s}\frac{\lambda E}{c_p(T_s - T_t)} \tag{8.24}$$

We will get back to this convective mass flux when inferring surface exchange properties and wind speeds in Section 8.5.

Step 5, the last step shown in the sequence of Fig. 8.1 refers to the frictional dissipation of raindrops as these fall from the height at which they were formed by condensation. As droplets fall, they convert their potential energy into the kinetic energy associated with the fall, which is then mostly dissipated by friction between the falling drops and the air. As described earlier, this frictional dissipation can contribute a significant share of about 1–2 W m^{-2} to the overall frictional dissipation (Pauluis et al. 2000; Lorenz and Renno 2002; Pauluis and Dias 2012). This frictional dissipation results in entropy production associated with precipitation, and is associated with about $\sigma_{prec} = G_{lift}/T_s = 1$ mW m^{-2} K^{-1} when using the values derived here. One should note that what we have not accounted for here is that a fraction of the drag by raindrops acts to force the surrounding air into a downdraft, so that a fraction of the potential energy reduction is converted into the kinetic energy of the downdraft. Note that this last step also yields the potential energy of water at the land surface, which drives the continental discharge by river flow to the oceans. As water flow on land covers a fraction of the total Earth surface and the differences in height are not as large as within the atmosphere, the frictional dissipation by water flow on land is thus of notably smaller magnitude.

In summary, the five steps shown in Fig. 8.1 describe how hydrologic cycling results from the combination of a latent heat flux from the surface, its condensation at height, and the generation of motion. It illustrates how thermodynamic disequilibrium is maintained in terms of the unsaturated air near the surface, which results from the combined action of generating motion and dehumidifying the atmosphere out of the latent heat flux. Hydrologic cycling then results in three irreversible terms that produce entropy: evaporation into unsaturated air of the boundary layer, σ_{evap}, the frictional dissipation of motion associated with moist convection, σ_{fric}, and the frictional dissipation of falling raindrops, σ_{prec}. The term σ_{evap} represents an unavoidable loss term from which no work is generated. It is unavoidable because water needs to first evaporate into unsaturated air before it can be released again upon condensation. This term thus maintains the evaporative flux. The sum of these three terms yields the total entropy production associated with hydrologic cycling, which is constrained by the entropy exchange associated with the evaporative flux at the surface and the radiative cooling at the cloud top:

$$\sigma_{tot} = \sigma_{evap} + \sigma_{fric} + \sigma_{prec} = \lambda E \cdot \left(\frac{1}{T_t} - \frac{1}{T_s} \right) \qquad (8.25)$$

Using the numbers from earlier, the contribution by the three terms is about $\sigma_{evap} =$ 7.3 mW m^{-2} K^{-1}, $\sigma_{fric} = 25$ mW m^{-2} K^{-1}, and $\sigma_{prec} = 1.0$ mW m^{-2} K^{-1}, which approximately sum up to a total entropy production $\sigma_{tot} = 35.2$ mW m^{-2} K^{-1} which is given by the boundary conditions of the system. The boundary conditions are described by the evaporative flux at the surface λE, as well as the temperature difference across the system, $T_s - T_t$. These boundary conditions thus play a critical role in setting the magnitude of hydrologic cycling in the Earth system. They are not static, but depend on the energy balances as well as the entropy exchange of the Earth system. To derive the thermodynamic limit on hydrologic cycling that can take place within the Earth system, we thus need to link these processes to the planetary energy and entropy balances in the next step.

8.5 Limits to hydrologic cycling

The limit to hydrologic cycling is now set by the limit of how much power can maximally be generated out of the condensational heating λP, in combination with the driving gradient, $T_b - T_t$. In other words, we ask for the maximum power by moist convection G_m, for a given radiative forcing of the system, which is comparable to the limit derived in Section 6.6, except that it deals more specifically with the individual processes involved in hydrologic cycling. The radiative forcing acts as a constraint to how much radiative and convective exchange can take place in the system and thus limits the magnitude of λP. The energy balances, in combination with the linearization of radiation as done in the chapter on radiation, then can provide a simple expression for the temperature difference, which depends on the latent heat flux. The thermodynamic limit to hydrologic cycling is thus set by the same setup as in the previous examples, in which the energy balance describes the trade-off between a heat flux (here, λP, which balances λE in steady state) with the temperature difference (here, $T_b - T_t$). We can thus estimate this limit by the components we already derived in the previous examples: the expression for the moist convective power G_m, is given by eqn. 8.17, while the surface energy balance was already described in Section 6.6. The maximum power limit on hydrologic cycling results from the combination of these two elements, as before.

When the expression for moist convective power G_m, is combined with eqn. 8.23, which expresses $T_b - T_t$ in relation to the overall temperature difference between the surface and the cloud-top temperature, $T_s - T_t$, we obtain

$$G_m = \lambda E \cdot \frac{s}{s+\gamma} \cdot \frac{T_s - T_t}{T_t} \tag{8.26}$$

The task to maximize this power is almost identical to the maximum power limit for a radiative heating gradient that was already derived in Section 6.6, with λE playing the role of the generic heat flux J and T_t represented by T_a (recall that we neglect the sensible heat flux from the surface in this chapter). The only difference to this previous example is the factor $s/(s+\gamma)$, which depends on temperature T_s. We can nevertheless use the formulation of the surface energy balance from Section 6.6 to express $T_s - T_t$ as a function of the latent heat flux, λE (as in eqn. 6.33):

$$T_s - T_t = \frac{R_s - R_{l,0} - \lambda E}{k_r} \tag{8.27}$$

With these two equations, eqns. 8.26 and 8.27, we can see the trade-off that shapes the maximum power limit: with greater values of λE, the power G_m increases as it is directly proportional to λE, but the temperature difference $T_s - T_t$ decreases with λE, thus shaping the maximum power limit. If we neglect the temperature dependence of the factor $s/(s+\gamma)$ in the maximization for simplicity, we arrive at the same expression for the maximum power limit as in Section 6.6, so that

$$\lambda E_{\text{opt}} \approx \frac{R_s - R_{l,0}}{2} \tag{8.28}$$

which yields a maximum moist convective power of

$$G_{m,\max} \approx \frac{s}{s+\gamma} \frac{(R_s - R_{l,0})^2}{4k_r T_t} \tag{8.29}$$

Note how the factor $s/(s+\gamma)$ reduces the estimate in comparison to the limit derived in Section 6.6. The other fraction, $\gamma/(s+\gamma)$, here describes the potential that is lost through the irreversible evaporation into unsaturated air at the surface.

Using values of the present day, these expressions yield an optimum evaporation rate of about $E_{\text{opt}} = 4.7$ mm d^{-1} and a maximum power of $G_{m,\max} = 12$ W m^{-2}. This optimum evaporative flux is of the right magnitude, but somewhat too high compared to observations-based estimates of the global mean evaporation rate of about 2.7 mm d^{-1}. The power that is lost by the irreversible entropy production by surface evaporation corresponds to about 7 W m^{-2}, which is somewhat higher than the estimate derived in the previous section. This overestimation is to be expected because of the bias in the radiation parameterization, since we considered an open water surface for the entire Earth's surface, and since we neglected the sensible heat flux. Note that the maximum power that can be derived from vertical convection is nevertheless about a factor of 5 larger than what is associated with the large-scale atmospheric circulation (cf. Section 7.4). This substantially greater magnitude simply reflects the fact that heat fluxes in the vertical are generally much larger than horizontal heat transport.

From the maximum power state we can infer more properties of the system. The starting point is the rate of mass exchange, J_m, within the system, which is inferred from the optimum latent heat flux by eqn. 8.24. This rate tells us how fast mass is circulated, transporting latent and sensible heat, yet as the system is in steady state, these heat fluxes also balance the rate by which water vapor is exchanged at the surface. As this surface exchange not only exchanges water vapor, but also momentum, we can infer the magnitude of surface exchange and near-surface wind speed associated with the maximum power state.

We first note that the formulation of the latent heat flux at the surface in terms of the difference in specific humidity, as formulated by eqn. 8.18, is of the same form as the exchange of momentum that was described by $F_{\text{fric}} = \rho C_d v^2$ of eqn. 7.8 in Section 7.3, with ρ being the air density, C_d the drag coefficient, and v the near-surface wind speed. When we assume that both properties, moisture and momentum, are exchanged at roughly the same rate (which should be the case if moist convection is the dominant process that drives the exchange, in contrast to the large-scale circulation), then the rate of mass exchange is related to the drag coefficient by

$$J_m = \rho C_d v \tag{8.30}$$

Hence, if we know the wind speed v, we can infer the drag coefficient from the rate of mass exchange and the wind speed by

$$C_d = \frac{J_m}{\rho v} \tag{8.31}$$

The wind speed is inferred from the kinetic energy balance of the system, which in steady state is given by $G_{m,\text{max}} = D$ (see also Section 7.3). This procedure essentially follows the approach taken by Emanuel (Emanuel 1986; Bister and Emanuel 1998; Emanuel 2007), who inferred the potential intensity of hurricanes from wind speeds that were derived by equating surface frictional dissipation with the Carnot limit. In our case, we deal with a convective boundary layer, in which the frictional dissipation is typically relatively uniformly spread across its depth, z_{bl} (Stull 1989). When we consider the rate of frictional dissipation near the surface, described by a drag force of the form of F_{fric} and referring to the wind speed v_{10}, at 10 m height, then the frictional dissipation is given by

$$\rho C_d v_{10}^3 = G_{m,\text{max}} \cdot \frac{z_{10}}{z_{bl}} \tag{8.32}$$

where $z_{10} = 10$ m. Using the expression for C_d from earlier, this then yields an expression for the wind speed v_{10}, of the form

$$v_{10} = \left(\frac{G_{m,\text{max}}}{J_m} \cdot \frac{z_{10}}{z_{bl}} \right)^{1/2} \tag{8.33}$$

This expression for the velocity can then be taken together with J_m to infer the drag coefficient by using eqn. 8.31.

To illustrate these expressions, we use the estimates for the maximum power limit of $\lambda E_{\text{opt}} = 135$ W m^{-2} and $G_{m,\text{max}} = 12$ W m^{-2} that were inferred earlier. The mass exchange rate J_m, given by eqn. 8.24, is about $J_m = 5.9 \cdot 10^{-3}$ kg m^{-2} s^{-1}. When we use a typical boundary layer height of $z_{bl} = 1500$ m (Stevens 2005), this then yields a wind velocity of $v_{10} = 3.7$ m s^{-1} and an exchange coefficient of $C_d = 1.3 \cdot 10^{-3}$. These values are well within the observed magnitude, with typical wind speeds over the ocean of $v_{10} \sim 10$ m s^{-1} and the exchange coefficient in the order of

$C_d \sim 10^{-3}$. Yet, it should also be noted that here it was assumed that the exchange coefficients for water vapor and momentum are equal, while in observations they are not necessarily equal yet still of similar magnitude (Andreas 2011). Also, the height of the boundary layer was taken from observations, but this property should also be connected to the dynamics of the system and linked to the temperature differences. The good agreement of these estimates with observations nevertheless suggests that much of the dynamical properties of the system are shaped by the maximum power limit.

Note that this is quite a different approach to derive the drag coefficient and near-surface wind velocity from how friction is commonly described. In the common approach, the drag coefficient is specified in relation to the von Karman constant and surface roughness in the context of the semi-empirical Monin–Obukhov similarity theory as well as certain stability functions (Stull 1989). The agreement of the approach taken here with the common approach is, however, not surprising nor contradictory. Kleidon et al. (2006) showed with climate model simulations that a state of maximum entropy production associated with atmospheric heat transport was associated with a value of the empirical von Karman constant that matched the observed value. What this suggests is that the common, semi-empirical drag parameterizations of surface exchange may in fact reflect this tendency of atmospheric turbulence to operate near its thermodynamic limit.

These estimates are, of course, rather simplified. We did not, for instance, consider temporal and spatial variations in E and P that are associated with lateral moisture transport by the atmospheric circulation that shape geographic precipitation patterns and with temporally highly concentrated precipitation events. Also, on land, the availability of water can be a limiting factor, the sensible heat flux plays a more important role, evaporation rates are affected by the presence of vegetation, and the diurnal cycle affects the energy partitioning. These aspects are revisited later in Chapter 10 that specifically deals with the conditions on land.

8.6 Sensitivity of hydrologic cycling

We next evaluate how this thermodynamic limit on hydrologic cycling changes with a change in radiative forcing. Such changes in the hydrologic cycle are commonly evaluated in the context of global climate change (Allen and Ingram 2002; Held and Soden 2006), with the focus being on how much mean precipitation increases with a certain increase in surface temperature. This change is referred to as hydrologic sensitivity, and is typically expressed in relative terms as $1/P{\cdot}dP/dT_s$. There are also phenomena that are closely related to moist convection for which such a sensitivity is documented, specifically for hurricanes (Emanuel 2005, 2013) and for lightning associated with thunderstorms (Romps et al. 2014). Even though hurricanes and thunderstorms are more regional and sporadic phenomena, they nevertheless provide some indication for the factors that control the sensitivity of hydrologic cycling to a change in radiative forcing. The expressions of the previous section allow us to evaluate these sensitivities and test the extent to which these are captured by the sensitivity of the thermodynamic limit on hydrologic cycling.

The typical starting point in describing hydrologic sensitivity is to note that warmer air can hold more moisture, which should yield an increase in evaporation. This increase in the saturation vapor pressure is about 6.5 % K^{-1} and can be seen in Fig. 8.3b. The simulated increase in the hydrologic cycle in climate models is, however, typically lower at about 2.2 % K^{-1} (Allen and Ingram 2002), which at first seems surprising. When we look at the thermodynamic limit derived in the previous section, the optimum evaporation rate, E_{opt}, described by eqn. 8.28, only depends on the radiative properties, so it is not surprising that the relative sensitivity would be different from the saturation vapor pressure curve. To evaluate this sensitivity, the change in E_{opt}, needs to be related to a change in surface temperature, which in turn is described by eqn. 8.27. The surface temperature is, in turn, changed by a change in R_s or a change in τ. Since we expressed the turbulent fluxes entirely by the latent heat flux here, the resulting sensitivities closely relate to those derived in Section 6.7, but are here interpreted in terms of hydrologic changes. As we neglected the sensible heat flux, which plays an important role in the surface energy balance on land, this aspect needs to be accounted for when dealing with the hydrologic sensitivity on land and non-tropical regions.

We can evaluate how E_{opt} changes with a given increase in surface temperature by $\Delta T_s = 1$ K from present-day conditions. If this increase is entirely caused by an increase in optical depth, it requires an increase by $\Delta\tau = (32/3) \cdot \Delta T_s/T_r \approx 0.04$ to cause a warming by ΔT_s (cf. eqn. 6.37). The increase results in a decrease of net terrestrial radiation at the surface of $\Delta R_{l,\text{net}} = -3.76$ W m^{-2}, which is compensated for by an increase in the latent heat flux. In relative terms, this corresponds to a relative increase of $1/E_{\text{opt}} \cdot \Delta E_{\text{opt}}/\Delta T_s \approx 2.8$ % K^{-1}, which is close to the typical sensitivity of 2.2 % K^{-1} reported from climate models. In principle, such a warming could also be caused by an increase in absorption of solar radiation ΔR_s, which, interestingly, results in a different response. The difference originates from the fact that R_s is connected to the radiative temperature, which changes with a change in ΔR_s. Using the linearization of radiative emission as before, this change in radiative temperature is given by $\Delta T_r = \Delta R_s/k_r$, while the surface temperature changes by $\Delta T_s = (1 + (3/32)\tau - (1/8)R_{s,a}/R_{s,toa}) \cdot \Delta T_r$. Hence, a warming of $\Delta T_s = 1$ K requires an increase in solar radiation of $\Delta R_s = 3.3$ W m^{-2}. This increase in solar radiation also results in an increase in net terrestrial radiation of $\Delta R_{l,net} = 1.44$ W m^{-2} because the downward terrestrial flux not only depends on τ, but also on the overall absorption of solar radiation. The warming caused by solar radiation then results in a comparatively weaker increase of $1/E_{opt} \cdot \Delta E_{opt}/\Delta T_s \approx 1.4$ % K^{-1}. Note how the type of radiative change actually matters for the sensitivity of evaporation to warming, an aspect that was previously noted in climate model simulations on solar geoengineering and global warming (Bala et al. 2008; Kleidon and Renner 2013a). It is thus relatively unrelated to the relative change in saturation vapor pressure, but rather shaped by the type of radiative change, as this alters the driving temperature difference in a different way. What this difference expresses is that in the end, the hydrologic sensitivity is determined by the change in the whole system, including fluxes and temperature differences. However, a more complete assessment requires the consideration of the sensible heat flux as well, particularly in non-tropical areas where temperatures are colder and the latent heat flux plays a less important role and on land, where water availability may limit evaporation.

In addition to the hydrologic sensitivity, we can also infer how the power involved in moist convection, $G_{m,max}$, changes with surface warming. This property is not commonly evaluated in climate models, but there is some indication of this sensitivity in two regional phenomena that are related to moist convection. The first phenomena are hurricanes. Emanuel (1987; 1999) developed a thermodynamic theory for hurricanes and described their intensity by a power dissipation index (Bister and Emanuel 1998). This index is derived by equating the power derived from the latent heat flux, using the Carnot limit of a dissipative heat engine, with the frictional dissipation at the surface. Using observed values for the drag coefficient and exchange coefficient for enthalpy, this kinetic energy balance can be solved for near-surface velocity. Their approach is similar to how velocity was inferred in the previous section, with the notable difference that in hurricanes, it is assumed that frictional dissipation takes place primarily near the surface, and not uniformly throughout the boundary layer. With this thermodynamic theory of hurricanes, Emanuel (1987) predicts a substantial increase of 40–50% in hurricane intensity with a global warming of 3 K, or about 15% K^{-1}, which was later confirmed by observations in recent decades (Emanuel 2005) and more recent climate model simulations (Emanuel 2013).

A second phenomenon that is directly related to moist convection are lightning strikes from thunderstorms. Romps et al. (2014) related the number of lightning strikes observed by satellites to atmospheric variables in an approach that is motivated by a thermodynamic expression. They combined precipitation with the convective available potential energy (CAPE) and a conversion efficiency, which has the units of power, and showed a high correlation between this thermodynamic expression with observations of lightning. They further showed that in climate model simulations of global warming, this expression increases by 12 ± 5 % K^{-1}, which is of similar magnitude as the sensitivity of hurricanes to surface warming.

Even though both examples relate to regional phenomena and are thus not directly comparable to the global mean considerations made here, they nevertheless indicate that the sensitivity of moist convection exceeds the hydrologic sensitivity of the evaporative flux. Both examples relate to the sensitivity by which moist convective power changes with temperature. This sensitivity can be evaluated here by considering the sensitivity of $G_{m,\max}$ with temperature, as given by eqn. 8.29. In this expression, the first term, $s/(s+\gamma)$, increases with surface temperature, while the second term increases quadratically with a change in optical depth that causes the surface warming. When evaluated numerically for global mean conditions at present day, $G_{m,\max}$ increases by about 7.6% with warming caused by a stronger greenhouse effect, and about 4% with warming caused by solar radiation. This enhanced sensitivity can be easily attributed to the two terms in the Carnot limit: as both the latent heat flux as well as the surface temperature increase with the optical depth, both terms in the Carnot limit increase with a stronger greenhouse effect, thus resulting in this more pronounced sensitivity. In the case of the change in absorbed solar radiation, the latent heat flux increases as well, but the temperature difference $T_s - T_t$ increases less because the radiative temperature also increases with more overall absorption. While this sensitivity of 7.6% K^{-1} with greenhouse warming is less than the estimates given earlier, it nevertheless points out that moist power is more sensitive to a change in radiative forcing than the flux or temperature difference alone. This is easily explained by the form of the Carnot limit, which combines both aspects and thus results in an enhanced sensitivity.

The last aspect to explore is the sensitivity of convective mass exchange to a change in radiative forcing, which is described by the sensitivity of J_m to surface temperature, T_s. It has long been stated that the convective mass flux should slow down with global warming (Betts and Ridgway 1989). With the expressions at maximum power state, this flux is given by $J_m = (s+\gamma)/s \cdot (k_r/c_p)$, so that its change in the formulation used here is determined by the change in the factor $(s+\gamma)/s$. This factor decreases with warmer temperatures, so that the convective mass flux decreases with -1.8 % K^{-1} for both forms of surface warming (because the factor $(s+\gamma)/s$ only depends on surface temperature). This sensitivity relates very well to the sensitivity of the convective mass flux over tropical oceans in climate model simulations of global warming of about -1.4 % K^{-1} (Bony et al. 2013). In the estimate derived from maximum power, this reduction relates to a greater effectiveness of heat transport at warmer temperatures, as warmer air can transport latent heat more effectively. Note, however, that in the derivation of this chapter, the sensible heat flux is neglected, which is justifiable for a tropical ocean surface, but less adequate for temperate regions and land. If the sensible heat flux is accounted for, this sensitivity is amplified (Kleidon and Renner 2013a).

To sum up, the sensitivities related to hydrologic cycling that are estimated from the maximum power limit are largely consistent with what has been reported. These sensitivities relate to how the overall system responds to a change in radiative forcing, with notable differences depending on whether this radiative change relates to solar or terrestrial radiation. They demonstrate the importance of considering the sensitivities in the context of the whole thermodynamic system, in which temperatures, fluxes, and the maximum power limit all react jointly to a change in radiative forcing. What has not been considered yet are the feedbacks of hydrologic cycling to the radiative forcing, an aspect that is dealt with next.

8.7 Thermodynamic consequences of hydrologic cycling

At the end of this chapter, we want to get back to the linkages of hydrologic cycling to the functioning of the Earth system as a whole. We focus on two consequences, along the lines shown in the overview figure (Fig. 1.5) in the first chapter: first, hydrologic cycling feeds back to the radiative forcing by altering the radiative properties of the Earth system, and second, hydrologic cycling affects the convective mass exchange and thereby, the transport of geochemical constituents, an aspect dealt with in the following chapter.

In terms of radiative effects, hydrologic cycling alters the absorption of radiation within the atmosphere due to the presence of water vapor and affects its reflectivity through effects of clouds and ice. Water vapor absorbs radiation very well at wavelengths starting at about $\lambda \approx 700$ nm in the near infrared and at longer wavelengths. As solar radiation contains a sizable fraction of radiation in these wavelengths, water vapor absorbs about 12% of the total absorbed solar radiation for a mean total column water vapor mass of about 30 kg m^{-2} (McDonald 1960; Manabe and Moeller 1961). Water vapor plays an even more pronounced role in absorbing terrestrial radiation in the atmosphere, and represents the most important greenhouse gas in Earth's atmosphere at present day. With greater amounts of water vapor in the atmosphere, there is thus an increased absorption of both solar and terrestrial radiation within the atmosphere, thus reducing surface solar radiation and enhancing the atmospheric greenhouse effect. The total amount of water vapor in the atmosphere is, however, not a fixed quantity but results from the combined effects of dehumidification of the atmosphere by moist convection and the ability of air to hold water vapor, both of which depend on temperature and, ultimately, the strength of the radiative forcing of the system.

These radiative effects of hydrologic cycling on the planetary forcing are further enhanced by the presence of clouds and ice. Clouds strongly reflect solar radiation, thus increasing the overall albedo of the surface-atmosphere system (with values of cloud albedo given in Table 6.2). They strongly enhance the greenhouse effect. Their presence hence alters the radiative forcing of the surface such that solar radiation is reduced while the greenhouse effect is enhanced. Ramanathan et al. (1989) estimated that at the global scale, clouds reduce shortwave radiation by about 45 W m^{-2} while enhancing the greenhouse effect by about 31 W m^{-2}. Clouds thus appear to cool the Earth's surface at the global scale by reducing the net radiative forcing at the surface by about 14 W m^{-2}. There are, however, large regional variations in these two effects and their sensitivity to changes in radiative forcing are uncertain (Stevens and Bony 2013).

Another aspect of how clouds alter the radiative forcing and that is likely to alter the maximum power limit deals with their spatial coverage. Clouds form as an inevitable consequence of hydrologic cycling as it represents the part of the cycle in which water vapor condenses and this necessarily results in the formation of droplets which are associated with the radiative effects of clouds. Yet, this requirement does not necessarily require that condensation takes place in a spatially or temporally uniform manner. In fact, when we think of convective clouds, also in how these were illustrated in Fig. 8.1, they typically only cover a certain fraction of the system, with the remaining fraction being clear sky. This fraction is not fixed by the condition that evaporation and precipitation balance each other in steady state, yet this fraction affects the radiative exchange to space.

There are three basic effects that may result in some form of optimum cloud cover that maximizes the power of moist convection. First, a greater cloud fraction results in greater reflection of solar radiation to space, thereby reducing the radiative heating of the surface and thus the latent heat flux. Second, a greater cloud fraction increases the optical depth of the atmosphere, as clouds typically have a strong greenhouse effect. A greater optical depth reduces surface cooling by net exchange of terrestrial radiation, which enhances the net radiative heating of the surface and thus the latent heat flux. The combination of the first and second effect can partly offset each other, with the net effect on the radiative forcing of the surface depending on how sensitive the planetary albedo and the optical depth are to changes in cloud cover. The third effect is relevant to the thermodynamics of the system and deals with the entropy exchange by emission to space. In the steady state that we consider, net absorption of solar radiation balances the net emission of terrestrial radiation to space. If the region is not entirely covered by clouds, then there is a certain rate of latent heat transport from the clear sky fraction to the cloudy fraction. This heat transport generates an imbalance in the emission to space of the two fractions, yet still satisfying the overall steady state. One may speculate that these factors could result in some optimum state of cloud cover that constitutes a means to further maximize the power of the moist heat engine, thus altering the thermodynamic limit of hydrologic cycling by the spatiotemporal arrangement of where and when condensation occurs. These factors would need to be considered in more detail.

An aspect that we have not dealt with so far is the presence of snow and ice, which represents another means by which water affects the radiative forcing of the Earth system. Snow and ice cover are associated with high values of surface albedo (Table 6.2), so that the presence of solid water reduces the absorption of solar radiation, thereby reducing the radiative forcing of the surface. At present, the effect of snow and ice cover at the planetary scale is relatively small as surface temperatures in most regions are well above the freezing point of water. The effect of snow and ice cover naturally becomes more important at colder temperatures that can either result from a weaker solar forcing or a weaker greenhouse effect. It is quite likely that such cases occurred in the Earth's history during "Snowball Earth" episodes (e.g., Hyde et al. 2000), in which large parts of the planetary surface were believed to be covered by ice or snow. Such a state would, obviously, thermodynamically be rather different in terms of the radiative forcing of the Earth system as well as the much reduced intensity of hydrologic cycling (Pierrehumbert 2002).

The other major consequence of hydrologic cycling is that it modulates the magnitude of convective mass exchange. So far, we dealt with this flux as the mass flux that exchanges heat between the thermal "reservoirs" that drive the moist heat engine. This mass exchange was also used here to infer the magnitude of the drag coefficient as well as the near-surface wind velocity. Yet, this convective mass exchange also transports other constituents between the surface and the atmosphere and thereby provides the means for large-scale geochemical cycling in the Earth system. The magnitude of this mass exchange increases with the radiative forcing of the system, but is also affected by the intensity of hydrologic cycling as water vapor represents an efficient means to transport large amounts of latent heat. When convection transports energy more efficiently by warm, moist air, then a given heat flux is associated with less convective mass exchange than in the absence of water vapor. In this way, water slows down this convective mass exchange, and thereby the cycling of other geochemical constituents between the surface and the atmosphere. This effect is looked at in more detail in the following chapters on geochemical cycling and on the processes on land. At the planetary scale, this tight linkage between the radiative forcing, convective mass exchange, and the ability of moist air to efficiently transport heat may also result in an optimum state at which convective mass exchange is maximized at intermediate temperatures. At the end of this book, this optimum state is evaluated in more detail and related to planetary evolution in Section 12.4.

8.8 Hydrologic cycling in the Earth system

In this chapter, hydrologic cycling was described as the consequence of a heat engine that is driven indirectly by the radiative forcing through condensational heating that generates convective motion, and that is maintained in a state of thermodynamic disequilibrium reflected in the presence of unsaturated air. At first sight, it would appear that the application of thermodynamics to hydrologic cycling would primarily apply to the phase transitions of water, an area that has a long history of its thermodynamic formulation. Yet, to maintain a hydrologic cycle in disequilibrium, the atmosphere needs to be continuously dehumidified, and this requires motion in order to exchange the moistened air at the surface with drier air from aloft. Thus, by constraining the generation of motion, thermodynamics plays an equally important role in shaping hydrologic cycling.

In contrast to the heat engines described in the previous chapters, the moist heat engine described here is not directly driven by the radiative heating of the surface, but rather indirectly by the condensational heating when water vapor is lifted to the height at which it reaches saturation. Through energy balance constraints, condensational heating is nevertheless strongly related to the radiative heating of the surface, although there are some irreversible losses associated with the evaporation into unsaturated air. The combination of the Carnot limit with energy balances resulted in a formulation very similar to the maximum power limit of radiative heating that was derived in Section 6.6. This limit results in global-scale estimates that are in reasonable agreement with observations in terms of the magnitude of evaporation and precipitation as well as the drag coefficient, near-surface wind speeds, and the sensitivities to changes in radiative forcing. Of particular relevance is the finding that the moist convective power reacts much more strongly to surface warming than the fluxes of the hydrologic cycle. This aspect can easily be explained by the maximum power limit, as it combines the sensitivity of the flux and the gradient, resulting in the enhanced sensitivity of moist convective power to surface warming. Even though several aspects have been neglected, notably other irreversible losses associated with water vapor (Goody 2000), the linkage with large-scale motion and large-scale moisture transport, and spatiotemporal variations in evaporation and precipitation, overall, it is quite remarkable how many first-order estimates can be derived from the maximum power limit of moist convection in steady state that are broadly consistent with previous studies given the simplicity of the formulations.

To summarize the work associated with the hydrologic cycle, the global estimates of this chapter as well as the linkages to other Earth system processes are summarized in Fig. 8.4. To start, using the evaporative flux of 88 W m^{-2} based on observations (Stephens et al. 2012), an equivalent of about 2 W m^{-2} cannot be used for work as it is lost by irreversible evaporation at the surface. The work output

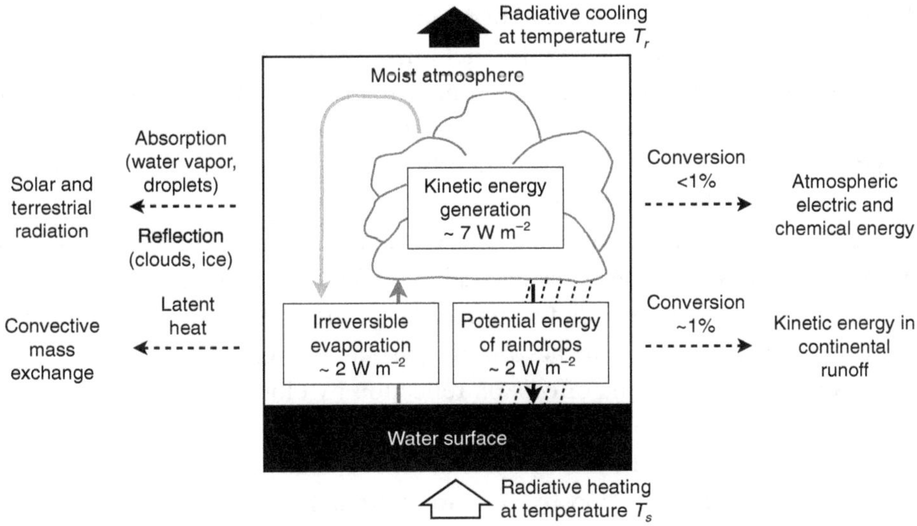

Figure 8.4 Summary of the global estimates of power involved in hydrologic cycling from this chapter using an estimate for evaporation based on observations. Note that the number given for evaporation represents the lost potential to perform work. Also shown are the consequences for the radiative forcing and mass exchange on the left and conversions into other forms of energy on the right.

of the moist heat engine results in the generation of motion that is constrained to be less than about 7 W m^{-2}. The work performed by moist convection is mostly dissipated by frictional dissipation and by the generation of potential energy of about 1–2 W m^{-2} associated with water droplets when water vapor condenses within the atmosphere. This potential energy is then converted into the kinetic energy of the falling raindrops, which is then dissipated into heat by friction.

Some of this work is converted further into other forms. As moist convection is associated with thunderstorms and lightning, some of the power in moist convection is converted into the electric energy of lightning and the global atmospheric circuit (Bering et al. 1998) and into the chemical energy of some compounds, particularly nitrous oxides that are generated by lightning (Schumann and Huntrieser 2007). This fraction is, however, likely to be small with a magnitude of less than 1% of the moist convective power (Borucki and Chameides 1984; Romps et al. 2014). Also note that a certain fraction of the potential energy associated with falling raindrops is not dissipated, but drives runoff processes on land when precipitation is intercepted at the land surface at an elevation above the sea level. With a mean continental elevation of about 800 m and a mean annual continental discharge of $45.5 \cdot 10^{12}$ m^3 yr^{-1} (Oki and Kanae 2006), this involves a depletion of potential energy of about 0.08 W m^{-2} of land surface which is also in the order of about 1%.

In summary, the work involved in hydrologic cycling entails mostly the generation and dissipation of motion and the frictional dissipation of raindrops. However, this does not mean that these small terms are irrelevant. After all, lightning charges the atmospheric electric circuit, and continental runoff enables sediment transport on land. It simply states that these contributions do not reflect major energetic expenses for the hydrologic cycle.

To place hydrologic cycling back into the Earth system context, we note that it strongly interacts with the planetary radiative forcing as the different phases of water substantially alter the radiative properties of the Earth system. These include the direct effects on altering the absorption of solar and terrestrial radiation within the atmosphere as well as the effects on reflection by clouds, snow, and ice cover. These effects feed back to the planetary forcing and quite substantially affect the thermodynamic state of the Earth system. These linkages were, however, only explored qualitatively here, mostly because a suitable, simple description of the thermodynamic constraints on cloud cover could not be formulated here. The other effect that hydrologic cycling has at the planetary scale relates to its effects on mass exchange associated with motion. This effect originates from the substantial latent heat that is involved in the phase changes of water and which alters the ability of a convective mass flux to transport heat and thus the effectiveness in depleting its thermodynamic driving gradient. As the maximum power limit sets the magnitude of the convective heat flux, the transport of moist air requires a lower mass flux. This effect is rather important in shaping the mass exchange between the surface and the atmosphere, and which we revisit in Chapter 10 on land processes.

Hydrologic cycling represents one dominant set of processes that deal with mass conversions within the Earth system and that shape the thermodynamic state of the planet. As a next step, we deal with geochemical cycling, which in terms of its thermodynamic formulation and its connection to motion is quite similar to hydrologic cycling in some respects.

9

Geochemical cycling

9.1 Drivers of chemical disequilibrium in the Earth system

We have now dealt with the thermodynamics of mass cycling associated with water in relation to the planetary forcing. The next step in describing the thermodynamic view of the Earth system shown in Fig. 1.5 is to extend and connect the geochemical cycling of mass to this view. As we will see in this chapter, the thermodynamic formulation of geochemical cycling is similar to hydrologic cycling, except that the connections of geochemical cycling to energy fluxes and radiative effects is more subtle. This formulation then allows us to understand how and by which processes chemical energy is generated within the Earth system. We can then understand how chemical disequilibrium is maintained in the Earth system, how biotic activity as a specific geochemical process fits into this description, by how much biotic activity contributes to the maintenance of chemical disequilibrium, and how geochemical cycling interacts with the thermodynamic state of the Earth system.

This thermodynamic description of geochemical cycling in the Earth system relates back to one of the motivations of the introduction, in which it was described that the Earth's atmosphere reflects a notable state of chemical disequilibrium, reflected mostly in the simultaneous presence of methane, CH_4, and oxygen, O_2 (Lovelock 1965). If left alone, methane would react with oxygen to form carbon dioxide and water, so it requires a continuous exchange of these compounds between the atmosphere and other compartments of the Earth system to maintain the simultaneous presence of these chemical compounds. This remarkable state of chemical disequilibrium has long been recognized as being a unique signature of the Earth system when compared to other planetary atmospheres of the solar system. It possibly serves as a fundamental indicator of a planet with life (Lovelock 1965; Hitchcock and Lovelock 1967; Lovelock 1975), it motivated the development of the Gaia hypothesis (Lovelock and Margulis 1974), and is indicative of geochemical cycling through the Earth's atmosphere as exchange fluxes of chemical compounds are needed to maintain this disequilibrium state.

This chapter deals with geochemical cycling as a thermodynamic phenomenon of the Earth system that reflects the maintenance of geochemical reactions in a state of disequilibrium at the planetary scale. The description of geochemical cycling starts with a thermodynamic formulation of chemical reactions and with relating the concentrations of chemical compounds to states of chemical disequilibrium. It is well established that chemical reactions act to minimize the Gibbs free energy of the involved compounds, much like evaporation into unsaturated air acts to minimize the free energy of a system composed of water in its liquid and vapor state. The depletion of Gibbs free energy is associated with an exothermic release or endothermic uptake of heat, thus altering the thermal entropy of the system. The state of chemical equilibrium in which the Gibbs free energy is at a minimum thus corresponds to a state of maximum entropy, similar to how the different phases of water reach their equilibrium in a state of saturation. The thermodynamic formulation of chemical reactions is equivalent to the formulation of phase transitions, except that we deal with multiple compounds rather than multiple phases of a single compound. In an isolated system, we would expect the chemical composition of the system to eventually reach a state of chemical equilibrium, in which no further chemical conversions or changes in forms of energy would take place. This raises the question of how a state of chemical disequilibrium can be maintained within the Earth system.

Ultimately, chemical disequilibrium in the Earth system is being generated and maintained by mechanisms that convert the energy of the planetary forcing into Gibbs free energy associated with different chemical compounds. The resulting chemical disequilibrium is then described by the distance of the Gibbs free energy within the system to its minimum value, ΔG. The changes in this disequilibrium are related to a number of processes that generate Gibbs free energy at rates G_i, and the rates of chemical reactions that dissipate this free energy D_j, which depend on chemical rate constants.

There are several mechanisms within the Earth system which generate Gibbs free energy, with five major mechanisms being summarized in Fig. 9.1. The first mechanism is the photochemical generation of chemical energy by which the absorption of highly energetic parts of solar radiation alters molecules, specifically by photodissociation due to absorption of ultraviolet radiation (see also Section 6.3.3). This mechanism is responsible for the generation and maintenance of short-lived ozone, O_3, and the atmospheric chemistry that is driven by its products, specifically the OH radical. This mechanism is part of the absorption of solar radiation within the atmosphere so that solar radiation acts as a direct source for this generation of chemical energy.

There are two mechanisms that generate chemical free energy that are driven by hydrologic cycling and which are thus indirectly related to solar radiation. First, the

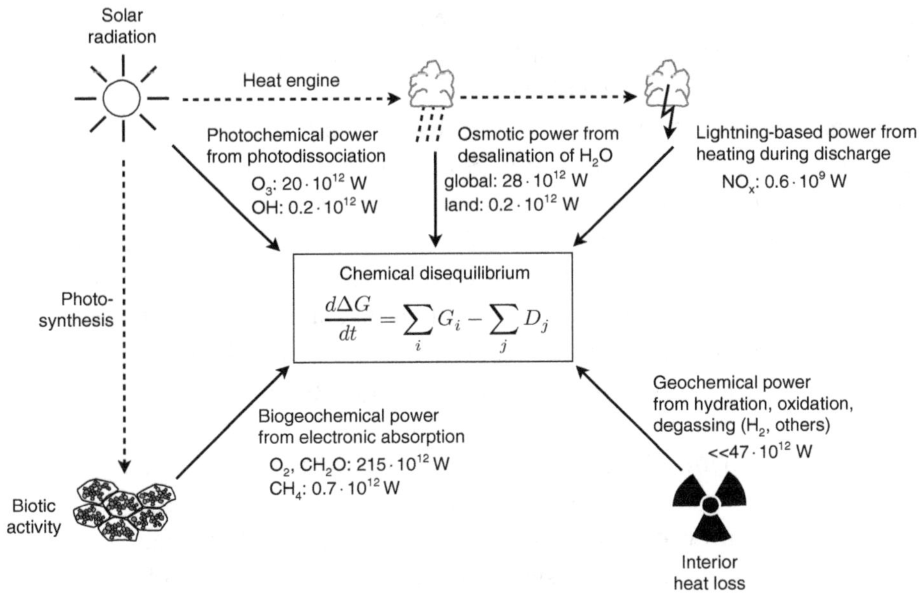

Figure 9.1 Overview of the major processes that generate chemical free energy, ΔG, and chemical disequilibrium within the Earth system, the main chemical compounds involved in disequilibrium, and how these processes ultimately are linked to the energy sources of solar radiation and the interior heat loss. The provided estimates are derived in the chapter. The rates G_i and D_j represent different generation and dissipation rates of chemical free energy.

evaporation of seawater results in desalination, so that the salts of seawater remain in the ocean when water evaporates. This results in atmospheric moisture being free of dissolved minerals, that is, it is desalinated. When this moisture precipitates over land and comes in contact with the continental crust, it dissolves minerals, thereby chemically weathering the continental surface. Chemical weathering processes are relatively slow, but they play an important role on long, geologic time scales over which they can slowly alter the chemical composition of the continental crust as well as the atmosphere. As this mechanism relates to changes in the osmotic potential of water, we refer to the free energy generation by this process as osmotic power. The energy to desalinate seawater and thus drive this form of chemical disequilibrium is reflected in a small increase in the value of the latent heat of vaporization compared to its value for vaporization of distilled water. The second mechanism by which hydrologic cycling generates chemical energy is through lightning. The high temperatures caused by lightning strikes cause a shift in the chemical equilibrium for the gas mixture of nitrogen and oxygen so that a number of new oxides are formed, mostly nitrous oxides such as NO and NO_2, which are referred to as NO_x. The chemical energy that is formed by lightning relates to the operation of the moist heat engine, described in the previous section, which in turn is driven by the radiative forcing of the surface–atmosphere system. The chemical energy generated by this process is referred to as lightning-based power.

Another broad mechanism of chemical free energy generation relates to the internal dynamics of the solid Earth. Some aspects of chemical disequilibrium associated with the Earth's interior may reflect leftovers from the formation of the Earth, since chemical reactions in the solid phase can take an extremely long time to deplete this disequilibrium. This would constitute a form of disequilibrium associated with the initial conditions of the system that simply have not reached their states of equilibrium because time was not sufficient. Here, however, we focus on the processes that actively generate chemical disequilibrium. As we will see in more detail, given sufficient time, the chemical compounds in a system will eventually reach a state of chemical equilibrium at given environmental conditions, particularly temperature and pressure. Such a state of equilibrium is equivalent to the saturation of air with water vapor of a system composed of liquid and water vapor. Just as the saturation vapor pressure, the state of chemical equilibrium changes when the temperature, pressure, or other environmental conditions changes. Such a change occurs when solids are moved by motion in the Earth's interior and brought to the conditions at the surface, or when material from the surface is subducted into the Earth's interior. The generation of chemical disequilibrium by geologic processes is thus closely linked to the rate by which mass is transported in the Earth's interior. The contact of solids to water and air results in degassing, e.g., of carbon dioxide, CO_2, oxidation, and chemical conversions that generate chemical free energy, e.g., associated with the release of molecular hydrogen, H_2. The chemical energy that is being generated by this change in environmental conditions contributes to the overall heat transport associated with the motion that generated the disequilibrium in the first place. The chemical energy thus constitutes a form of latent heat, just as in the case of hydrologic cycling, and the magnitude by which chemical free energy is generated is thus constrained by the differential heating that causes the convective motion in the interior, again in direct correspondence of how the differential radiative heating constrains the intensity of the hydrologic cycle. This rate by which chemical free energy is formed by geologic processes is referred to as geochemical power in Fig. 9.1.

Last, but not least, biotic activity affects the geochemical disequilibrium of the Earth system. We can broadly distinguish between two different effects associated with different types of life, in terms of biotic activity consuming geochemical free energy from the Earth system, as in the case of chemotropic and heterotrophic life forms, and in terms of biotic activity that generates geochemical free energy, as is the case of photosynthesis. As we will see, the free energy generated by photosynthetic life represents a major contribution to the generation of chemical free energy in the Earth system. Photosynthesis is a photochemical process that draws its energy from the absorption of solar radiation. In contrast to the photochemistry involved in ozone, photosynthesis utilizes solar radiation at longer wavelengths by electronic absorption, so that the absorbed photons are less energetic and the chemical conversion operates at a more controlled level. Photosynthesis generates chemical free energy out of solar radiation in the form of carbohydrates that are then used to produce biomass. This production of biomass and photosynthesizing tissues requires other chemical compounds, particularly carbon dioxide, water, nitrogen, phosphorus, and sulfur, among others. As these compounds are drawn from the environment, biotic activity substantially affects the rates by which these chemical compounds are cycled within the Earth system, resulting in a strong modulation of geochemical cycling by biotic activity. The overall generation of chemical energy by biotic activity is referred to as biogeochemical power.

The requirements for these chemical compounds to form biomass constitutes a major constraint on biotic activity. Patterns in marine productivity, for instance, show that these are greatest in areas of strong oceanic mixing and coastal upwelling (Field et al. 1998) rather than areas of greatest solar radiation. This is typically referred to as a nutrient limitation on biotic productivity. Yet it is important to note that it is not directly the availability of nutrients, but rather the rate by which these are transported that represents the limiting factor. As this rate of mass exchange and transport is constrained by thermodynamics, as described in the previous chapters, they play a critical role in shaping biotic activity on Earth. There are, however, other means by which the biosphere can modulate this constraint. This is accomplished by food webs in which biomass is consumed and converted by consumers, resulting in an accelerated cycling of nutrients within the ecosystem. There are hypotheses in ecology that suggest that this internal cycling of nutrients within ecosystems represents a means to enhance and perhaps even maximize biomass production (Lotka 1922a,b; Odum 1969; Ulanowicz and Hannon 1987; Volk 1998). By this internal cycling of chemical compounds within the biosphere, the contribution of biotic activity to the overall generation of chemical free energy within the Earth system can thus be further enhanced.

These processes of generating chemical free energy span a remarkable range of time scales, from the fast processes involved in ozone generation and depletion to the very slow processes of rock weathering and the interior that slowly changes the chemical composition of the surface and the atmosphere on geologic time scales. The generation rates of chemical free energy have therefore likely changed quite substantially over the evolution of the Earth system. They most likely altered the state of chemical disequilibrium, although the disequilibrium is not just shaped by the magnitude of the generation rates, but also by how fast the chemical kinetics deplete the disequilibrium state.

The resulting state of chemical disequilibrium, however, is not the main factor that feeds back to the overall thermodynamic functioning of the Earth system. Geochemical cycling typically does not involve substantial amounts of energy and its effects do not alter the radiative forcing as drastically and immediately as the different phases of water. The more relevant feedbacks associated with geochemical cycling relate most strongly to the effects that these have on the atmospheric composition, which alters the radiative properties of the system. The presence of ozone, for instance, absorbs shortwave radiation in the stratosphere, resulting in radiative heating. The cycling of carbon dioxide and methane shape their atmospheric concentration, and since both are important greenhouse gases, the intensity of geochemical cycling is linked to the magnitude of the atmospheric greenhouse effect. Additionally, terrestrial vegetation on land is associated with substantial biomass, which typically lowers the surface albedo, thereby enhancing the absorption of solar radiation and radiative heating at the surface. When the intensity of geochemical cycling changes, it results in changes in the radiative forcing of the system, which in turn affects the intensity of mass exchange between the surface and the atmosphere, as described in the previous chapters. As chemical reactions require a supply of reactants and a removal of products to be sustained, and this exchange is accomplished by the mass exchange within the system, there is a two-way interaction between geochemical cycling, radiative forcing, and mass exchange within the Earth system.

This chapter first describes the thermodynamics of chemical reactions, the state of chemical equilibrium, and how chemical disequilibrium is maintained. This is then applied to the abiotic generation processes within the atmosphere, describing the three generation processes shown in Fig. 9.1 in greater detail and providing estimates of their magnitudes. This is followed by a treatment of how interior processes result in chemical disequilibrium and thus in the generation of chemical free energy. The thermodynamics of photosynthesis is then described as the major form by which biotic activity generates geochemical free energy, and how internal cycling of geochemical compounds alleviates nutrient limitations. The feedbacks of geochemical cycling on the Earth system are then explored by highlighting the connections with mass exchange within the system. The description of geochemical cycling focuses on a qualitative account of how thermodynamics applies to chemical reactions, as the specifics depend quite strongly on the actual chemical reaction being involved. This chapter closes with a brief summary, in which the magnitudes of the different generation processes of chemical free energy are compared, and the connections to other processes of the Earth system are highlighted.

9.2 Thermodynamics of chemical reactions

Chemical reactions relate to conversions between molecular binding energies, thermal energy, and pressures. Similar to phase transitions of water vapor, the reconfiguration of atoms and molecules into different arrangements can alter the overall binding energy, with the difference in binding energy being released or consumed as thermal energy or resulting in pressure changes. As the total entropy of the system is composed of the molar entropies of the compounds in the system as well as the thermal energy and uncompensated heat, this rearrangement in an isolated system represents the manifestation of the second law in maximizing the total entropy of the system and resulting in a state of chemical equilibrium. In this section, this direction to chemical equilibrium is illustrated with a very simple chemical system. We show the correspondence to the phase transitions of water, and describe how chemical reactions are described in thermodynamics. We draw on the simple example given in Simoncini et al. (2013), which is highly simplified and is used here to illustrate chemical reactions in general. For a more detailed treatment, the reader is referred to specialized textbooks on thermodynamics (e.g., Kondepudi and Prigogine 1998) or chemistry (Atkins and de Paula 2010).

We consider an isolated system that is made up of two hypothetical compounds, A and B, which chemically react to form a compound AB at a fixed, given temperature T and pressure p. This reaction is expressed by

$$A + B \rightleftharpoons AB + \Delta H_r \tag{9.1}$$

where ΔH_r is the change in enthalpy of the reaction, which relates to the difference in the Gibbs free energies of formation between the reactants and products. The change in the Gibbs free energy by the reaction ΔG_r, is then given by the change in enthalpy ΔH_r, as well as the change in entropy ΔS_r, of the reaction

$$\Delta G_r = \Delta H_r - T\Delta S_r \tag{9.2}$$

These properties of the reaction are determined from the molecular properties of the compounds involved in the reaction and are listed in chemistry textbooks (e.g., Kondepudi and Prigogine 1998; Atkins and de Paula 2010).

In this reaction, the compounds A and B are referred to as the reactants, while the compound AB is the product of the reaction. This reaction is exothermic if it releases heat, that is, $\Delta H_r > 0$, while it is endothermic if it takes up heat, $\Delta H_r < 0$. Its state is described by the concentrations, or densities, of the three compounds, abbreviated by $[A]$, $[B]$, and $[AB]$.

To illustrate the progression to an equilibrium state, we consider three different states with different concentrations of A, B, and AB as illustrated in Fig. 9.2. Note

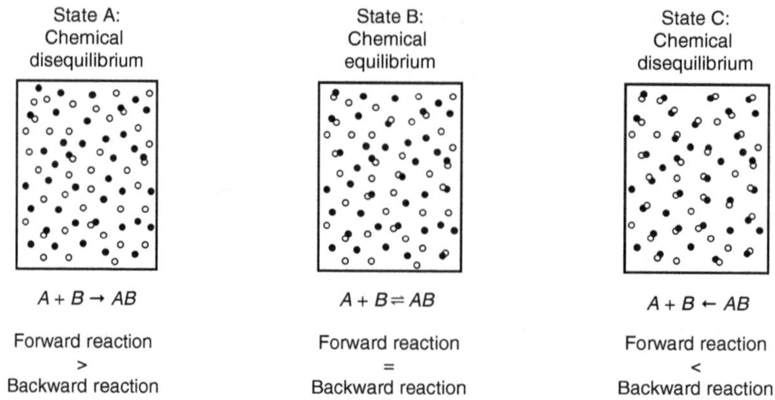

Figure 9.2 Illustration of three states with different concentrations of hypothetical compounds A (white circles), B (black circles), and AB (combined) and their relation to chemical equilibrium. Note the similarity of this example to saturation of water vapor and the description of phase transitions in Fig. 8.2.

that while the total mass of the system remains unchanged in the three states, the number of molecules changes because two molecules A and B are converted into a single molecule AB in the reaction.

State A in Fig. 9.2 represents a state with high concentrations of the reactants, that is, $[A]$ and $[B]$ are large, and a low concentration of the product, that is, $[AB]$ is small. The forward reaction that forms compound AB out of A and B is proportional to the concentrations of the compounds $[A]$ and $[B]$. More generally, the rate is proportional to the so-called activity of the compounds, which correspond to the partial pressure in case of an ideal gas or the mole fraction in a solution, but "concentration" is used to keep the description in the following less technical. This rate, $J_{m,f}$, can be expressed as

$$J_{m,f} = k_f[A][B] \tag{9.3}$$

with k_f being a rate constant of the forward reaction by which A and B is converted into AB. The reverse, or backward, reaction that dissociates AB back into its parts A and B, is proportional to the concentration of the compound $[AB]$. Its rate, $J_{m,r}$, is expressed as

$$J_{m,r} = k_r[AB] \tag{9.4}$$

with k_r being the rate constant of the reverse reaction. With these two rates, we can then express the mass balances of the system that describe the change of the concentrations in time

$$\frac{d[A]}{dt} = J_{m,A} - J_{m,f} + J_{m,r}$$
$$\frac{d[B]}{dt} = J_{m,B} - J_{m,f} + J_{m,r}$$
$$\frac{d[AB]}{dt} = J_{m,AB} + J_{m,f} - J_{m,r} \tag{9.5}$$

with $J_{m,A}$, $J_{m,B}$, and $J_{m,AB}$ being mass exchange fluxes across the system boundary that are included here for generality, but in the case of the isolated system considered here are zero. These dynamics that depend on the masses of the chemical reactants and products is generally referred to in chemistry as the law of mass action.

In state A in Fig. 9.2, the forward reaction is higher than the backward reaction, $J_{m,f} > J_{m,r}$, so that the system progresses towards higher concentrations of $[AB]$ while lowering the concentrations $[A]$ and $[B]$ in equal proportions. The net rate by which the reactants are converted into products is referred to as the reaction velocity v, and is given by

$$v = J_{m,f} - J_{m,r} = k_f[A][B] - k_r[AB] \tag{9.6}$$

In state A, this reaction velocity is $v > 0$. The reaction velocity depends on the concentrations of the compounds, but also on the forward and backward rate constants, k_f and k_r, which are specific to the particular reaction. The rate constants thus shape the speed by which a reaction proceeds towards equilibrium.

State C in Fig. 9.2 shows the reverse case in which the concentration of the product is greater than the concentrations of the reactants. This results in the backward reaction being greater than the forward reaction, $J_{m,r} > J_{m,f}$, a negative reaction velocity, $v < 0$, and in a decrease in the concentration of the product $[AB]$ while increasing the concentrations of the reactants, $[A]$ and $[B]$. In both cases, the reactions would shift the concentration to a more balanced state in which the concentrations are such that the forward reaction and the backward reaction are in balance.

The final state is the state of chemical equilibrium, shown as state B in Fig. 9.2. In this state, the forward and backward reactions balance each other, $J_{m,f} = J_{m,r}$, the reaction velocity is $v = 0$, so that there is no net change in the concentrations of reactants and products in time. This equilibrium state is characterized by an equilibrium constant, K_{eq}, given by

$$K_{\text{eq}} = \frac{[A][B]}{[AB]} = \frac{k_r}{k_f} = e^{-\frac{\Delta G_r}{RT}} \tag{9.7}$$

where the latter expression provides the linkage of the equilibrium state with the Gibbs free energy of the reaction, ΔG_r, as well as the temperature T, at which this equilibrium takes place. Note how the equilibrium constant only depends on temperature as an environmental condition of the system, an important thermodynamic result. The equilibrium constant implies that with greater releases of energy associated with the reaction and/or at lower temperatures, the equilibrium state is characterized by a higher concentration of products and lower concentrations of reactants. As we will see, this effect of temperature on the equilibrium state of the system is a critical factor to understand how chemical free energy is generated within the Earth system.

Before we continue with the thermodynamic description of this example, let us briefly highlight the equivalence of a chemical reaction to the formulation of hydrologic cycling in Section 8.2. Hydrologic cycling can be seen as a simple chemical reaction in which liquid water forms the reactant and water vapor the product. It can be expressed as $H_2O_l \rightleftharpoons H_2O_v + \Delta H_r$, where ΔH_r is the latent heat of vaporization (with $\Delta H_r < 0$ as the evaporation process takes up heat). In this reaction, evaporation represents the forward reaction, while condensation represents the backward reaction. The reaction velocity relates to $E - C$, and the state of chemical equilibrium is represented by the saturation state of water vapor. There is thus a direct correspondence of the different states of disequilibrium with respect to the water vapor concentration in Fig. 8.2 and the states of chemical disequilibrium that are shown in Fig. 9.2.

The thermodynamics of chemical reactions are expressed in terms of the chemical potential, just like for the evaporation process. As described earlier (eqn. 8.1), the chemical potential of a substance is defined by the change in Gibbs free energy of the system with respect to a change in its mass, evaluated at constant pressure and temperature. This results in an expression for the chemical potential of a substance with concentration c that is identical to the formulation of water vapor, as given by eqn. 8.9

$$\mu(p, T, c) = \mu_0(p, T) + RT \ln(c) \tag{9.8}$$

where $\mu_0(p, T)$ is the chemical potential at a reference state that is independent of the concentration and is given by the standard molar Gibbs free energy of formation of the compound. Related to the aforementioned example, each of the compounds involved in the chemical reaction thus has a chemical potential, μ_A, μ_B, and μ_{AB}.

When the chemical reaction progresses, it changes the concentrations of the compounds as well as the chemical potentials. As these relate to the concentrations, the progress of the chemical reaction is expressed in terms of the affinity of the reaction, A. The affinity is defined by the chemical potentials of the products subtracted from the chemical potentials of the reactants. For the example here, the affinity is described by

$$A = \mu_A + \mu_B - \mu_{AB} \tag{9.9}$$

The progress of the reaction towards equilibrium is then described by the extent of the reaction, ξ. The extent of the reaction describes how much mass of the compounds is changed. In the example here, it is described by

$$d\xi = -dN_A = -dN_B = dN_{AB} \tag{9.10}$$

where dN_A, dN_B, and dN_{AB} are the changes in the concentrations expressed in moles and relate to the changes in $[A]$, $[B]$, and $[AB]$ that were used earlier. More generally, the factors in the extent of the reaction are given by the stoichiometric numbers of the reaction, which in our case are simple because all components change in equal proportions. The use of the extent of the reaction simplifies the description of the mass balances, because these changes occur in fixed proportions as the reaction advances. The reaction velocity (eqn. 9.6) can be expressed directly in terms of how the extent of the reaction changes in time:

$$v = \frac{1}{V}\frac{d\xi}{dt} \tag{9.11}$$

We can link the formulation of the chemical reaction in terms of its affinity and the extent of reaction back to the changes in Gibbs free energy as the reaction advances to the state of chemical equilibrium. As the affinity relates to the chemical potentials of the involved compounds, which in turn express the change in the Gibbs free energy of the substances as the concentrations change, the dissipation of the Gibbs free energy in the system is described by

$$D = A \cdot \frac{d\xi}{dt} = V \cdot A \cdot v \tag{9.12}$$

When applied qualitatively to our example, the concentrations of A and B are high compared to AB in state A, resulting in comparatively large numbers of N_A and N_B compared to N_{AB}. As our example is set up to attain an equilibrium state at some intermediate concentrations of the compound AB, the chemical potential of this compound needs to be smaller than that of the compounds A and B. The affinity of the reaction in state A is thus comparatively high. The change in the extent of the reaction is positive, indicating the direction of the chemical reaction, which is to deplete A and B, with $dN_A < 0$, $dN_B < 0$, and $dN_{AB} > 0$. The product of the two terms is thus positive, that is, the Gibbs free energy is dissipated with $D > 0$. This dissipation by the chemical reaction is achieved because the reactants have a different chemical potential than the product of the reaction. The state of chemical equilibrium, state B, is then represented by the minimum of the Gibbs free energy of the system, which is reflected by some intermediate concentrations of the compounds and associated chemical potentials. Formally, this chemical equilibrium state is described by $\partial G/\partial\xi = 0$. Note that in state C, which has a concentration of AB that is higher than the equilibrium concentration, the signs of both terms is reversed, thus also resulting in a dissipation of the Gibbs free energy, as is to be expected.

As the chemical reaction proceeds to a state of chemical equilibrium, the reduction in Gibbs free energy is associated with an overall increase in entropy of the system. This increase is described by the entropy production associated with the chemical reaction, σ_{chem}, which is expressed by the affinity and the change in the extent of the reaction:

$$\sigma_{chem} = \frac{A}{T}\frac{d\xi}{dt} = \frac{D}{T} \tag{9.13}$$

This entropy production is thus directly related to the reduction in the Gibbs free energy of the system. When the state of chemical equilibrium is reached, the entropy production by the chemical reaction thus vanishes ($\sigma_{chem} = 0$).

Once the state of chemical equilibrium is reached, any external perturbation of this state can be described in general by Le Châtelier's principle, named after the French chemist Henry Louis Le Châtelier (1850–1936). This principle states that (as cited in Kondepudi and Prigogine 1998):

> "any system in chemical equilibrium undergoes, as a result of a variation in one of the factors governing the equation, a compensating change in a direction such that, had this change occurred alone, it would have produced a variation of the factors considered in the *opposite* direction."

This principle formulates how chemical reactions are driven by the direction imposed by the second law. It is central to understanding how Earth system processes that are external to chemical reactions result in the generation of chemical disequilibrium, and it allows us to understand this at a general and qualitative level.

Two examples of the application of the principle of Le Châtelier are shown in Fig. 9.3. The first example deals with an external change in the concentrations of reactants and products (Fig. 9.3a). This is, for instance, accomplished by mass exchange (i.e., non-vanishing fluxes of $J_{m,A}$, $J_{m,B}$, and $J_{m,AB}$ in the mass balances, eqn. 9.5), which removes products from the reaction site and replenishes reactants, or the reverse. An externally caused increase in the concentrations of the reactants, indicated by $+[A]$ and $+[B]$ in the figure, alters the concentrations and thereby directly, the extent of the reaction, resulting in some change $\Delta\xi$. Furthermore, since the chemical potentials depend on the concentrations as well, the affinity of the reaction changes as well by some change ΔA. The response of the system is then such that it aims to compensate for this external change. The compensation is accomplished by the chemical reaction proceeding towards chemical equilibrium by converting the reactants A and B into the product AB. This description is equivalent to the dynamics described earlier, except that its starting point is the state of chemical equilibrium and disequilibrium is generated by mass exchange.

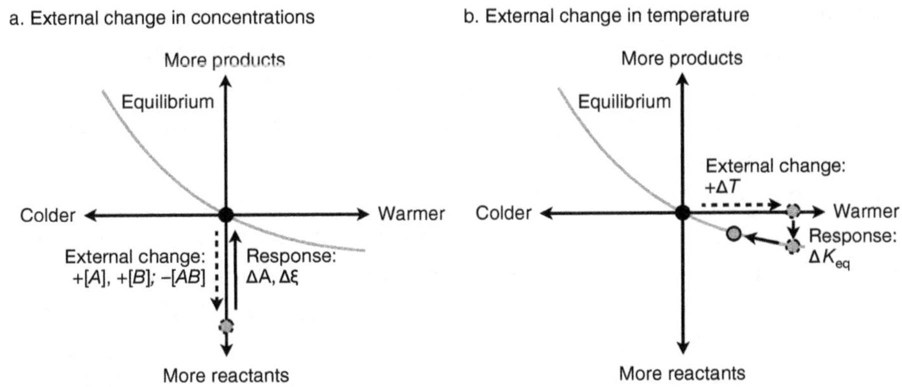

Figure 9.3 Schematic illustration of Le Châtelier's principle for (a) a change in the concentrations of reactants, $[A]$ and $[B]$, and products, $[AB]$, and (b) for a change in temperature, ΔT. The external change is indicated by dashed arrows, while the response is shown by solid arrows. The state of chemical equilibrium is indicated by the black circle at the origin, while the perturbed states are shown by the grey circles. The grey line shows the equilibrium concentrations following eqn. 9.7 qualitatively.

The second example shown in Fig. 9.3b is an externally caused increase in the temperature of the system, indicated by $+\Delta T$. The principle of Le Châtelier states that the system would respond to this change by aiming to reduce this external cause for warming. We can see how this response is manifested in the equations given earlier. The increase in temperature results in a change in the equilibrium constant, ΔK_{eq} (eqn. 9.7), shifting the state of chemical equilibrium towards a higher concentration of reactants and away from products. The concentrations $[A]$, $[B]$, and $[AB]$ in the system that were in chemical equilibrium before the warming occurred are thus no longer in equilibrium, but have an excess concentration of $[AB]$, as in state C in Fig. 9.2. The resulting chemical reaction would thus convert the product AB into its reactants A and B by taking up heat ΔH_r, from the environment. This uptake of heat is the effect that aims to compensate for the external warming of the system. It, however, does not necessarily compensate the external change completely, so that the final state likely remains warmer with an altered state of chemical equilibrium, as indicated by the grey circle in Fig. 9.3b.

There are also other external changes that one can consider using Le Châtelier's principle. For instance, we can consider the consequences of a system that is brought into contact with water, which results in hydrating reactions, or in contact with oxygen, resulting in oxidizing reactions. This principle also works for the phases of water for which we already established the equivalence. If, for instance, a system in equilibrium is heated, the saturation vapor pressure of the air is raised, resulting in enhanced evaporation and latent cooling of the system, which opposes the external heating of the system. At a more general level, these examples allows us to understand how chemical disequilibrium is caused in a system by external heating and cooling as well as by transport processes that alter the concentrations of reactants and products.

The extent of chemical disequilibrium, or the chemical free energy ΔG, in the system is then determined by the generation rate of this form of disequilibrium (which is a form of power), and also by the rate of its dissipation. As the dissipation rate depends on the rate constants of the reactions (cf. eqn. 9.12), the extent of chemical disequilibrium that is maintained in steady state depends on both the generation rate as well as the chemical kinetics that deplete it. A high level of chemical disequilibrium can thus represent either a high generation rate, or a low rate constant. This is equivalent to, e.g., kinetic energy as a form of disequilibrium, in which a high wind speed can reflect a strong generation rate or weak frictional dissipation. This is an important distinction for the interpretation of chemical disequilibrium, because disequilibrium does not reflect on the power by which this state is being maintained.

Le Châtelier's principle allows us to evaluate chemical reactions and geochemical cycling of the Earth system in the context of re-establishing chemical equilibrium within the system without much knowledge of the particular chemical reactions involved. It thus allows us to formulate what we would expect a chemical system to do that is based on the second law of thermodynamics, and it allows us to relate these dynamics to the uptake and release of thermal energy, which provides the critical link to our thermodynamic view of the whole Earth system. As the magnitude of heat fluxes and transport processes are thermodynamically constrained, as described in the previous chapters, these heat fluxes thus also inform us about the constraints by which Earth system processes can generate chemical disequilibrium within the Earth system and how these should react and feed back to change.

9.3 Atmospheric processes

We first focus on the processes within the atmosphere that generate chemical disequilibrium, which refer to the photochemical and lightning-based power shown in Fig. 9.1, while osmotic power is dealt with in the next section. Most of the following chemistry is basic knowledge in atmospheric chemistry (e.g., the textbook by Hobbs 2000), yet the rates of energy that are associated with the chemical conversions are rarely quantified. The provided estimates of the associated energy conversion rates are thus uncertain and should be taken with a note of caution.

The most dominant source of chemical free energy generation and thus chemical disequilibrium within the atmosphere are chemical reactions driven by photodissociation, or more generally by photolysis, of atmospheric compounds by the absorption of ultraviolet radiation. Most importantly, it is the photolysis of molecular oxygen, O_2, in the stratosphere that results in the production of energetically excited oxygen atoms and short-lived ozone, O_3. These chemical reactions are known as the Chapman reactions. In the first of these reactions, molecular oxygen is dissociated by the absorption of shortwave radiation with wavelengths between 185 and 220 nm to produce oxygen atoms:

$$O_2 + h\nu \rightarrow O + O \tag{9.14}$$

where $h\nu$ represents the energy of the absorbed photon. The oxygen atoms then quickly react to form ozone by

$$O + O_2 + M \rightarrow O_3 + M \tag{9.15}$$

where M is a third molecule, either N_2 or O_2. Much of the ozone, however, is rapidly destroyed, mostly by photolysis of ozone

$$O_3 + h\nu \rightarrow O_2 + O \tag{9.16}$$

which involves absorption of solar radiation with wavelengths of 305 to 320 nm. The oxygen atom formed by this reaction then reacts with ozone to form molecular oxygen again

$$O + O_3 \rightarrow O_2 + O_2 \tag{9.17}$$

These destruction reactions increase strongly with ozone concentrations so that ozone concentrations are comparatively low. Note how this state of chemical disequilibrium that is associated with the presence of oxygen atoms and ozone is strongly shaped by the absorption of solar radiation. Photolysis is not just the process that generates this chemical disequilibrium, it also contributes to its rapid dissipation. This state of chemical disequilibrium is thus strongly driven purely by radiative processes.

In terms of the energy that is involved in generating this disequilibrium, Cicerone (1987) estimated that globally, the generation of ozone utilizes about $20 \cdot 10^{12}$ W or about 0.04 W m^{-2} of solar radiation. These reactions describe the major pathway by which ozone is generated in the stratosphere, so that this estimate likely describes the magnitude of chemical free energy generation by photochemistry within the atmosphere. It should, however, be noted that there are also other pathways by which ozone is generated, for instance in the troposphere, and there are also other photochemical reactions, e.g., the photolysis of NO_2, that generate chemical free energy, so that this number is probably too low.

A small fraction of about 1% of the photolysis of ozone (Hobbs 2000) initiates much of stratospheric chemistry by the generation of the hydroxyl radical, OH.

The OH radical is formed when an oxygen atom from the photolysis of ozone reacts with water or methane. This radical is short-lived and has very low concentrations in the atmosphere, but nevertheless plays a central role in the oxidation of other compounds in the atmosphere, such as methane, sulfur, or nitrogen compounds. Even though only a small fraction of the photolysis is used, this contribution is very important for providing the oxidizing capacity of the atmosphere and therefore for atmospheric chemistry (Thompson 1992). Using the estimate of $20 \cdot 10^{12}$ W associated with ozone production and the 1% estimate of OH production yields a global estimate in the order of about $0.2 \cdot 10^{12}$ W or $4 \cdot 10^{-4}$ W m^{-2} associated with the generation of the OH radical for atmospheric chemistry. As the OH radical is involved in the oxidation of compounds and thus brings atmospheric constituents into equilibrium with the oxygen concentration of the atmosphere, it is likely to play a more important role in depleting, rather than enhancing, the overall chemical disequilibrium within the atmosphere.

Another process within the atmosphere that generates chemical disequilibrium is lightning. The average energy release of a single lightning stroke has been estimated to be somewhere between $4 \cdot 10^8$ J (Borucki and Chameides 1984) and $7 \cdot 10^9$ J (Price et al. 1997), resulting in substantial local heating and high temperatures. These high temperatures result in the thermal dissociation of molecular oxygen and a shift in the chemical equilibrium of oxygen and nitrogen compounds, so that lightning forms nitrous oxide, NO, and N_2O out of N_2 and O_2. While at the high temperatures during lighting that can exceed 10000 K, this chemical equilibrium is rapidly established, it takes much longer to reach chemical equilibrium at colder temperatures below 1000 K when air cools down. Thus, the rapid and intense heating associated with lightning discharges causes a chemical equilibrium between oxygen and nitrogen that involves NO and N_2O at high temperatures for a very short time, resulting in a disequilibrium in these compounds when the air is cooled down by mixing with the surrounding air of the atmosphere. The generation of chemical free energy by lightning is thus comparable to the second case described in Fig. 9.3b in the context of Le Châtelier's principle. As lightning is mostly caused by moist convection, this energy is related to the power involved in hydrologic cycling.

Borucki and Chameides (1984) further estimated that the total energy release by lightning of about $40 \cdot 10^9$ W or $8 \cdot 10^{-5}$ W m^{-2} produces nitrous oxide at a rate of $2.6 \cdot 10^9$ kg N yr^{-1}. Using the standard thermodynamic properties of NO, N_2, and O_2, these numbers correspond to an efficiency of about 1.5% in converting the released energy during lightning into the chemical free energy associated with NO, so that this would translate into a global generation rate of about $0.6 \cdot 10^9$ W or $1.2 \cdot 10^{-6}$ W m^{-2}. A more recent estimate of the global NO production rate by Schumann and Huntrieser (2007) is about a factor of 2 higher, which suggests that these generation rates may be twice as high as well. The contribution of chemical free energy generation in the atmosphere by lightning would nevertheless seem much smaller than that by photochemistry.

9.4 Hydrospheric processes

Chemical disequilibrium is also caused directly by hydrologic cycling involving the evaporation and condensation of water. During the evaporation of seawater at the surface, it is only the water that is being evaporated and not the salts and gases that are dissolved in the water. When water vapor condenses in the atmosphere and forms liquid water again, this water is pure, distilled water. This water is then in disequilibrium with the atmospheric composition and, once it precipitates on land, in disequilibrium with the mineral composition of the solid surface of the Earth. It can thus drive chemical changes in the atmosphere and in the solid Earth by dissolving compounds in water. This disequilibrium is depleted by the diffusion of atmospheric compounds into the water droplets and the chemical weathering of the continental crust on land.

The magnitude of the power involved in desalination can be estimated by considering the energetic changes of the opposite process when the salts of seawater are dissolved in pure water (Isaacs and Schmitt 1980). The dissolution of salts results in a change in the osmotic pressure of water, which is essentially a chemical potential. In weak solutions, that is, a solvent such as water with a low concentration of solutes, the osmotic pressure π, is directly related to the concentration of the substances and expressed by the van't Hoff equation (Kondepudi and Prigogine 1998)

$$\pi = i R T \tag{9.18}$$

with i being the van't Hoff factor which describes how many ions are generated out of the dissolved substance, R is the ideal gas constant, and T is the temperature of the solution. Note how this equation is practically identical to the ideal gas law for solute molecules occupying a volume equal to the volume of a solution. Applied to salts in seawater, in which there are two ions for each dissolved mole of salts, that is, $i = 2$, the mean osmotic pressure is about $\pi = 4.8 \cdot 10^3$ J mol^{-1}. The molar concentration of sea salts is derived from the mean sea salt concentration of about 35.2 g kg^{-1} and a mean molar weight of 70.6 g mol^{-1} (Schlesinger 1997) to yield a concentration of 515 mol m^{-3}. Taken together and applied to the total flux of ocean evaporation of $361 \cdot 10^{12}$ m^3 yr^{-1} (Peixoto and Oort 1992), this yields an estimate of the associated power of $28 \cdot 10^{12}$ W or 0.077 W m^{-2} per unit area of ocean surface with the desalination of seawater, an estimate consistent with previous estimates (Isaacs and Schmitt 1980; Kleidon 2010). As noted already in Chapter 8, this value slightly adds to the latent heat of vaporization. As it adds only about $2.5 \cdot 10^3$ J kg^{-1}, this contribution is three orders of magnitude less than the latent heat of vaporization so that it does not noticeably affect the estimates associated with hydrologic cycling.

The power of desalination has several consequences, including physical consequences. Evaporation results in the remaining seawater becoming saltier and thus heavier, hence contributing to the generation of negative buoyancy and the generation of thermohaline motion in the ocean (which relates to Chapter 7). Most of this chemical free energy is, however, dissipated when precipitation mixes the desalinated water with seawater. Thus, this form of chemical free energy also contributes little to generate chemical disequilibrium for atmospheric chemistry.

The part of the evaporated ocean water that is transported and precipitates on land to form continental river discharge, which is about 10% of ocean evaporation or $33 \cdot 10^{12}$ m^3 yr^{-1} of water (Peixoto and Oort 1992), is able to perform chemical work by dissolving rock-based minerals of the continental crust, resulting in chemical weathering reactions. At most, chemical weathering brings the precipitated water into a state of saturation, which reflects a chemical equilibrium between the water with the minerals at the interface of the continental crust. The assumption that water on land reaches this level of saturation and then exports dissolved minerals of the continental crust to the ocean by river discharge appears to be a reasonable assumption to estimate the chemical weathering rates (Arens and Kleidon 2008). Chemical weathering is of particular relevance for the Earth system with respect to the weathering of silicate rocks, as this releases calcium, which then combines with carbonates in the ocean and remove carbon dioxide from the atmosphere on geologic time scales (Walker et al. 1981). An estimate of the upper limit on the work involved in chemical weathering can be based on the global estimate for osmotic power given earlier. To do so, we first note that most of the osmotic power relates to sodium chloride, which constitutes about 94% of the dissolved compounds in seawater. When the estimate is reduced to the flux of river discharge and to the minerals other than sodium chloride, we obtain an upper limit of the power involved in chemical weathering on land of $(33/361) \cdot 6\% \cdot 28 \cdot 10^{12}$ W, which yields a power of $0.2 \cdot 10^{12}$ W or $1.1 \cdot 10^{-3}$ W m^{-2} per unit area of land surface (Fig. 9.1).

There are several other factors affecting actual chemical weathering rates on land, particularly at the local to regional scale, such as soil development and the lithology of the region (Kump et al. 2000). Also, at these scales, chemical weathering reactions are often treated as being kinetically limited, while the approach taken here focuses on the process that causes the disequilibrium by the evaporation of seawater and establishes the limit that is set by the saturation state of chemical equilibrium.

9.5 Geologic processes

Processes in the Earth's interior are related to a number of forms of chemical disequilibrium that affect the atmospheric composition as well as the composition of the crust. The disequilibrium and the associated chemical free energy is generated primarily by interior motion that is reflected by mantle convection, plate tectonics, subduction, and uplift of crust. Interior motion exchanges mass between vastly different conditions of the interior and the surface, and these conditions are associated with different states of chemical equilibrium. What we do here is to use Le Châtelier's principle to generally formulate the chemical changes that occur as a result of interior motion without specifying the substances that are involved. The focus in the following is thus not on the particular chemical kinetics that take place within the interior Earth, but rather on the broad picture of how chemical free energy is being generated.

We first focus on the difference in temperature between the interior and the surface of the solid Earth, and how mass exchange between these temperatures causes chemical disequilibrium. This case is illustrated in Fig. 9.4 in terms of the hypothetical reaction from earlier, in which two reactants, A and B, form a product AB at the prevailing conditions of the surface. We assume that this reaction is exothermic and releases heat ΔH_r, as the reaction proceeds to chemical equilibrium. As the temperature of the surface is relatively cold, in chemical equilibrium

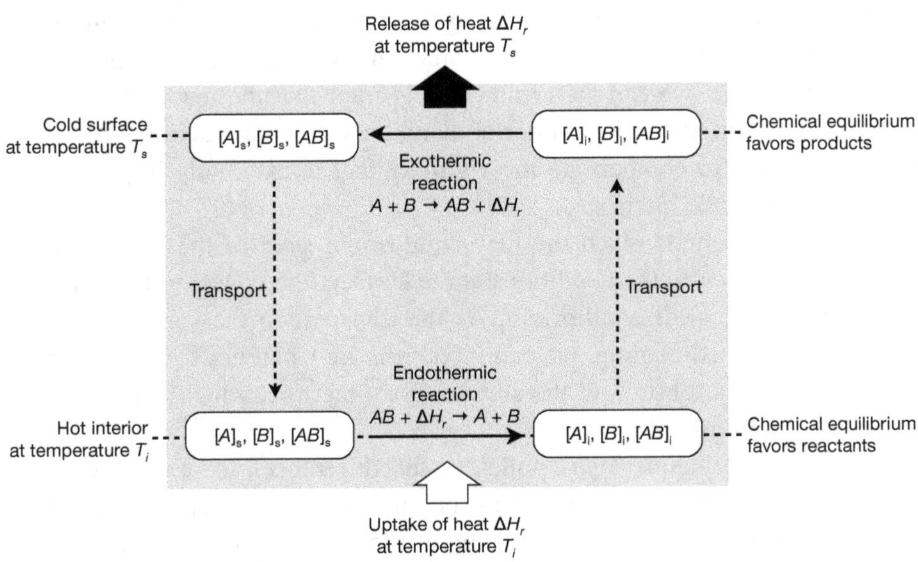

Figure 9.4 Schematic diagram to illustrate how chemical disequilibrium is generated by interior motion, using the hypothetical reaction $A + B \leftrightharpoons AB$ as an example.

the concentration of the product, $[AB]$, is comparatively high compared to the concentrations of the reactants, $[A]$ and $[B]$. This can be seen by the expression of the equilibrium constant, which for the surface conditions is given by

$$K_{eq,s} = \frac{[A]_s[B]_s}{[AB]_s} = e^{-\frac{\Delta G_r}{RT_s}} \tag{9.19}$$

where the index s denotes the values of the properties of the surface at cold conditions with temperature T_s.

When the product of this reaction is then transported into the interior by subduction, it is brought to warmer conditions. These warmer conditions alter the value of the equilibrium constant, and thus the state of chemical equilibrium. The concentrations of the reactants and products that were in equilibrium at surface conditions, $[A]_s$, $[B]_s$, and $[AB]_s$, are thus no longer in equilibrium with each other. The concentrations in chemical equilibrium are given by a different equilibrium constant that is evaluated at the hotter temperature of the interior, T_i

$$K_{eq,i} = \frac{[A]_i[B]_i}{[AB]_i} = e^{-\frac{\Delta G_r}{RT_i}} \tag{9.20}$$

where the index i denotes the values of the properties at the hot, interior conditions. Since the interior is warmer than the surface, $T_i > T_s$, the equilibrium constant at the hotter temperature is less than the value at the surface, $K_{eq,i} > K_{eq,s}$, so that the concentrations of the reactants are greater in equilibrium at the high temperature T_i than at the surface temperature, T_s. When the reactants and products are brought to the warmer conditions, this represents an external change of warming. From Le Châtelier's principle, we can then immediately infer that the reaction proceeds to the new equilibrium state of the interior as an endothermic reaction that takes up heat, ΔH_r, aiming to compensate the warming that is externally imposed by the mass transport into the interior.

When the compounds reach the new equilibrium state of the interior and are then transported back to the conditions at the surface, the concentrations are again in a state of chemical disequilibrium. As the transport to the surface resulted in an externally imposed cooling, we can infer from Le Châtelier's principle that the reaction towards equilibrium at the surface is exothermic, which was the starting point of the description of Fig. 9.4.

Le Châtelier's principle also applies to the differences in other environmental conditions between the interior and the surface. The pressure in the interior is substantially higher in the interior than at the surface. When solid material is transported to the surface, it tends to degas, e.g., carbon dioxide, which aims to compensate for the externally imposed reduction in pressure, while the opposite process of permeation takes place when the material is brought to the high pressure of the interior. Another difference in conditions is set by the presence of oxygen at the surface. As solid material is transported to the surface and is brought in contact with oxygen, it oxidizes. This uptake of oxygen from the atmosphere is aimed to compensate the external change imposed by bringing the material into an oxic environment. When the oxidized material is transported into the interior, the opposite reaction takes place to counter the external change towards an anoxic environment, resulting in reduction of the compounds.

The last example of Le Châtelier's principle described here deals with the presence of water at the seafloor. When solid material is transported to the surface at mid-oceanic ridges and brought in contact with water, the resulting chemical reaction aims to compensate for this external change of wetting the material. This results in a broad class of chemical reactions that is referred to as hydration which consume water from the environment. If we consider a hypothetical mineral A, this reaction has the general shape of

$$2A + 2H_2O \rightarrow 2A(OH) + H_2 + \Delta H_r \qquad (9.21)$$

which releases molecular hydrogen as a gas to the atmosphere as well as heat. Such hydrogen release is observed at mid-oceanic ridges at the seafloor in which the hot material from the interior reaches the surface and gets in contact with seawater. As the hydrated material is subducted into the interior, the opposite, endothermic dehydrating reactions take place, which again aim to compensate for the externally imposed change by transport to the dry environment of the interior.

While there may well be a myriad chemical reactions involved in depleting chemical disequilibrium in the Earth's interior and at the surface, the overall magnitude of the generation of chemical free energy is nevertheless constrained. The reactions have in common that they generally proceed endothermically in the interior and exothermically at the surface, which is simply the result of the environmental conditions with much higher temperatures in the interior and much colder temperatures at the surface. The different composition of the material that is lifted from the subducted material thus transports energy, just as hydrologic cycling transports latent heat. Since the geothermal heat flux is estimated to be about $47 * 10^{12}$ W (Davies and Davies 2010) or 0.09 W m^{-2}, this flux sets an upper bound to the overall energy that can be transported as a result of chemical reactions (see also, e.g., Rosing et al. 2006). This release thus sets an upper bound on the generation of chemical disequilibrium and chemical free energy generation by interior processes. In fact, the overall energy transport by chemical reactions needs to be less than this upper bound. This is because of a fundamental trade-off between the energy that is being transported per unit mass and the magnitude of transport: the more geochemical energy being transported by interior motion, the less thermal energy is available to drive the interior heat engine because the overall rate of heat transport is constrained by the geothermal heat flux. We encountered this trade-off before in the description of the hydrologic cycle in terms of how convective mass exchange declines with a warmer global surface temperature in Section 8.6. Since the mechanical power generated by the interior heat engine yields interior motion, plate tectonics, subduction, and alike, which require this power to be sustained, the overall rate by which geochemical cycling in the interior transports heat is likely to be considerably less than the geothermal heat flux, although the magnitude is difficult to quantify.

While this description of chemical free energy generation by interior processes is necessarily general and leaves out much of the complexities involved, it nevertheless provides a broad picture of how thermodynamics shapes these mechanisms, how these interact with other processes, and how it constrains the overall magnitude. It emphasizes the importance of transport by interior motion as a physical requirement to generate chemical disequilibrium, and highlights the importance of the trade-off between convective heat transport and transport in the form of chemical compounds, which then limits the magnitude of the overall generation of chemical free energy. Le Châtelier's principle then places the resulting geochemical reactions in the thermodynamic context of re-establishing thermodynamic equilibrium and resisting the externally imposed change in environmental conditions. The effects of degassing, oxidation, and hydration at the interface to the ocean and atmosphere then form the basis for establishing the interaction between the inner Earth and the atmosphere.

9.6 Biotic processes

A particular set of geochemical transformations are represented by biotic processes. Biotic activity is driven by various forms of metabolisms that provide the energy supply to sustain life. As already noted by Schrödinger (1944), life is a complex form of a dissipative process that maintains its order by consuming low entropy food and exporting high entropy waste. This entropy exchange allows for the maintenance of a state of disequilibrium in the living cell.

Linking this requirement of any living organism to the context of geochemical cycling, the metabolic activity requires material exchange of chemical compounds with the abiotic environment. This material exchange is necessary not only to construct living tissues, but also to accomplish the supply and transport of chemical energy to drive the chemical reactions involved in the metabolism of living cells and to dispose their waste products. Depending on the means by which organisms derive their energy, forms of life are broadly categorized into chemotrophs, organisms that derive their energy from inorganic forms of chemical compounds of the environment; phototrophs, which derive their energy from sunlight and transform chemical

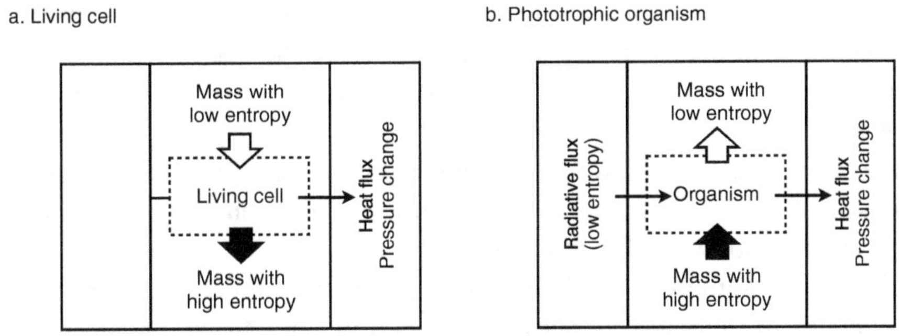

Figure 9.5 Mass and entropy exchange for (a) Schrödinger's general description of life, which describes chemotrophic and heterotrophic forms of life, and (b) the description of phototrophic organisms.

compounds to store the energy; and heterotrophs, which derive their energy from organic chemical compounds of the environment.

The goal of this section is not to go into the specifics of the various forms of metabolisms and forms of life, which are tremendously diverse (Amend and Shock 2001), but rather to focus on how biotic activity is embedded in the functioning of the Earth system so that we can estimate thermodynamic limits of its activity. We thus focus not on single organisms, but rather on the activity of the biosphere which is the sum of all living organisms. As was already alluded to, transport plays a critical role because – using Schrödinger's view of life as a dissipative process – it accomplishes the supply of low entropy food and removal of high entropy waste products. Schrödinger's view of a living cell is illustrated in Fig. 9.5a: the center of the diagram shows the living cell that sustains its metabolic activity by the uptake of mass from the environment with low entropy, or more precisely, it takes up the reactants with a certain associated amount of chemical free energy, and releases mass with high entropy or low chemical free energy – the products of the metabolic reactions – into the environment. This reaction takes place exothermically, so that it releases thermal energy into the environment and changes its pressure, which, taken together, constitutes a flux of enthalpy. This view applies to chemotrophic and heterotrophic forms of life. Note how these life forms act to deplete chemical disequilibrium in the environment, in contrast to Lovelock's notion of chemical disequilibrium as a sign of life that was used at the beginning of this chapter.

Figure 9.5b shows the case of phototrophic life. Here, the situation is reversed as the chemical free energy is not taken up from the environment, but rather taken from the absorption of low entropy solar radiation. By doing so, phototrophic life can alter the chemical composition and generate a state of disequilibrium within the environment. This is in fact opposite to Schrödinger's view of the living cell, because the chemical free energy is generated from the low entropy source provided by the absorption of solar radiation rather than from a source of chemical free energy from the environment.

While these illustrations simply show that the metabolic activity of organisms are consistent with thermodynamics and follow the second law, thermodynamic limits apply here in two different ways. Thermodynamics constrains the conversion of radiation into some form of free energy, as we have already seen in Chapter 6, which applies to the photochemical conversion of phototrophs. Such limits have been explored for a long time regarding the maximum efficiency of photosynthesis. Another constraint that thermodynamics imposes regards the rate by which reactants and products are exchanged with the environment. This type of limit was already described in the previous chapters with respect to motion and mass exchange and earlier in this chapter, for the geochemical cycling within the interior Earth, so it should not come as a surprise that such transport limitations may play a decisive role in limiting biotic activity, not only at the scale of individual organisms, but also for the biosphere at the planetary scale.

Chemotrophic life forms are typically found at or near sites where chemical compounds from the interior reach the surface. This is the case at mid-oceanic ridges on the seafloor (Schrenk et al. 2013), which form as a result of seafloor spreading due to plate tectonics. As hot material from the interior reaches the surface, chemical reactions such as the hydration of the newly formed crust release molecular H_2 among other chemical compounds, while the hot surface drives convective motion within hydrothermal vent systems. These compounds supply chemical free energy to drive metabolic reactions at continuous rates. As such sites likely existed throughout most of Earth's history, provide comparatively high rates of chemical free energy, and were likely more active in the past, these sites are viewed as the environment in which life first emerged (Russell and Hall 1997; Nisbet and Sleep 2001; Martin et al. 2014). The activity of a chemotropic biosphere is, however, limited by the supply rate of chemical free energy that is provided from the geologic source. As we estimated this rate to be a fraction of the geothermal heat flux, and mid-oceanic ridges represent only a fraction of the surface area of the Earth, the overall dissipative activity of chemotrophs at such sites should be well below the $47 \cdot 10^{12}$ W of the geothermal heat flux. More realistic estimates for the carbon fixation at hydrothermal vents are in the range of $0.2 - 2.0 \cdot 10^{12}$ mol C yr^{-1} (Marais 2000), which would translate into about $0.03 \cdot 10^{12}$ W or 0.014% of the power generated by photosynthesis.

Phototrophic life circumvents the limited energy supply by performing photosynthesis, which draws its energy directly from the low entropy of solar radiation and incorporates this energy into compounds taken up from the environment.

This is the main process by which biotic activity generates the energy to sustain the metabolic activity of the biosphere. With the absorption of solar radiation, photosynthesis converts carbon dioxide and water into carbohydrates and molecular oxygen, which can be written in short as

$$CO_2 + H_2O + h\nu \rightarrow CH_2O + O_2 \tag{9.22}$$

where CH_2O is used to represent the carbohydrates, and $h\nu$ stands for the use of solar photons in the reaction, with a minimum requirement of eight photons, with four photons of wavelengths 680 and 700 nm each, for each molecule of carbon dioxide. It involves a complex biochemical conversion chain that effectively removes protons from water and adds it to the carbon dioxide molecule, rather than transferring the carbon atom to the water molecule as it may appear in the way the chemical reaction is written (Rabinowitch and Govindjee 1969; Walker 1992). Note that even though photosynthesis appears like a photochemical reaction similar to ozone chemistry in the atmosphere, it is quite different as it uses wavelengths that are much less energetic and which are absorbed by electronic absorption. This allows photosynthesis to proceed in a highly controlled way.

Energetically, photosynthesis utilizes about 1370 kJ of radiative energy per mole of fixed carbon associated with the eight photons, and yields about 479 kJ of chemical free energy that is bound in glycose. This energy conversion proceeds with an efficiency of 479 kJ/1370 kJ = 35%, and was found by Hill and Rich (1983) to operate near the maximum thermodynamic efficiency at low light intensities. However, only about 48.5% of the solar radiation can be used by photosynthesis, the so-called fraction of photosynthetically active radiation. This fraction already includes mechanisms by which plants downgrade photons of shorter wavelengths to the wavelengths used in photosynthesis, although this conversion is necessarily associated with energetic losses. When these losses are combined, the maximum thermodynamic efficiency of photosynthesis to convert solar radiation is placed at around 12%. The evaluation of this thermodynamic limit to radiation use by photosynthesis spans a long history (Duysens 1958; Mortimer and Mazo 1961; Ross 1966; Knox 1969; Radmer and Kok 1977; Landsberg and Tonge 1980; Bolton and Hall 1991) so that this maximum efficiency of photosynthesis appears robust. From observations, however, it is also long known that the maximum efficiency of photosynthesis in natural environments is estimated to be much lower with less than 3% (Monteith 1972, 1978; Zhu et al. 2008; Amthor 2010), even when water and nutrients do not limit plant productivity as in well-fertilized crops. This apparent discrepancy between the actual and theoretically possible photosynthetic efficiency is a field of active research as it is seen as a key aspect that could result in substantially greater agricultural yields (Zhu et al. 2008; Amthor 2010), an aspect that we revisit in Chapter 11 when we deal with human activity and food supply.

Given that photosynthesis has likely been present on Earth for more than 3.5 billion years and photosynthetic life had a long time to evolve and adapt, this discrepancy between theory and observation of the photosynthetic efficiency is rather surprising. The use of the thermodynamic limit in the previous examples of motion and hydrologic cycling yielded estimates that compare well with the observed magnitude of the fluxes, so why would photosynthetic life also not evolve to its thermodynamic limit? One possible explanation for this discrepancy is that there is another environmental process (apart from water and nutrients) that is thermodynamically limited and that has not been taken into account in the quantification of the limit. One process that could potentially play this role is the exchange of carbon dioxide across the interface between the photosynthesizing tissues and the atmosphere (or ocean), which is tied to the rate of mass exchange that was described in Sections 7.5 and 8.5. While in the ocean, this mass exchange would relate to the intensity of mixing of the surface ocean, which we did not explore in great detail, on land, this mass exchange relates to the convective mass exchange between the surface and the atmosphere. As the rate of this mass exchange is closely related to absorbed solar radiation, such a transport limitation may thus be disguised as an apparent limitation by the utilization of light. Such a light-limited rate of photosynthesis is often used to estimate rates of photosynthesis in natural environments, which is then modified by other limiting factors (e.g., Monsi and Saeki 1953; Monteith 1972; Field et al. 1998). This process will be dealt with in greater detail when describing thermodynamic limits over land, so that we follow up on this hypothesis in Chapter 10.

To obtain global estimates of the chemical energy generated by photosynthesis, we could take the 3% observed maximum efficiency and apply it to the absorbed solar radiation at the surface of $R_{s,s} = 165$ W m^{-2}. This yields an estimate of $2500 \cdot 10^{12}$ W globally, or 5 W m^{-2} per unit area of chemical energy generated. When the 479 kJ per mole of fixed carbon is used to derive the associated uptake of carbon dioxide from this energy rate, this yields a global rate of carbon uptake of $2.0 \cdot 10^{18}$ gC yr^{-1}, which is about a factor of ten too large. Observations estimate the annual mean uptake by the marine and terrestrial photosynthesis to be about $50 \cdot 10^{15}$ and $120 \cdot 10^{15}$ gC yr^{-1} respectively (Ciais et al. 2013), yielding a global total flux of about $170 \cdot 10^{15}$ gC yr^{-1}. These observed estimates correspond to generation rates of chemical energy of $63 \cdot 10^{12}$ W or 0.2 W m^{-2} per unit ocean surface for photosynthesis by the marine biota, $152 \cdot 10^{12}$ W or 1.1 W m^{-2} per unit land surface for the terrestrial biota, or a global total of $215 \cdot 10^{12}$ W or 0.4 W m^{-2}.

These numbers illustrate two important points about photosynthesis. Firstly, they show that there are other processes that generally limit photosynthesis more strongly than the availability of light. In the ocean, the low marine productivity is attributed to a strong limitation by the availability of nutrients. Nutrient availability is, in turn, strongly related to oceanic mixing and upwelling of deep ocean water that is enhanced with nutrients due to decomposition, and to river basin discharge to a lesser extent. Hence, high concentrations of phytoplankton are found in areas of coastal upwelling along the western coastlines of continents in the subtropics, and in areas of strong ocean mixing due to mid-latitude storm activity. The photosynthetic activity of the marine biosphere is thus predominantly shaped by the mixing and transport in the oceanic environment, and not by the availability of solar radiation. On land, photosynthetic activity correlates more strongly with solar radiation, but is substantially affected by the availability of water due to the tight coupling between the uptake of carbon dioxide and evaporative water loss by leaves. As there are large regions in which water availability is limited by precipitation, such as in the subtropical desert regions, terrestrial photosynthesis is substantially reduced by the lack of water. This brings us to the second important point. Even though the mean rate on land is also much lower than the 5 W m^{-2} rate inferred from a 3% maximum efficiency, photosynthesis on land is nevertheless much less constrained than in the ocean, with its mean rate per unit area being about a factor of five higher than for the ocean. The conditions on land thus seem more conducive to perform high rates of photosynthesis.

In summary, biotic activity in the form of photosynthesis generates a remarkably large amount of chemical free energy. At the global scale, the $215 \cdot 10^{12}$ W or 0.4 W m^{-2} that is associated with photosynthetic activity is notably higher by an order of magnitude than the estimates of atmospheric photochemistry, lightning, the hydrologic cycle or interior processes shown in Fig. 9.1. This highlights the importance of biotic activity in shaping the geochemical environment of the Earth, even though it operates well below the limit by which solar radiation could be converted into chemical free energy.

9.7 Biospheric activity

The chemical energy generated by photosynthesis from solar radiation, CO_2, and H_2O feeds the activity of the biosphere. This activity can broadly be classified into the producers, the phototrophic life forms that generate the chemical free energy, and the consumers, heterotrophic life forms that consume chemical free energy in organic compounds stored in living and dead organic material (Fig. 9.6). The activity of the biosphere is then represented by the overall dynamics of its producers and consumers.

The dynamics of the biosphere involves the interweaving of various interactions. Consumers obtain their chemical free energy from dead biomass, but also from the biomass of producers or other consumers. They thus feed on the availability of biomass, yet as consumers provide digestive systems highly suited for

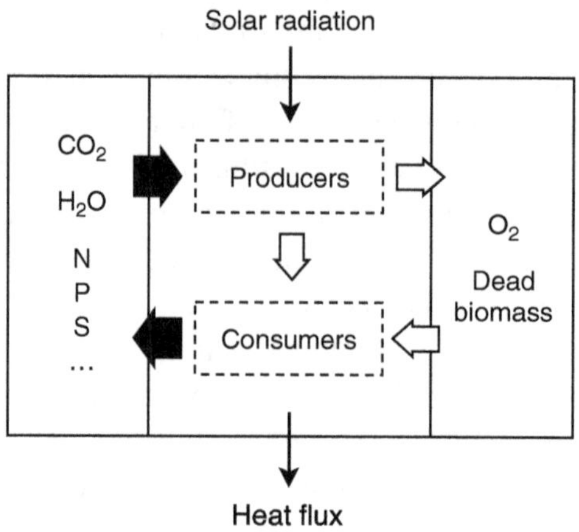

Figure 9.6 The activity of the biosphere is represented by the sum of all producers and consumers, which through the material requirements to form biomass convert chemical compounds of the Earth system. The activity is driven by the chemical free energy generation resulting from photosynthesis out of solar radiation, which is in the end released back to the Earth system in the form of a heat flux.

decomposition and the ability to transport, they modulate the speed and location by which the chemical compounds embedded in the biomass are released back into the environment, thereby making them available for the uptake by producers.

The chemical elements involved in biotic activity relate mostly to six chemical elements of which biomass is composed: carbon, oxygen, and hydrogen that are directly involved in the chemical reaction of photosynthesis as well as nitrogen, phosphorus, and sulfur as nutrients. Biomass production is thus associated with the uptake of these chemical compounds from the physical environment in certain stoichiometric ratios. This uptake can be associated with substantial energetic costs, for instance, associated with the fixation of molecular nitrogen from the atmosphere or with the weathering of rocks to obtain phosphorus. The further conversion of biomass by consumers releases some of these compounds back into the physical environment. The activity of the consumers can thereby feed back to the activity of producers, resulting in complex networks of interactions that are described as food webs. By affecting the activity of the producers, these interactions can feed back to the overall generation of chemical free energy by the photosynthetic activity of the producers, and thereby to the overall magnitude of biotic activity. In all of this, energy and mass are being converted and chemical compounds are exchanged between the biosphere and the physical environment, thereby affecting the magnitude of geochemical cycling within the Earth system.

When biotic activity is not seen at the level of individuals, but rather at the aggregated scale of the biosphere, it is a dissipative activity which converts solar radiation into thermal energy, as shown by the energy exchange at the top and bottom of Fig. 9.6. The biosphere thus represents a dissipative system that is driven primarily by photosynthesis. As we have seen in the previous chapters, dissipative systems have mechanisms by which free energy is generated and dissipated, which applied to the biosphere, is represented by the generation of chemical free energy by the producers and the dissipation by producers as well as consumers, which use the chemical energy to maintain their metabolism and growth. This dissipation of chemical free energy by the metabolic reactions and the associated gas exchange of oxygen and carbon dioxide is referred to as respiration.

Lotka in his seminal work (Lotka 1922a,b) has argued that the biosphere would evolve in time to reach the limit of maximum power, and this concept has been further explored over decades in ecosystem science, for instance by Odum and Pinkerton (1955), Odum (1969), Odum (1988), Schneider and Kay (1994a) and Yen et al. (2014). This evolution towards maximum power is what we would expect of complex dissipative systems in general in the context that we evaluated in the past chapters and in terms of the generalized dynamics that we explored in Chapter 5. We would thus expect the activity of the biosphere to evolve in the same way as dissipative systems evolve in general up to the point at which they reach their thermodynamic limit imposed by the boundary conditions. The activity of the biosphere would then be predictable by these limits set by the boundary conditions. This, of course, raises the questions as to what these limits are and how they are being reached. We have seen earlier that it is not the thermodynamic limit of radiative conversion which limits photosynthetic activity and speculated on mass exchange of carbon dioxide being the limiting factor for limiting photosynthetic activity. Yet, nutrient availability plays an important role as well. This availability can to some extent be altered by the internal dynamics of consumers within the biosphere, which we illustrate in the following example.

To illustrate the importance of consumers in shaping the overall availability of nutrients, possibly even to a level in which these no longer limit biotic activity, we consider an example that includes the biota as well as their essential nutrients within a single system. This combination of living organisms with their immediate environment is formally defined as an ecosystem (Odum 1969). We consider such an ecosystem with a fixed, given amount of nutrients. These nutrients are distributed among an inorganic reservoir of the environment as well as the organic reservoirs associated with the biomass of producers and consumers. We further consider the total biomass in this ecosystem in a mean steady state, so that the production of biomass is balanced by its decomposition and consumption by consumers, with the latter being described by the heterotrophic respiration of the ecosystem. The dissipative activity of the ecosystem is then described by the rate by which biomass is produced, which is referred to as the net primary productivity of the ecosystem.

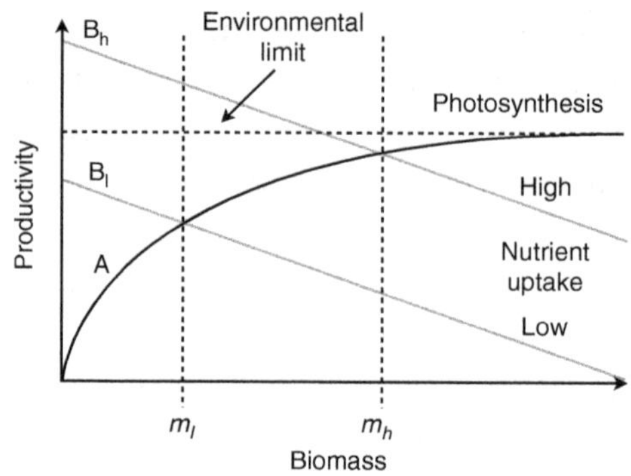

Figure 9.7 Illustration of how net primary productivity is a function of biomass. Line A represents the increase of the photosynthetic rate with biomass which approaches an environmentally limited rate (horizontal dashed line). Lines B_l and B_h represent two cases of nutrient uptake due to contrasting nutrient availability in the environment. Productivity is constrained to the minimum of the limitations imposed by photosynthetic activity (line A) and nutrient uptake (lines B_l and B_h). The vertical dotted lines mark the maximum productivity that can be attained given the limitations imposed by photosynthesis and nutrient availability.

In the following, we evaluate the dissipative activity of the ecosystem by its productivity, although one should note that it does not include the autotrophic respiration associated with the metabolic activity of the producers. As the net primary productivity is typically about 50% of the gross primary productivity by photosynthesis, this contribution by the producers could simply be added to the following considerations. Also note that consumers typically represent less than 10% of the biomass within an ecosystem on land (Odum 1988), so that most biomass represents the producers of an ecosystem.

We next consider how different levels of steady-state biomass affect the productivity of the ecosystem. This sensitivity is illustrated in Fig. 9.7. A greater biomass is typically associated with a greater amount of photosynthesizing tissues, so that the photosynthetic rate should be higher with greater biomass. This is shown in Fig. 9.7 by line A. The photosynthetic rate should eventually saturate against a limit imposed by the environment, which is indicated by the horizontal dashed line. With a greater biomass, however, there are more nutrients incorporated in the biomass, so that the availability of nutrients in the inorganic reservoir is reduced. The rate by which new biomass can be produced should then be reduced; also less nutrients are available with greater biomass. This is shown by lines B in Fig. 9.7. The biomass at which these two limitations match each other mark the state at which the productivity of the ecosystem is at a maximum. If the biomass is lower than this optimum biomass, the productivity is limited by the photosynthetic activity due to low biomass. At the other side of this maximum, it is the availability of nutrients that limits productivity. Note also that if the total stock of nutrients in the ecosystem is enhanced, which can be achieved, for instance, by the allocation of chemical free energy to the fixation of nitrogen from the atmosphere or by the enhancement of rock weathering to release phosphorus, the level of maximum productivity shifts to a greater value and gets closer to the limit imposed by the environment on photosynthesis. This can be seen in Fig. 9.7 by comparing the two cases of nutrient availability in the ecosystem, with line B_l representing a lower nutrient stock while line B_h represents a higher stock. The case with a greater nutrient stock associated with a greater biomass m_h and the productivity is closer to the environmental limit indicated by the horizontal dashed line.

These considerations demonstrate that there is a state of maximum productivity associated with a fixed stock of nutrients within an ecosystem, and that this maximum approaches the environmental limit with a greater nutrient stock. One mechanism by which an ecosystem can achieve this state of maximum productivity is by the presence of a certain level of consumers within the ecosystem. The notion that ecosystems actually adjust to this maximum state has been formulated by the optimum grazing hypothesis and was tested in grassland and aquatic ecosystems which provides some evidence for its applicability (McNaughton 1979; Carpenter and Kitchell 1984; Williamson et al. 1989). These studies evaluated the extent to which the presence of consumers, for instance in the form of large grazers in grasslands, can enhance nutrient cycling and thus the productivity of the ecosystem. The presence of grazers, in turn, is affected by the abundance of carnivores, resulting in multiple trophic levels within an ecosystem that shape the intensity of biomass consumption and internal nutrient cycling, and thus the dissipative activity of the ecosystem. Similar kinds of maxima of biotic activity have been identified in complex metabolic networks (Meysman and Bruers 2007; Vallino 2010; Unrean and Srienc 2011).

In the limiting case, the internal dynamics of an ecosystem with consumers at different trophic levels can result in a state of maximum productivity that does not appear to be affected by the availability of nutrients but only reflects the environmental limit described by the horizontal dashed line in Fig. 9.7. This state is associated with a high degree of nutrient cycling within the ecosystem. This enhancement in nutrient cycling is well established in ecosystem theory (Odum 1969) and was quantitatively demonstrated by Volk (1998). Volk introduced the notion of cycling ratios, which relate the influx of nutrients into the producers to the abiotic rate by which a nutrient is made available. Using estimates based on observations, he showed that these ratios are large, with a range of 500–1300 for nitrogen cycling and a value of 46 for phosphorus cycling (Volk 1998). These high ratios demonstrate the importance of internal nutrient cycling so that the resulting levels of productivity are no longer a direct consequence of nutrient limitations.

The environmental limit of productivity is then likely to be related to the physical processes of the environment, possibly related to the hypothesis formulated earlier that it is the physical exchange of mass that limits producers and thus the activity of the biosphere. This potential transport limitation as well as the limitation imposed by the presence of water on land would then act as the primary limitations for the productivity of the biosphere on land. The maximization of productivity within ecosystems can then explain why the large-scale distribution of vegetation types, such as forests, grasslands, and deserts, as well as their productivity can be predicted very well by purely physical variables of solar radiation, precipitation, and temperature that represent the environmental conditions. The close correspondence between climatic properties and vegetation types has long been observed and recognized, and it is represented by climate classification schemes, such as the Köppen climate classification (Köppen 1923). This perspective would link the activity of the terrestrial biosphere to the thermodynamic limits of motion and the planetary radiative forcing as these relate to the mass exchange between the biosphere and its physical environment.

At the end of this section it may also be noted that some of the biotic activity is associated with the emission of trace gases such as methane, CH_4, or nitrous oxide, N_2O. These compounds are generally created upon incomplete decomposition. For instance, the incomplete decomposition of organic material in the absence of oxygen in wetlands or termite mounds results in a reaction that is different from the reverse reaction of eqn. 9.22 and has the form

$$2CH_2O \rightarrow CO_2 + CH_4 + \Delta H_r \tag{9.23}$$

When this methane is released to the atmosphere, it is then in disequilibrium with the high concentration of atmospheric oxygen and can be oxidized by

$$CH_4 + 2O_2 \rightarrow CO_2 + 2H_2O + \Delta H_r \tag{9.24}$$

However, note that this methane release effectively represents a "leakage" of chemical free energy into the atmosphere due to incomplete decomposition. Simoncini et al. (2013) estimated the power involved in maintaining this equilibrium to be $6.7 \cdot 10^{11}$ W, which represents less than 0.3% of the power involved in photosynthesis. If the biosphere had the ability to prevent this leakage, this chemical free energy could be used for additional metabolic activity. The relevance of this aspect is that it relates back to the interpretation of chemical disequilibrium in the atmosphere as a sign of a habitable planet. The cause for this biotic methane release reflects a small degree of inefficiency in the biosphere that would be prevented if the means were available. Hence, the interpretation of chemical disequilibrium within the atmosphere as a sign of a habitable planet seems questionable.

9.8 Thermodynamic consequences of geochemical cycling

So far, we have dealt with the various means by which processes generate chemical free energy and drive chemical disequilibrium in the Earth system. As a next step, we want to ask how these chemical conversions feed back to the functioning of the Earth system, specifically regarding the effects on its thermodynamic drivers and the overall thermodynamic state. We distinguish the effects between the direct effects associated with energy release or uptake of a chemical reaction from the indirect effects associated with changes in the material properties of the physical environment. Most relevant here are the indirect effects in terms of changes in the radiative properties of the atmosphere, the conditions for condensation of water vapor in the atmosphere, the melting characteristics of the crust, and the heat capacity of the mass that is being transported as it affects the associated rate of mass exchange. These changes in the physical properties affect the radiative forcing, phase transitions, heat fluxes, mass exchange, and thereby the thermodynamic functioning of the Earth system.

In the following description of feedbacks, the focus is set on the effects of chemical reactions on mass exchange as one of the main processes that feed back to the conditions that maintain chemical reactions. From chemical kinetics, one may think that the temperature at which the reaction takes place would be a more direct control of a chemical reaction. Yet the examples provided here as well as the example of hydrologic cycling in the previous chapter have highlighted that mass exchange plays a central role that maintains chemical reactions by the supply of reactants and removal of products. This focus on mass exchange thus emphasizes the strong degree of coupling of chemical reactions and disequilibrium states with the broader thermodynamic setting of the Earth system.

In terms of the direct effects of a chemical reaction that takes place due to disequilibrium at a local scale, the exothermic heat release of this reaction can result in the generation of a convective flux that can result in a positive feedback to the chemical reaction. This feedback is schematically shown in Fig. 9.8a. We consider an exothermic chemical reaction that takes place at a certain temperature. The release of the thermal energy associated with the reaction attempts to counteract the progress of the chemical reaction, following Le Châtelier's principle. The warmer temperature would then act to slow down the chemical reaction, which is indicated by the dashed arrow from the reaction temperature to the chemical reaction. The reaction temperature is, however, also affected by other heat fluxes. In particular, the temperature difference between the reaction site and the environment can provide the driving gradient to generate a convective heat flux. This convective heat flux acts to reduce the reaction temperature, which is indicated by the dashed arrow in the figure and which would form an immediate negative feedback on the reaction temperature (which can again be interpreted by Le Châtelier's principle).

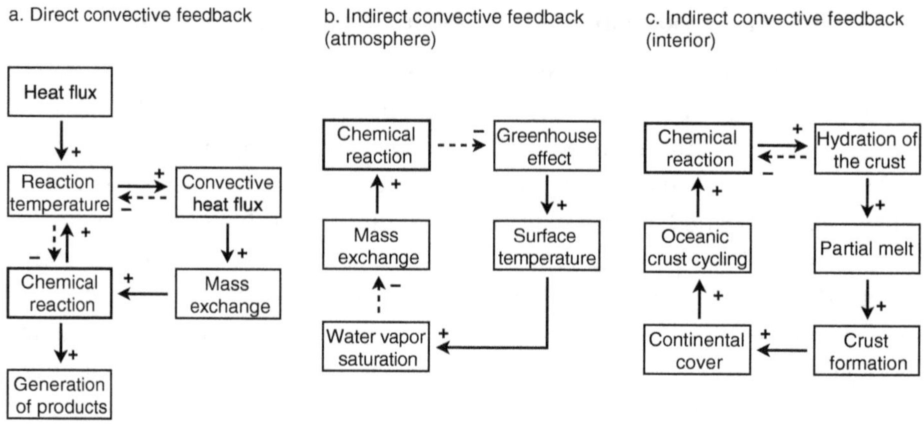

Figure 9.8 Positive feedbacks between a chemical reaction and convection due to (a) the direct release of thermal energy, (b) indirect effects on the radiative forcing, and (c) indirect effects on material properties and continental insulation.

Yet, convection is also associated with convective mass exchange, and this mass exchange can supply reactants to and remove products from the reaction site. As this exchange maintains or even enhances the local disequilibrium in the concentrations of reactants and products at the reaction site, this would act to enhance the progress of the chemical reaction, thus forming a positive convective feedback on the rate of an exothermic reaction, while the convective cooling may act as a regulator on the temperature at the reaction site.

The following two examples illustrate cases in which such a convective feedback may play a decisive role for the chemical reaction. The first example is the case of hydrothermal vents at mid-oceanic ridges at the seafloor. The geothermal heat flux from the interior is not distributed uniformly at the surface, but substantially enhanced at mid-oceanic ridges at the seafloor, sites where the oceanic plates diverge and hot material from the interior reaches the surface of the solid Earth. The contact of the new crust with the ocean water results in hydration reactions that can result in substantial heat release. For such systems, Jupp and Schultz (2000, 2004) found that convection acts as a regulator of the temperature found at hydrothermal vents, while Simoncini et al. (2011) explored the feedbacks that convection has on the generation of H_2 as an energy source for chemotrophic life at these sites. These sites are seen as the sites at which life likely first emerged on Earth (Russell and Hall 1997; Martin et al. 2014). They represent favorable conditions with a steady flux of chemical free energy and the convective feedback may play a critical role in enhancing these conditions by providing higher rates of chemical free energy as well as moderating the temperatures at these sites. The second example where such a convective feedback plays an important role is in the case of wildfires (Clark et al. 1996). Wildfires are driven by the combustion of biomass, which proceeds highly exothermically and which can generate strong convection. Convection in turn feeds back to the fire as it exchanges products with reactants, particularly oxygen. However, both examples are necessarily of local extent, because chemical reactions at the global scale do not release substantial rates of heat compared to the radiative heating due to the absorption of solar radiation. Hence, the indirect effects of geochemical cycling are likely to play a greater role for the thermodynamic state of the Earth system at the planetary scale than the direct effect of heat uptake or release.

The indirect effects of chemical disequilibrium and the chemical reactions that it results in can also form a positive feedback on mass exchange within the atmosphere, which is shown in Fig. 9.8b. By the conversion of reactants into products, the chemical reaction alters the chemical composition within the system. This, in turn, can affect the radiative properties of the system and, for instance, reduce the greenhouse forcing. In Section 8.6, we noted that the mass exchange is actually enhanced in the presence of a weaker greenhouse effect due to the colder temperature and the lower value of the slope of the saturation vapor pressure curve. This enhanced mass exchange could then enhance the exchange of reactants and products, thus forming a positive feedback. This feedback would be much more subtle than the direct feedback discussed earlier, yet it would operate at the planetary scale.

The changes in atmospheric composition throughout the Earth's history may serve as an example for this indirect feedback on chemical reactions. Such changes in atmospheric composition affect the radiative forcing of the surface through changes in the greenhouse effect. The chemical composition of the atmosphere has likely changed substantially through the history of the Earth (Kasting and Catling 2003), from an initially strong greenhouse effect of the early Earth's atmosphere to a comparatively weak greenhouse effect in the recent past, with glacial cycles over the last two million years. This change in the atmospheric greenhouse effect has likely been caused primarily by changes in the atmospheric concentration of carbon dioxide. These changes relate to the effects of biotic activity on carbon cycling, directly by photosynthetic activity and deposition of organic material, and indirectly by its effects on chemical weathering of silicate rocks (Schwartzman and Volk 1989). The feedback would then be established by linking mass exchange back to biotic activity, which we speculated earlier to possibly play a major role in setting the limits to biotic activity.

There are, of course, various other forms of chemical alterations that take place within the Earth system. For instance, chemical compounds provide and alter cloud condensation nuclei within the atmosphere, which in turn affect optical properties of clouds and provide the surfaces at which condensation can take place. The maintenance of stratospheric ozone concentrations is associated with the heating of the atmosphere from above, which alters the radiative forcing for convection but also reduces the damaging effects of photodissociation from taking place near the surface. We did not deal at all with ocean chemistry here, which is important, for instance, for buffering carbon dioxide in the dissolved form. Yet, the point that is emphasized here by the feedbacks shown in Fig. 9.8 is that mass exchange could well play a critical link in maintaining a chemical reaction in the environmental setting of the Earth system, and this aspect results in quite a different interpretation of what the relevant feedbacks may be between geochemical cycling and the Earth system.

At the end of this section, we want to illustrate a similar, potential feedback between chemical reactions at the surface and convection in the interior by changes in the material properties of the Earth's crust, as shown in Fig. 9.8c. A major change in the Earth's history was the chemical differentiation into the heavier oceanic and lighter continental crust that led to the rise and growth of continental surfaces. While estimates on how continents evolved through time show a high degree of uncertainty, all estimates agree that the young Earth likely had no continental crust, continental coverage increased through time, and possibly saturated near current level (Taylor and McLennan 1996; Cawood et al. 2013). Continental crust is made of granite-like minerals and their formation has been linked to the presence of water on Earth (Campbell and Taylor 1983). Granites are formed through a complex geochemical transformation that includes a partial melt in the interior, which allows the lighter minerals to separate, rise to the surface, and form continental crust.

It has been observed that there is a coincidence of growth periods of continental crust and changes in photosynthetic activity in the geologic record. Rosing et al. (2006) suggested that the extra energy made available by photosynthesis may have contributed substantially to the formation of continental crust. An important component in the formation of granitic melts are dehydration reactions in the interior, that is, the opposite reaction to eqn. 9.21. These reactions result in a substantial lowering of the temperature at which basaltic crust melts (Rosing et al. 2006). Once the crust melts and sufficient granitic rocks accumulate, their buoyancy can separate the lighter material from the basaltic crust upon subduction. One way by which an effect of photosynthesis on these processes can be understood is by considering the hydration reaction described by eqn. 9.21 in the presence of oxygen that results from photosynthesis. Then, the reaction could change to

$$4A + 2H_2O + O_2 \rightarrow 4A(OH) + \Delta H_r \tag{9.25}$$

The difference from the previous case is that for the same amount of water, the rate of hydration in the presence of oxygen would be doubled. This enhanced hydration could then contribute to the lowering of the melting temperature of the subducted crust, as described earlier, so that the products of photosynthetic activity could contribute to the formation of continental crust.

To work our way towards a potential positive feedback, we note that the presence of continental crust plays an important role in that it effectively "floats" on the oceanic crust and acts to insulate the surface, resulting in warmer temperatures in the mantle below. In this context, Lenardic et al. (2005) suggest that the presence of a partial surface cover of continental crust may in fact result in an overall enhancement of heat loss from the interior through the effect of the warmer temperatures on the viscosity of the mantle material. By using a conceptual model and laboratory experiments with a highly viscous fluid, the authors suggested that the lower viscosity would increase the turnover of oceanic crust, resulting in a greater heat loss through the oceanic lithosphere. Even though the heat loss through the continental crust would be reduced, the overall effect would still enhance interior heat loss, with possibly even an optimum surface coverage of continental crust that would maximize the turnover of oceanic crust and the associated heat loss from the interior.

When these two effects are combined – the enhanced formation of continental crust due to photosynthetic activity and the enhanced heat loss and convective mass transport due to the insulating effects of continental crust – we may get a positive feedback that is similar to the other examples described earlier. Photosynthetic activity, taking place at the interface between the solid interior and the atmosphere–ocean system, would cause material changes that result in the formation and coverage of the surface by continental crust. The insulating effect of continental crust could then feed back by enhancing convective heat loss and material exchange with the solid Earth, thereby supplying rock-based nutrients to surface life at a greater rate. The overall effect of these chemical alterations would then possibly result in a maximized rate of interior cooling of the planet, as suggested by Lenardic et al. (2005). Although both mechanisms as well as the potential feedback that may result from these are not well established, the causal linkages are further supported by a recent modeling study by Hoening et al. (2014) who resolved these processes more specifically and at greater detail.

To sum up, the outlined feedback loops described here shared the focus of a thermodynamic Earth system perspective in which chemical reactions were linked to their consequences for mass exchange which sustain their rates. This general treatment also applies to biotic activity as a specific set of chemical reactions that operates at the interface between the atmosphere and the solid Earth. While the effects outlined here are not broadly accepted and may need to be refined, they nevertheless embraced this critical link between geochemistry and mass exchange, rather than the more common view that geochemical reactions are primarily constrained by their kinetics.

9.9 Geochemical cycling in the Earth system

In this chapter, we described the physical mechanisms by which chemical free energy and thus disequilibrium is generated, which form the basis for geochemical transformations and cycling. We described five major mechanisms that generated chemical free energy, with the global estimates given in Fig. 9.1: photochemical energy generated by photodissociation as a result of absorption of solar radiation, osmotic energy generated by the desalination of seawater upon evaporation, lightning-based chemical energy generated by moist convection, geochemical energy generated by interior processes, and biogeochemical energy generated by photosynthesis. While photochemistry plays a somewhat special role as it is driven by absorption processes, geochemistry can be understood in general terms as the outcome of Le Châtelier's principle. This principle is a manifestation of the second law and describes that chemical compounds react in a way to resist externally imposed change. A key process that imposes such external change is the transport and exchange of mass, linking geochemical disequilibrium to motion. Photosynthesis as the major process by which life creates energy to sustain metabolic activity combines the highly controlled absorption of non-destructive solar radiation with the requirement for mass exchange of chemical compounds that are involved in photosynthesis and the formation of biomass. The resulting chemical disequilibrium in the Earth system reflects the balance between the generation of chemical free energy and its dissipation, identical to the maintenance of disequilibrium associated with the kinetic energy of planetary motion (Chapter 7) or the disequilibrium of unsaturated air associated with hydrologic cycling (Chapter 8).

In the summary of the global estimates shown in Fig. 9.1, we note how much biotic activity contributes to chemical free energy generation within the Earth system. Its magnitude is likely at least an order of magnitude greater than any other abiotic process within the Earth system. Yet, the chemical disequilibrium that this is mostly associated with is the disequilibrium between the reduced carbon compounds stored in biomass, dead organic material stored in the soil and geologic reservoirs, and the large amounts of molecular oxygen in the Earth's atmosphere, that is, chemical disequilibrium that extends across the different reservoirs of the Earth system. This major form of disequilibrium has relatively little to do with the chemical disequilibrium within the atmosphere that is reflected in the coexistence of atmospheric methane and oxygen as was proposed by Lovelock (1965). As described here, methane is produced by the biosphere as a byproduct of incomplete respiration in anoxic environments, such as wetlands, while most respiration takes place in oxic environments. Hence, the chemical disequilibrium within the atmosphere seems to be little indication of the overall biotic activity in the Earth system. Furthermore, it should also be noted that chemical disequilibrium is also not indicative of its generation rate (Simoncini et al. 2013), because the extent of chemical disequilibrium also depends on the rate by which it is depleted. This should come, in fact, as no surprise. We have seen in the previous chapters that the disequilibrium reflected in wind speeds does not indicate how much power there is in the generation, or that relative humidity reflects the power of the hydrologic cycle. While chemical disequilibrium certainly reflects that there are chemical reactions taking place within a system, we cannot use chemical disequilibrium on its own to infer the rates by which chemical free energy is generated and dissipated.

To illustrate this discrepancy between chemical disequilibrium and its generation rate, we can use the level of atmospheric oxygen throughout Earth's history as an example. As just described, the major form of chemical disequilibrium is between the oxic atmosphere and the reduced carbon compounds at or below the surface. Geological reconstructions suggest that the level of atmospheric oxygen was maintained relatively stable at high concentrations over the last 400 million years (Holland 2006), suggesting that the associated chemical disequilibrium probably did not change over this time period. Yet, during this period, the colonization of land and evolution of higher plants took place, which most likely led to a substantial increase in terrestrial productivity, culminating in present-day terrestrial productivity being considerably greater than the marine counterpart. Even though the chemical disequilibrium remained stable during this time period, biotic activity and related biogeochemical cycling has likely substantially increased. This example substantiates the point that chemical disequilibrium contains little indication of the magnitude of biotic activity and thus the habitability of the planet, which is contrary to Lovelock's (1965) initial suggestion.

When placing chemical disequilibrium in the context of the Earth system, the process that repeatedly plays a key role in driving chemical reactions and disequilibrium is the rate of mass exchange. We saw earlier that it drives the generation of chemical free energy in the interior, and may play a key role in sustaining biotic activity. This rate of mass exchange is related to convective heat transfer, which in turn is thermodynamically constrained by the planetary forcing, thus placing mass exchange in a planetary thermodynamic context and linking it to the planetary forcing. It was also described in this chapter how and why mass exchange is greater at mid-oceanic ridges, sites that provide the favorable conditions for the emergence of chemotrophic life. Describing favorable conditions for life in terms of its mass exchange may thus form a novel and different approach to characterize the habitability of the planetary environment.

Before we get back to this general discussion of habitability at the end of the book, the next two chapters in a sense also deal with specific aspects of habitability. We first deal with the conditions on land, as these form the environment where photosynthetic activity per unit area reaches the highest values on Earth, which could be seen as highly habitable conditions. While a thermodynamic description of land processes are valuable in its own right, it can also further inform us about the thermodynamic characteristics of habitability.

10
Land

10.1 The thermodynamic setting of land

The next step in describing the thermodynamics of the Earth system deals with the conditions on land. The land surface – in contrast to the oceanic surface – has a particular relevance in the Earth system. While it covers less than a third of the Earth's surface, it harbors a disproportionately large share of the biosphere, culminating in the lush and highly productive tropical rainforests that show among the highest levels of photosynthetic activity on the planet. These high levels of photosynthetic activity on land are achieved by large, complex, and highly organized vascular plants rather than by small and comparatively primitive microorganisms that are the primary producers of the oceans. On land, the high rates of photosynthetic activity are associated with a physical imprint on the characteristics of the surface. Note how different the land surface covered by rainforest is compared to an ocean surface, which is almost entirely described purely by its physical state, as exemplified by Fig. 10.1. On land, forest canopies provide dark and heterogeneous surfaces which absorb solar radiation while their root systems reach deep into the soil where they are able to extract water and transport it into the canopies to sustain an evaporative flux into the atmosphere. By dominating the absorption of solar radiation and evaporation, forests shape the partitioning of the surface energy balance. Biotic effects on the functioning of the Earth system are thus particularly strong at the land surface. In this chapter, we want to understand the conditions that allow for and favor these strong biotic effects and how these feed back to biotic activity from the insights gained so far from thermodynamics.

We have seen in the previous chapters that the partitioning of absorbed solar radiation into radiative and convective cooling in the surface energy balance is constrained by the maximum power limit as it results from a close interaction of the convective heat flux with its driving temperature difference between the surface and the atmosphere. Hence, the effects of forests on the surface energy

Figure 10.1 Illustration of the contrast in the physical appearance between (a) a densely vegetated land surface, showing a tropical rainforest in Singapore, and (b) an ocean surface near the Danish coast that is described almost entirely by its physical state.

balance extend further into the atmosphere and have the potential to alter the thermodynamic limit of the surface–atmosphere system. As convection drives the mass exchange between the surface and the atmosphere, this potentially feeds back on the gas exchange between vegetation and the atmosphere and thus on the level of biotic activity, as was hypothesized in the previous chapter. Overall, these effects illustrate that the land surface provides a thermodynamic setting that is quite different from an ocean surface.

In this chapter, we focus on the factors that shape thermodynamic limits of land surface functioning that are particularly different from an ocean surface and how these relate to biotic activity. One, almost obvious difference is the difference in water availability between land and ocean. Since in many regions on land, water availability is limited, the surface energy partitioning and convective heat fluxes described in Chapter 8 need to be modified to account for water limitation. While this aspect is described here as well, the main focus in this chapter is on three major differences that relate to variations in the forcing and storage effects with respect to the radiative heating and cooling, the wetting by precipitation and drying by evaporation, and the carbon uptake and release by biotic activity. As we will see in this chapter, these storage effects affect the thermodynamic limits and convective mass exchange that was described in the previous chapters and thereby result in unique differences in the thermodynamic functioning of the land surface that are then further shaped by biotic effects on these exchanges.

The importance of the buffering effect of storage changes is illustrated in Fig. 10.2 for radiative fluxes and heat storage. The figure shows two extreme cases of heat storage effects on the energy balance partitioning of a system that is heated

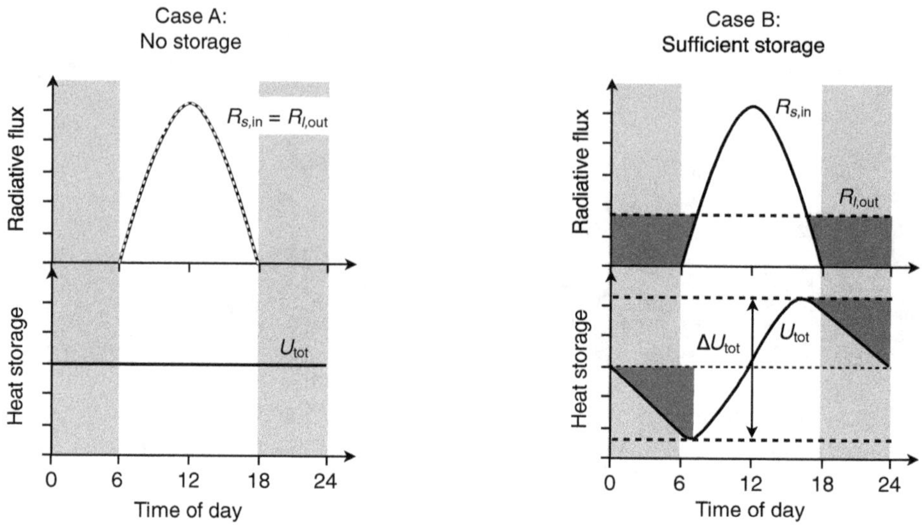

Figure 10.2 Illustration of the effect of heat storage changes on the radiative fluxes over a day, where the areas shaded by light grey denote night. Case A on the left shows a steady state in which solar radiative heating by a flux $R_{s,in}$ is instantaneously balanced by the emission of radiation $R_{l,out}$, while case B on the right shows a state in which sufficient heat storage distributes the emission of radiation over the whole length of the day to accomplish a steady steady. While case A is associated with no changes in heat storage so that heat storage U_{tot} remains unchanged, case B requires a sufficient ability to store and release heat within the system of some magnitude ΔU_{tot}. The periods over which heat is released from storage to sustain $R_{l,out}$ are indicated by the dark grey areas.

by absorption of solar radiation $R_{s,in}$ that varies through the course of the day and is cooled by the emission of radiation, $R_{l,out}$. In case A, the steady state is instantaneously achieved, so that $R_{l,out} = R_{s,in}$ at every time during the course of the day. To maintain this steady state, no change in heat storage within the system, U_{tot}, takes place. However, since there is no emission of radiation during the night, this case is associated with a nighttime temperature of 0 K. This case is not observed on Earth. Case B represents the other extreme, in which the balancing of the radiative fluxes is achieved in the mean taken over the day, so that $R_{l,out} = R_{s,in,avg}$, where the "avg" is used to denote the average of $R_{s,in}$ over the day. This case requires a change in heat storage within the system to balance out the difference between $R_{s,in}$ and $R_{l,out}$ and, in particular, to sustain $R_{l,out}$ during the night. This case is more representative for the Earth system, where diurnal temperature differences are relatively small compared to the mean surface temperature. The pertinent question regarding heat storage is how the surface–atmosphere system accommodates such heat storage changes for land and ocean and how it affects thermodynamic limits.

This brings us to the first difference between land and ocean. Solar radiation is absorbed differently over land, where it is absorbed *at* the surface, while in the oceans, it is absorbed in the top 20 to 50 meters *below* the surface. This aspect simply reflects the difference in the optical properties of water and the land surface, with water being transparent while the land surface is not. Hence, solar radiation penetrates the water column of the ocean relatively deeply before it is absorbed, while on land, it is absorbed highly concentrated at the surface, thus providing a strong heating source. This difference in how solar radiation is absorbed has a substantial consequence for how the diurnal variations in solar radiative heating between day and night are buffered by the system. For oceans, these diurnal variations are buffered by the surface ocean and result in only a small diurnal variation in temperature due to the large heat capacity of the absorbing water body (Oke 1987). On land, however, these variations are typically buffered by heat storage changes in the lower atmosphere that are reflected in the pronounced development of a convective boundary layer on land. This is because the soil conductivity on land is rather small, so that only a small fraction of the solar heating can penetrate into the soil. It thus heats the surface and generates buoyancy, resulting in a convective heat flux that heats the lower atmosphere. This convective motion is not based on a heat engine driven by the condensational heating of moisture as described in Chapter 8, but rather by the direct solar heating of the surface. As it does not involve condensational heating, we refer to this heat engine as a dry heat engine, which then results in dry convection and a sensible heat flux. The resulting heat storage changes in the lower atmosphere then need to be accounted for in the derivation of the Carnot limit as this heat storage change inside the heat engine modifies the limit quite substantially. It modifies the expression for the maximum power limit and the partitioning of the surface energy balance because of the different way by which the diurnal variation in solar radiation is buffered by the land–atmosphere system. The same considerations also apply to seasonal variations, but we focus here on the diurnal variation in solar radiation as this represents the most pronounced variability in the forcing with pronounced effects on thermodynamic limits.

The second major difference on land relates to the availability of water at the surface. For the ocean surface, water availability is obviously not a limiting factor, but on land, the presence of water is shaped by the amount and temporal variations in moisture transport by atmospheric motion and precipitation as well as the ability of the land surface to store and release water back into the atmosphere. The surface energy balance was dealt with before when discussing the limits of hydrologic cycling in Chapter 8, where it was assumed that water is not a limiting factor, that the convective heat transport from the surface to the atmosphere was entirely due to the latent heat flux, and that motion resulted from a moist heat engine driven by condensational heating. While these assumptions are quite reasonable for the tropical ocean, on land these do not hold. The strong solar heating of the land surface during the day heats the near-surface air, creates buoyancy, and initiates convective motion and thus passively transports the evaporated vapor. As evaporation involves substantial latent heat, the evaporative flux contributes to the ability of the convective flux to transport heat, which affects both the cooling of the surface and the rate of convective mass exchange. These interactions are further complicated by the temporal mismatch in precipitation and evaporation, with precipitation typically occurring during short, wet episodes, while evaporation taking place over prolonged dry episodes in which condensation and the associated latent heat may not necessarily be released within the atmosphere. These relate to storage effects in the soil, which can store precipitated water for dry periods, and to storage effects in the atmosphere, which accumulate evaporated moisture until it condenses and precipitates. These storage effects of the soil and the atmosphere are similar to the effects of heat storage shown in Fig. 10.2, except that these deal with mass storages of water and vapor, and with short wet periods and prolonged dry periods over which a steady state is accomplished in the mean water balance. The ability to evaporate or condense water thus affects surface conditions, the convective mass exchange with the atmosphere, the convective heat flux, as well as the maximum power that can be derived from solar radiative heating of the surface.

These differences in the buffering of the diurnal cycle of solar radiation, dry convection, and moisture effects result in a considerably different thermodynamic setting of the surface–atmosphere system on land. The third major difference relates to the presence of biotic activity in the form of vegetation that was already described at the beginning of this chapter. The land surface of some regions, such as the humid tropics which in their natural state are covered by dense rainforests, are so strongly shaped by biotic activity that the relevant questions seem to relate to the how and why biotic activity shapes the physical conditions rather than how the physical conditions shape biotic activity. Focussing on the thermodynamic setting of the land surface, we can identify two major effects of vegetation activity that substantially alter the thermodynamic forcing: the enhanced absorption of solar radiation by vegetated surfaces and the modulation of the evaporative flux into the atmosphere. These effects then feed back to the conditions in which photosynthesis takes place. The particular process that we focus on here that limits and is affected by surface conditions is the mass exchange of carbon dioxide (CO_2), between the photosynthesizing tissues of the vegetative cover and the atmosphere. This limitation to photosynthesis was already hypothesized in Section 9.6; it is speculative in that it is not considered in common descriptions of the environmental factors shaping photosynthetic rates on land, yet, as we will see, it provides a simple, physically consistent explanation for the low efficiency of photosynthesis in natural environments that is consistent with observed relationships. Vegetation faces a particular challenge to achieve a high rate of photosynthesis during the diurnal cycle with respect to this mass exchange, which deals with a similar storage effect as shown in Fig. 10.2 except that it deals with storage changes of CO_2. Photosynthesis is active during periods of daylight, which on average is half the length of a day, while respiration, the opposite process that converts carbohydrates back into CO_2 and that relates to the metabolic activity of the ecosystem, takes place over day and night. This diurnal mismatch in the process that takes up CO_2 from the atmosphere compared to the process that releases it back into the atmosphere requires that during periods of photosynthesis, CO_2, on average, needs to be taken up twice as fast during daylight compared to its release. It thus requires a sufficiently fast mass exchange between the vegetated surface and the atmosphere, so that the stock of CO_2 within the atmosphere is depleted during the day to sustain high photosynthetic rates. This mismatch in the carbon exchange between day and night is well known in ecosystem science and is referred to as the net ecosystem exchange of CO_2. This exchange is explored here as the limiting process for photosynthetic CO_2 uptake by relating it to the mass exchange associated with the maximum power limit of the convective flux over land. It then serves as the basis to describe the thermodynamic limit on carbon uptake as well as the feedbacks between biotic activity and the overlying atmosphere under the constraint of maximum power.

To sum up the scope of this chapter, we first describe thermodynamic limits for the land–atmosphere system with respect to the limit associated with dry convection and then explore the three major aspects of the forcing of the land surface–atmosphere system that substantially shape the thermodynamic limit to land surface functioning. These topics frame the next four sections of this chapter. Section 10.2 deals with the maximum power limit of dry convection, the next section with the effects of the diurnal variation in solar radiative forcing and associated heat storage changes on this limit. This is followed by a section on hydrologic effects on this limit, in terms of the general availability of water at the surface as well as temporal variations and storage effects. We then describe the limits of CO_2 exchange between the surface and the atmosphere as a potential limit for biotic activity on land. The last section then describes the interactions and feedbacks between vegetation, the surface, and the atmosphere. The chapter closes with an emphasis on the linkages of the processes on land to the Earth system and to the following chapter on human activity.

10.2 Limits to dry convection on land

To derive a thermodynamic limit for dry convection, that is, convective motion that results directly from the radiative heating of the surface rather than condensational heating as in Chapter 8, we revisit the maximum power limit for a vertical difference in radiative heating and cooling from Section 6.6. To apply it to dry convection on land, we follow previous work (Kleidon and Renner 2013b; Kleidon et al. 2014b) and express the convective heat flux J that was used in Section 6.6 as the sum of the sensible and latent heat flux, H and λE, so that $J = H + \lambda E$. Both of these fluxes relate to the temperature difference between the surface and the atmosphere, and are expressed in a similar way as in Sections 7.5 and 8.4

$$H = c_p J_m \cdot (T_s - T_a) \tag{10.1}$$

and

$$\lambda E = \lambda J_m (q_s - q_a) \approx c_p J_m \frac{s}{\gamma} (T_s - T_a) = \frac{s}{\gamma} H \tag{10.2}$$

where c_p is the specific heat capacity of air, J_m is the rate of convective mass exchange, $q_s - q_a$ is the difference in specific humidity between the surface air and the atmosphere, s is the slope of the saturation vapor pressure curve, and γ is the psychrometric constant (cf. eqn. 8.12). In the approximation made for the latent heat flux, it is assumed that the difference $q_s - q_a$ can be expressed by the values at saturation at the respective temperatures, equivalent to the step made in the derivation of eqn. 8.18 in Section 8.4.

The difference to the previously derived limits is that we derive the thermodynamic limit on dry convection specifically by maximizing the power associated with the sensible heat flux. The expression for this power G_d, is derived here using the Carnot limit of a dissipative heat engine, the sensible heat flux, and the temperature difference between the surface and the atmosphere:

$$G_d = H \cdot \frac{T_s - T_a}{T_a} \tag{10.3}$$

When a simplified expression for net terrestrial radiation is used of the form $R_{l,\mathrm{net}} = k_r \cdot (T_s - T_a)$ as in Kleidon et al. (2014b) with a surface energy balance in steady state of the form

$$R_{s,s} - R_{l,\mathrm{net}} - H - \lambda E = 0 \tag{10.4}$$

then the maximization of power, $\partial G_d / \partial H = 0$, yields a partitioning of the surface energy balance of

$$R_{l,\mathrm{net}} = \frac{R_{s,s}}{2} \tag{10.5}$$

$$H_{\mathrm{opt}} = \frac{\gamma}{s + \gamma} \cdot \frac{R_{s,s}}{2} \tag{10.6}$$

$$\lambda E_{\mathrm{opt}} = \frac{s}{s + \gamma} \cdot \frac{R_{s,s}}{2} \tag{10.7}$$

These expressions are associated with an optimum temperature difference between the surface and the atmosphere given by

$$T_s - T_a = \frac{R_{s,s}}{2k_r} \tag{10.8}$$

which result in an expression for the maximum power $G_{d,\max}$, of

$$G_{d,\max} = \frac{\gamma}{s+\gamma} \cdot \frac{R_{s,s}^2}{2k_r T_a} \tag{10.9}$$

Furthermore, the ratio of the sensible to latent heat flux, known as the Bowen ratio B, is given by

$$B = \frac{H_{\text{opt}}}{\lambda E_{\text{opt}}} = \frac{\gamma}{s} \tag{10.10}$$

This expression, as well as the functional form for the latent heat flux given by eqn. 10.7, is very similar to the well-established concept of equilibrium evaporation (Schmidt 1915; Penman 1948; Priestley and Taylor 1972) and empirical evaporation formula of similar form (Makkink 1957; de Bruin and Lablans 1998; Xu and Singh 2000). The equilibrium evaporation rate is often described and used to infer the atmospheric demand or potential evaporation rate of the land surface, that is, the rate of evaporation in the absence of water limitation. The derivation of this rate here has a very similar interpretation. As it results from the maximization of dry convective power, it represents the thermodynamic limit for evaporation from locally driven convective exchange. It thus represents an atmospheric transport limitation, which is driven by radiative heating of the surface, rather than a direct energy limitation to which this rate is commonly attributed to.

The set of equations 10.5–10.9 describe the partitioning of the surface energy balance in the climatological mean under the assumption that the surface exchange is driven entirely by the radiative heating of the surface. It is entirely determined by the solar radiative forcing of the surface $R_{s,s}$, the atmospheric temperature T_a, which is determined from the total absorbed solar radiation $R_{s,\text{toa}}$, (see Section 6.6) and the surface temperature T_s, from which the value of k_r can be inferred. Using global means of these values with $R_{s,s} = 165$ W m^{-2} and $T_s = 288$ K yields an estimate for the maximum power of $G_{d,\max} = 4.0$ W m^{-2}, and optimum fluxes of sensible and latent heat of $H_{\text{opt}} = 31$ W m^{-2} and $\lambda E_{\text{opt}} = 52$ W m^{-2}. These values compare reasonably well with global estimates (Kleidon and Renner 2013b) as well as the geographic variation on land (Kleidon et al. 2014b), although the overall magnitude of turbulent fluxes is underestimated, particularly for forested regions. We will see later how this underestimation can be attributed to diurnal storage effects.

This is, of course, a highly simplified description. The use of the radiative temperature of the atmosphere likely overestimates the temperature difference and thus the power associated with dry convection. Furthermore, this approach does not take into account the effects of diurnal variations or water limitations, and it assumes a steady state between evaporation and precipitation. These aspects are dealt with in the following two sections. Also not considered here are effects of the large-scale atmospheric circulation on surface exchange, which in common approaches are accounted for by an explicit dependence of surface exchange on wind speeds as in the bulk formula described by eqn. 7.8. The omission of this effect of the large-scale circulation can be justified by noting that the power associated with the large-scale flow of about 2 W m^{-2}, of which about half results in the turbulent dissipation near the surface, is smaller than the maximum power associated with the sensible heat flux of 4 W m^{-2}. This suggests that surface exchange on land is more strongly driven by the radiative heating of the surface rather than by the large-scale atmospheric flow. This notion is not commonly acknowledged but justified by the greater power that can be derived from the much stronger vertical radiative forcing of the surface–atmosphere system.

10.3 Limits to dry convection with heat storage changes

Conditions on land are strongly shaped by the diurnal variation in solar radiation. The associated changes in heat storage that are required to buffer this variation affect the thermodynamics of the land surface-atmosphere system and the partitioning of the surface energy balance. We next evaluate the effect of these diurnal heat storage changes on the maximum power limit associated with dry convection. To do so, we revisit the derivation of the Carnot limit of a heat engine. We essentially follow the approach as in Section 4.2, except that we also consider a heat storage change within the engine, as shown in Fig. 10.3a. This altered formulation is then used to derive the maximum power limit. Heat storage changes are inferred from the diurnal solar forcing, and these estimates are then applied to the diurnal cycle.

We derive the Carnot limit with heat storage changes using the first and second law of thermodynamics for the engine. Compared to eqn. 4.1, the first law takes a slightly altered form because the change in heat content of the engine is accounted for by the term dU_e/dt on the left hand side:

$$\frac{dU_e}{dt} = J_{\text{in}} - J_{\text{out}} - G \tag{10.11}$$

The entropy budget for the engine is also different from eqn. 4.2 because it includes an explicit change of entropy of the engine dS_e/dt,

$$\frac{dS_e}{dt} = \frac{1}{T_e}\frac{dU_e}{dt} = \frac{J_{\text{in}}}{T_h} + \sigma - \frac{J_{\text{out}}}{T_c} \tag{10.12}$$

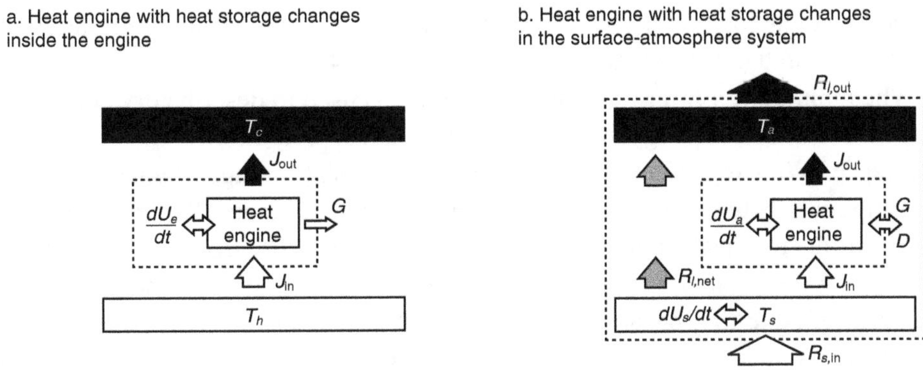

Figure 10.3 Schematic diagram of (a) a heat engine with heat storage changes, dU_e/dt, inside the engine. This figure is similar to Fig. 4.1, except that it specifically includes a term dU_e/dt that represents the change in heat storage. Diagram (b) shows a setup of the surface–atmosphere system which includes heat storage changes within the atmosphere dU_a/dt, and in the ground dU_s/dt, as well as net radiative exchange.

Here, the change in entropy of the engine is given by the change in heat content and the respective temperature T_e, of the engine, which is bound by the temperatures T_h and T_c. The maximum value of G_d is derived by assuming no dissipative loss within the engine; so we set $\sigma = 0$. This yields an altered Carnot limit of the form

$$G_d = J_{\text{in}} \cdot \frac{T_h - T_c}{T_h} - \frac{dU_e}{dt} \cdot \frac{T_e - T_c}{T_e} \tag{10.13}$$

If we assume that $T_e \approx T_h$, we obtain this limit in a simpler form of

$$G_d \approx \left(J_{\text{in}} - \frac{dU_e}{dt} \right) \cdot \frac{T_h - T_c}{T_h} \tag{10.14}$$

We refer to this limit as the Carnot limit of a cold heat engine. It is of the same form as the regular Carnot limit, except that the heat flux is reduced by the change in heat storage that occurs within the engine. This reduction originates from the lower entropy export of the engine when the heat content of the engine increases, thus lowering the Carnot limit. Note, however, that if $T_e \approx T_c$, then the extra term of the heat storage change vanishes, because the entropy increase in the engine due to storage changes is indistinguishable from entropy export. This then results in the Carnot limit as before. This case is, however, unlikely for the lower atmosphere, which maintains temperatures closer to the surface temperature rather than the radiative temperature of the atmosphere. The use of $T_e = T_h$ as in eqn. 10.14, however, results in an underestimation of the power as the temperature of the lower atmosphere is certainly lower than the surface temperature. When we further account for the dissipative heating within the engine, this only alters the temperature in the denominator to T_c, as in Section 4.4.

When this Carnot limit of a cold heat engine is used to describe dry convection, we use the surface energy balance, as in eqn. 10.4, to express the temperature difference. With regard to the representation of heat storage changes, we need to distinguish between changes in atmospheric heat storage, dU_a/dt, which take place within the heat engine so that $dU_e/dt = dU_a/dt$, from the heat storage changes in the ground dU_s/dt, which is represented by the ground heat flux at the surface. Note that because the ground heat flux is diffusive, it is generally of a small magnitude on land compared to the other heat fluxes. Both forms of heat storage changes contribute to the total, so that $dU_{\text{tot}}/dt = dU_a/dt + dU_s/dt$. Note also that atmospheric heat storage changes relate to direct changes associated with the sensible heat flux and indirect changes in latent form, which are associated with water vapor, and which are altered by evaporation and condensation. What we focus on in the following is the role of these heat storage changes in altering the maximum power limit for surface energy balance partitioning. While changes in the atmospheric heat storage affect the Carnot limit described by eqn. 10.14, changes in the ground heat storage act to reduce the effect of the solar radiative forcing to $R_{s,s} - dU_s/dt$, and would thus overall lower both the net terrestrial radiative flux as well as the heat fluxes.

With the effects of heat storage changes in the ground and in the atmosphere, we obtain an altered energy balance partitioning at the surface at a state of maximum power

$$R_{l,\text{net}} = \frac{R_{s,s} - dU_s/dt - dU_a/dt}{2} \tag{10.15}$$

$$H_{\text{opt}} = \frac{\gamma}{s+\gamma} \cdot \frac{R_{s,s} - dU_s/dt + dU_a/dt}{2} \tag{10.16}$$

$$\lambda E_{\text{opt}} = \frac{s}{s+\gamma} \cdot \frac{R_{s,s} - dU_s/dt + dU_a/dt}{2} \tag{10.17}$$

Note that the surface energy balance is altered in this case to $R_{s,s} - R_{l,\text{net}} - H_{\text{opt}} - \lambda E_{\text{opt}} - dU_s/dt = 0$, so that there is an accumulation (or release) of thermal energy below the surface, depending on the sign of dU_s/dt. The associated expressions for the optimum temperature difference and maximum power associated with dry convection, $G_{d,\max}$, are given by

$$T_s - T_a = \frac{R_{s,s} - dU_s/dt - dU_a/dt}{2k_r} \tag{10.18}$$

and

$$G_{d,\max} = \left(H_{\text{opt}} - \frac{dU_a}{dt}\right) \cdot \frac{R_{s,s} - dU_s/dt - dU_a/dt}{2k_r T_a} \tag{10.19}$$

The expressions for the maximum power limit are of similar form to the case without storage, yet show a clear effect of heat storage changes within the system. They show that heat storage changes affect the fluxes and that it matters *how* the change in heat storage is being accommodated by the system, as the terms dU_s/dt and dU_a/dt enter the expressions with different signs. Specifically, their effect is opposite to each other for the optimum sensible and latent heat fluxes as they enter H_{opt} and λE_{opt} with different signs, yet their effect on the maximum power is similar in that both act to reduce the power when the storage fluxes are positive.

In principle, these expressions associated with the altered maximum power limit require the same forcing as before, except that additionally the changes in heat storage need to be specified. When we approximate the diurnal cycle to be in a steady state, the heat storage terms vanish when integrated over the whole day since there is no net increase or decrease in storage in the steady state. This steady-state condition is represented in case B of Fig. 10.2. The diurnal variation of the solar radiative forcing sets the magnitude of the total change in heat storage, as indicated by ΔU_{tot} in Fig. 10.2. When we consider that solar radiation is constrained to about 12h, but the emission of radiation occurs over a 24h period, then half of the absorbed solar radiation needs to be stored to sustain nighttime emission, so that $\Delta U_{\mathrm{tot}} = R_{s,s}/2 \cdot \Delta t_{\mathrm{day}}$, with $\Delta t_{\mathrm{day}} = 24\mathrm{h}$. Note that this consideration assumes that there is sufficient heat storage within the system so that the emission of terrestrial radiation to space does not vary during the course of the day. This assumption is justified by the comparatively large heat capacity of the atmosphere, which was already introduced in Section 2.3 and which results in a comparatively large residence time, of about ninety days, for heat in the atmosphere.

This consideration allows us to assume that the radiative temperature of the atmosphere should remain approximately constant during day and night. As the emission of radiation from the system is then fixed, we can then infer that the total change in heat storage within the system is given by

$$\frac{dU_{\mathrm{tot}}}{dt} = \frac{dU_s}{dt} + \frac{dU_a}{dt} = R_{s,s} - R_{s,s,\mathrm{avg}} \tag{10.20}$$

where $R_{s,s,\mathrm{avg}}$ is the average of $R_{s,s}$ taken over the whole day. This expression can be graphically seen in case B of Fig. 10.2. With $R_{s,s}$ representing the term $R_{s,\mathrm{in}}$ and $R_{s,s,\mathrm{avg}}$ the term $R_{l,\mathrm{out}}$, this expression of heat storage change describes the increase in total heat storage during daytime at a rate $R_{s,s} - R_{s,s,\mathrm{avg}}$, and the decrease of heat storage during nighttime at a rate $R_{s,s,\mathrm{avg}}$. From this consideration, we also immediately see that the total heat storage change over the whole day vanishes because $R_{s,s,\mathrm{avg}}$ represents the diurnal average of the absorbed solar radiation, $R_{s,s}$.

With this steady-state constraint, the only aspect that is yet undetermined in our description is the relative distribution of total heat storage change between the

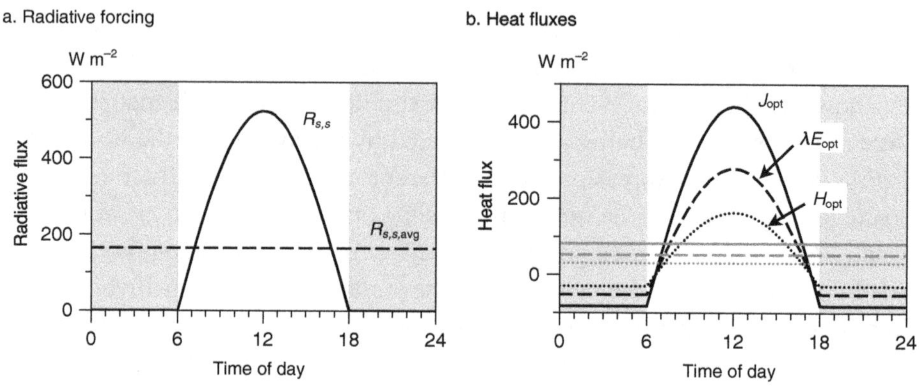

Figure 10.4 The effects of storage changes on the energy partitioning associated with the maximum power limit. Panel (a) shows the diurnal variation of the absorbed solar radiation, $R_{s,s}$, with areas shaded in light grey marking night. In (b), the optimum heat fluxes H_{opt} (dotted line), λE_{opt} (dashed line), and the sum, $J_{\text{opt}} = H_{\text{opt}} + \lambda E_{\text{opt}}$ (solid line) are shown for two cases of diurnal heat storage changes. The black lines show J_{opt}, H_{opt}, and λE_{opt} for the case A in which all diurnal heat storage changes take place in the atmosphere, whereas the grey lines show the case S in which the storage changes take place below the surface. The global mean value for average solar absorption at the surface and surface temperature is used in the figure.

surface and the atmosphere. To illustrate the effect of this distribution, we consider a mean diurnal cycle in which the absorbed solar radiation corresponds to the global mean value of $R_{s,s,\text{avg}} = 165$ W m^{-2} and the two extreme cases in which all heat storage change takes place at the surface (case S) or within the atmosphere (case A). The two cases are shown in Fig. 10.4. They considerably differ in the magnitude of the optimum heat fluxes as well as their diurnal variation. While case S shows a constant flux throughout the day with $H_{\text{opt}} + \lambda E_{\text{opt}} = R_{s,s,\text{avg}}/2$ as in Section 10.2, case A shows much greater heat fluxes that result in a greater mean of $3/4 \cdot R_{s,s,\text{avg}}$. These greater fluxes correspond very well in magnitude and temporal variations with observed values of the diurnal surface energy balance on land (Oke 1987). The enhancement by 50% originates from the earlier consideration by which on average, half the absorbed solar radiation is stored for nighttime emission, of which half contributes to the optimum heat fluxes at the maximum power limit.

Note that the nighttime optimum fluxes H_{opt} and λE_{opt} are negative. This outcome results from the sign change in dU_a/dt in the Carnot limit (eqn. 10.14) and the requirement to maintain a temperature difference $T_s - T_a$ during the night to generate power. It furthermore requires the cooling to occur solely from the atmosphere, which represents conditions that are probably quite rare on land.

These effects can also be seen directly in the expressions associated with the maximum power limit when the expression for dU_{tot}/dt given by eqn. 10.20 is used. For case S, $dU_s/dt = dU_{\text{tot}}/dt$ and $dU_a/dt = 0$, and eqns. 10.15–10.19 simply reproduce the expressions derived in Section 10.2. In this case, all diurnal heat storage changes take place below the surface, which is similar to an ocean surface, as described in the beginning of the chapter. Case A is given by $dU_s/dt = 0$ and $dU_a/dt = dU_{\text{tot}}/dt$. This case yields daytime optimum sensible and latent heat fluxes that are proportional to $R_{s,s} - R_{s,s,\text{avg}}/2$ and thus show a pronounced diurnal variation and considerably greater magnitudes, which is similar to a land surface. The stark difference in the convective heat fluxes over land and ocean seen in observations are thus reproduced by the maximum power limit given the different assumption of where the diurnal changes in heat storage take place.

These two cases are surely idealized and a couple of aspects are not included here. For instance, nighttime inversions often form near the ground, due to a partially transparent atmosphere. Also, the explicit dynamics of how temperatures change throughout the day, particularly of the lower atmosphere in terms of the growth of the atmospheric boundary layer (which is closely linked to the temperature T_e above), and the flux into the ground are not represented explicitly nor are other sources of variations such as the seasonal cycle. This simplified representation also results in a surface temperature that is uniform throughout the day, which can be seen by using dU_{tot}/dt (eqn. 10.20) in the expression of $T_s - T_a$ (eqn. 10.18), although observed surface temperatures typically show a diurnal variation. These deficiencies point out that a more detailed treatment would probably be needed to describe how T_e as well as the height of the boundary layer changes throughout the day.

The highly simplified description of the diurnal dynamics presented here nevertheless share key aspects with common descriptions of convective boundary layer dynamics. These descriptions explicitly formulate how the potential temperature and thus the heat content of the lower atmosphere changes during the day due to the sensible heat flux and entrainment processes (Tennekes 1973; McNaughton and Spriggs 1986; Stull 1989; Garratt 1992). These dynamics play a key role here as well, although they were not explicitly treated, but rather inferred from the thermodynamic constraint of maximum power of the sensible heat flux and the effect of heat storage changes as constrained by the diurnal variation of solar radiation. Hence, even though the formulation of boundary layer dynamics from the thermodynamic perspective taken here may look different from common approaches, it represents essentially the same dynamics yet generalizes the formulation by also including the constraints imposed by thermodynamics.

In summary, this example illustrates that the inclusion of heat storage changes in the maximum power limit substantially affects the outcome in terms of the magnitude, the mean, and temporal variations of the associated energy balance partitioning. We considered two contrasting cases of how the surface–atmosphere system accommodates the buffering of the strong diurnal variation in solar radiation. When the surface buffers these variations, similar to an ocean surface, the maximum power limit yields uniform conditions and fluxes. In contrast, when these variations are buffered by the lower atmosphere, which is more representative of the conditions on land, the maximum power limit yields greater fluxes with a strong diurnal cycle. This difference in buffering the diurnal variation of the solar radiative forcing can thus be seen as a major difference in the thermodynamic setting of the land surface–atmosphere system.

10.4 Hydrologic effects on land–atmosphere exchange

What we have not considered so far are factors that affect the ability of the land surface to evaporate water. There are two primary factors that are considered in the following. The first factor deals with the general availability of moisture at the surface. Some regions are generally limited by the availability of water, such as deserts or some tropical regions during the dry season. The lack of water reduces the latent heat flux and imposes a different constraint on surface energy balance partitioning than the passive ability of convecting air to transport moisture. This factor clearly affects not only surface energy balance partitioning, but also the power that can be derived from surface heating. The second factor deals with the steady-state condition reflected in $E = P$, while climate is typically shaped by a pronounced variability in the periods in which precipitation and evaporation occur. Precipitation events typically occur during comparatively short time periods with large fluxes, while evaporation is sustained over much longer periods with low fluxes. This has two consequences: first, evaporation can only be sustained during dry episodes when there is sufficient ability to store water, and second, the energy balance partitioning of the surface-atmosphere system is altered during dry episodes when no condensation occurs within the atmosphere. These factors are also affected by the presence of vegetation, which is dealt with later in this chapter. The goal of this section is to evaluate how these two factors alter the maximum power limit and the associated conditions at the land surface.

The central balance equation that describes the water availability for evaporation at the land surface is the soil water budget, which is an expression of the conservation of mass. This budget describes the temporal dynamics of water stored within the rooting zone of the soil, and in simple form can be described by

$$\frac{dW_s}{dt} = P - E - Q \qquad (10.21)$$

where W_s is the amount of soil water, P is the addition of water by precipitation, E is the removal of water by evaporation, and Q is the runoff and drainage of water that contributes to river discharge. Note that E includes a considerable contribution by the transpiration of plants, thus forming a flux of evapotranspiration. We refer to the total flux as evaporation for simplicity, as it nevertheless describes a phase transition and a mass flux of water vapor into the atmosphere. Plants also determine the depth of the rooting zone as this is the volume of the soil in which soil water can be taken up by plants, brought up into the canopy, and be evaporated into the atmosphere while the ability of bare soil to evaporate water is rather limited. The ability of the soil to store and release water relates to the capillary and adhesive forces in the soil matrix that are characterized by a chemical potential (Edlefsen and Anderson 1943). This potential is known as the matric potential in hydrology. Movement and redistribution of soil water is driven by differences in the soil water potential, which combines the matric potential with the gravitational potential of the water. A state of thermodynamic equilibrium for a given amount of soil water is reached when the Gibbs free energy associated with the soil water potential is minimized, which is referred to as the hydraulic equilibrium of the soil. Note that the soil matrix potential contributes to the difference in chemical potentials that drives the evaporation process, as described in Section 8.4, but because its contribution is small, it can easily be neglected. More details on the thermodynamics of soil hydrology may be found in Kleidon et al. (2009), Porada et al. (2011), and Zehe et al. (2013).

To bring these dynamics of soil water into a simple form, we consider the soil water budget in the climatological mean, which yields $P - E - Q = 0$. This budget sets a water limitation on the evaporation rate of the form $E = P - Q$. In the limiting case of no runoff, it yields a water limited rate of evaporation, which is simply described in the climatological mean by $E = P$ and which can be lower than the evaporation rate E_{opt} inferred from the limits of dry convection. Note that the limit $E = P$ implies sufficient ability to store water and thus may include vegetation effects through the depth of the rooting zone and the contribution by plant transpiration.

To evaluate the effects of this limitation on the evaporation rate, we refer to the actual evaporation rate by E_{lim}. We assume that it is proportional to the evaporation rate in the absence of water limitation, as expressed by eqn. 10.2, so that $E_{\text{lim}} = f_w E$ with f_w being a factor that describes the extent of water limitation with values between zero and one. In addition, we consider a case in which the evaporation rate is no longer matched by the precipitation within the system. This may represent a transient state in which water vapor accumulates in the atmosphere, or a case with a net import or export of water vapor from the system. The two energy balances of the land surface and the atmosphere are then expressed by

$$R_{s,s} - R_{l,\text{net}} - H - \lambda E_{\text{lim}} = 0 \tag{10.22}$$

and

$$R_{l,\text{net}} + H + \lambda P - R_{l,\text{out}} = 0 \tag{10.23}$$

Note the two modifications introduced here by using E_{lim} rather than E to account for the effect of water limitation on evaporation, and by describing the condensational heating within the atmosphere by P rather than E.

The energy partitioning associated with the maximum power limit can be derived as it was done earlier. When using the same formulations as in Section 10.2, the associated partitioning of the surface energy balance results in a very similar form as given by eqns. 10.6 and 10.7, except for a modification by the reduction factor f_w (Kleidon et al. 2014b)

$$H_{\text{opt}} = \frac{\gamma}{f_w s + \gamma} \cdot \frac{R_{s,s}}{2} = \frac{R_{s,s}}{2} - \lambda E_{\text{lim}} \tag{10.24}$$

and

$$\lambda E_{\text{opt}} = \frac{f_w s}{f_w s + \gamma} \cdot \frac{R_{s,s}}{2} = \lambda E_{\text{lim}} \tag{10.25}$$

The effect of water limitation on the maximum power limit is thus to merely shift the reduction in the latent heat flux to an enhancement in the sensible heat flux, while the trade-off with the temperature difference, $T_s - T_a$, is not directly affected. Because the sensible heat flux is enhanced in the presence of water limitation, the maximum power associated with dry convection is enhanced as well.

Another change that is introduced by these modifications is that the temperatures of the system are affected because the energy balance of the surface–atmosphere system is altered. Since $P = E_{\text{lim}}$ is no longer assumed, there is a net source or sink term in the atmospheric energy balance that is described by the rate $\lambda(P - E_{\text{lim}})$. This term either relates to the accumulation or release of latent heat associated with water vapor within the system or to a net transport of vapor in or out of the system. It alters the energy balance of the surface–atmosphere system to

$$R_{l,\text{out}} = R_{s,s} + R_{s,a} + \lambda(P - E_{\text{lim}}) \tag{10.26}$$

The associated change in radiative temperature can be derived by linearization of the change in T_a due to the term $\lambda(P - E_{\text{lim}})$. The modified radiative temperature is then described by a reference value, $T_{a,0}$, that is associated with $P = E_{\text{lim}}$ and a deviation ΔT_a due to the imbalances associated with $P - E_{\text{lim}}$

$$\Delta T_a = \frac{\Delta R_s + \lambda(P - E_{\text{lim}})}{k_r} \tag{10.27}$$

Here, the term ΔR_s refers to differences in solar absorption in both, $R_{s,s}$ and $R_{s,a}$, that are associated with the imbalance in $P - E_{\text{lim}}$ because episodes with $P > 0$ are associated with clouds which alter the absorption of solar radiation within the system. This difference in radiative temperature shifts the temperature difference to the surface, $T_s - T_a$, and results in a difference in surface temperature compared to the reference case in which $P = E_{\text{lim}}$

$$\Delta T_s = \Delta T_a + \frac{\Delta R_{s,s}}{2k_r} \tag{10.28}$$

where $\Delta R_{s,s}$ is the difference in absorption of solar radiation at the surface. This expression for ΔT_s describes how surface temperature is affected by evaporation during periods when it exceeds precipitation.

In periods with $P = 0$, evaporation results in a reduction of T_a and thus of T_s, an effect that is well known as evaporative cooling. Using the estimates of the global mean climatic state result in a magnitude of up to 20 K of this cooling effect, consistent with what has been estimated with extreme simulations with atmospheric climate models (Shukla and Mintz 1982; Kleidon et al. 2000). The interpretation here is, however, somewhat different from the common explanation of evaporative cooling because the cooling effect results from the temporal imbalance of precipitation and evaporation, modulated by variations in the radiative forcing due to clouds through the term $\Delta R_{s,s}$. This different interpretation results from a perspective in which evaporation forms only one component of the overall heat transport in the surface–atmosphere system. The overall heat transport is considered here under the water mass balance constraint and the limit of maximum power associated with dry convection. For the water balance in steady state, the evaporated water from the surface is balanced by condensation and precipitation in the atmosphere, with some net transport of water vapor in the climatic mean. The associated condensational heating results in the opposite effect on T_a (cf. eqn. 10.27). If no changes in the solar radiative forcing are considered between dry and wet episodes (i.e., $\Delta R_s = 0$ and $\Delta R_{s,s} = 0$), then these two effects cancel out. This compensating effect also explains why the expression for $T_s - T_a$ derived in Section 10.2 for the maximum power limit for dry convection in the climatic mean does not explicitly depend on evaporation (cf. eqn. 10.8). Yet, periods of condensation and precipitation are associated with clouds, which result in $\Delta R_s < 0$ and thus a cooling effect. It would thus seem that the overall cooling of the surface by evaporation results from the combination of these temporal variations in precipitation and evaporation with the associated radiative effects. This line of reasoning is consistent with recent idealized model simulations (Rochetin et al. 2014).

In summary, water availability on land has two pronounced effects on the maximum power limit of dry convection on land. First, a reduction of evaporation due to limited supply of water is compensated for by an increase in sensible heat flux, which directly enhances the power of dry convection within the system. In the simple formulation used here, this compensation does not affect the surface or radiative temperature in the climatic mean state. The second effect deals with temporal variations in precipitation and evaporation that result in short wet and prolonged dry episodes. These episodes are associated with a mismatch between precipitation and evaporation, which in turn is linked to storage changes and lateral transport of atmospheric moisture. This difference in $P - E$, in combination with differences in the radiative forcing, has a strong effect on temperatures and results in the well-known effect of evaporative cooling. Even though the formulations here are highly simplified, these insights nevertheless appear robust as they address first-order effects on the surface-atmosphere energy balances.

10.5 Convective mass exchange on land

Both effects introduced by the changes in diurnal heat storage and the availability of water at the surface, affect the sensible heat flux, dry convective power, and the mass exchange of dry air, water vapor, and other compounds between the surface and the atmosphere. The magnitude of this convective mass exchange was already inferred before from the ratio of the heat flux in relation to the driving temperature difference (cf. Sections 7.5 and 8.5). Here, the same approach is used to infer the convective mass exchange associated with dry convection, J_m, from eqn. 10.1:

$$J_m = \frac{H_{\text{opt}}}{c_p(T_s - T_a)} \tag{10.29}$$

The same expression for J_m can be derived from the latent heat flux. This expression is similar to the commonly used aerodynamic conductance for sensible heat (e.g., Garratt 1992), with the difference being that here the temperature difference that drives the heat engine, $T_s - T_a$, is used rather than the temperature difference between the surface and near-surface atmosphere. While $T_s - T_a$ is the more adequate choice for describing the overall mass exchange by the engine as it is this difference that drives the heat engine, one should note that the use of the radiative temperature of the atmosphere is an approximation that is likely to overestimate the magnitude of the temperature difference.

We can now evaluate how the two specific differences on land described in the previous sections affect the magnitude of convective mass exchange. The effects of diurnal heat storage changes in the atmosphere result in a pronounced diurnal variation of the optimum sensible heat flux, with the expression given by eqn. 10.16. In the simplified limiting case considered in Section 10.3, the temperature difference did not, however, change over the course of the day. Hence, the mass exchange varies predominantly with H_{opt}, and thus also shows a pronounced diurnal variation. Compared to diurnal mean conditions, the diurnal variation thus introduces a substantially elevated mass exchange during the course of the day. This

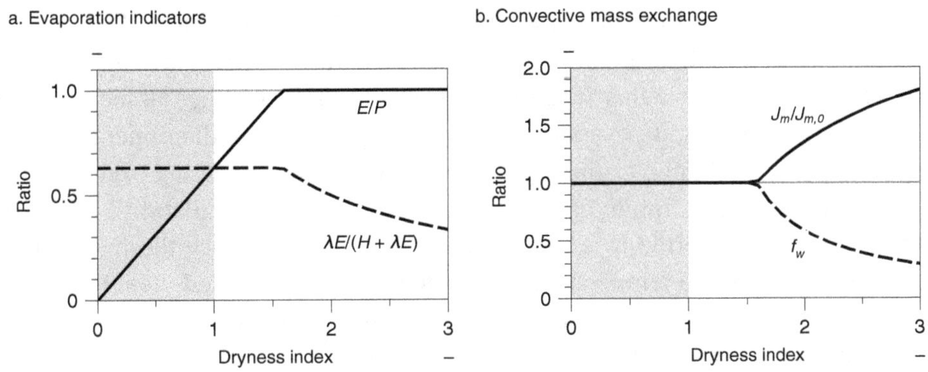

Figure 10.5 The effects of water limitation on evaporation shown as a function of Budyko's dryness index, defined by $R_{net}/\lambda P$. Panel (a) shows the ratio of evaporation E, to precipitation, P ("evaporative index"), and the ratio of the latent heat flux λE, to the total convective heat flux, $H + \lambda E$ ("evaporative fraction"). Panel (b) shows the wetness factor f_w as well as the enhanced convective mass exchange J_m, under dry conditions, shown in relation to the rate of mass exchange without water limitation $J_{m,0}$. The areas shaded grey denote the range in which $\lambda P > R_{net}$, which are commonly referred to as regions in which evaporation is energy limited. The graphs are plotted using global mean conditions.

enhancement is of equal proportion as for H in Fig. 10.4b when comparing the two cases of heat storage changes.

Water limitation also affects the magnitude of convective mass exchange. We saw in the previous section that in the maximum power limit, such a reduction in the ability of the land surface to evaporate is compensated by the sensible heat flux (cf. eqn. 10.24), yet this compensation does not affect the temperature difference $T_s - T_a$. Thus, this enhancement of the sensible heat flux results in an enhanced convective mass exchange as well. This enhancement is shown in Fig. 10.5 in the context of the well-known framework of Budyko (1974). This framework provides a hydrologic characterization of regions according to the relative magnitudes of net radiation and precipitation. The characterization is based on the radiative index of dryness, or dryness index, defined as the ratio of net radiation R_{net}, to the energy that would be required to evaporate all precipitation λP. When this ratio is less than one, precipitation provides more water than can potentially be evaporated by net radiation, while a ratio greater than one represents the case where net radiation is greater than the energy required to evaporate the water provided by precipitation. The effect of a limited water supply on evaporation is shown in Fig. 10.5a as a function of the dryness index. The figure shows evaporation in relation to precipitation (solid line), a ratio referred to as the evaporative index, for the limiting case where evaporation is described by $E_{lim} = E_{opt}$ up to the point when precipitation limits evaporation, which is then described by $E_{\text{lim}} = P$. The first case, $E_{\text{lim}} = E_{\text{opt}}$, is commonly referred to as an energy-limited regime for evaporation, although the interpretation here is that it is the convective mass exchange associated with dry convection that limits evaporation in this regime rather than energy. As the convective mass exchange is directly driven by the radiative heating of the surface, this case is, in the end, limited by energy rather than water availability. The other case, $E_{\text{lim}} = P$, is referred to as the water-limited regime. Since the figure shows E_{lim} for a given surface temperature, the factor $s/(s + \gamma)$ is fixed, which results in the linear increase in E/P with the dryness index for the first regime up to $E_{\text{lim}} = P$.

Note that observations typically fall below this idealized description of E, as some runoff still occurs in the transition to water-limited conditions. Also shown in Fig. 10.5a is the evaporative fraction, defined as the ratio of the latent heat flux to the total convective heat flux, $\lambda E/(H + \lambda E)$, which is used in micrometeorology to characterize surface energy balance partitioning. Since a fixed radiative forcing is used in the figure, the evaporative fraction is described by a constant value of $s/(s + \gamma)$ when water availability does not limit evaporation. In the water-limited case, this fraction decreases with a greater dryness index. This decrease of the evaporative fraction results from the maximum power limit for dry convection in the presence of water limitation. The limit fixes the fraction of convective heat fluxes regardless of water availability, so when the latent heat flux is reduced, this reduction is compensated for by a greater value of the sensible heat flux.

The increase in the sensible heat flux then results in an increase in convective mass exchange. Figure 10.5b shows this increase in relative terms in which J_m in the presence of water limitation (with H given by eqn. 10.24 with a value of $f_w < 1$) is related to the value of J_m in the absence of water limitation (i.e., for a value of $f_w = 1$, with the mass flux referred to as $J_{m,0}$). This increase in the convective mass exchange results from the lower ability of the sensible heat flux to transport heat for a given rate of mass exchange when compared to the latent heat flux. To accommodate a greater sensible heat flux hence requires a greater intensity of convective mass exchange. The implication of this hydrologic effect is that the ability of the land surface to evaporate water affects the magnitude of convective mass exchange.

The magnitude of mass exchange between the surface and the atmosphere is thus quite different over land compared to oceans. The maximum power limit of dry convection with the two effects of atmospheric heat storage and water availability that are specific to land act to enhance this rate and result in a pronounced diurnal variation. These different conditions for convective mass exchange then affect the rate by which the land surface exchanges geochemical compounds with the atmosphere.

10.6 Limits to terrestrial photosynthesis

The convective mass exchange just described applies to the exchange of CO_2 between the vegetative cover and the atmosphere as well. This exchange plays an important role in sustaining photosynthetic activity during the day. As already described in the introduction of this chapter, the challenge that photosynthesis faces is that it takes up CO_2 at a higher rate during daylight than by which CO_2 is released by respiration over the whole course of the day. Photosynthesis by terrestrial vegetation thus depletes the stock of atmospheric CO_2 during the day. This difference in the exchange fluxes of CO_2 is referred to as net ecosystem exchange and leaves a noticeable imprint in terms of diurnal variations in the concentration of CO_2 in the convective boundary layer of the lower atmosphere. This indirect manifestation of photosynthetic activity on land has been used to infer the gross carbon uptake by photosynthesis (Desjardins et al. 1982; Wofsy et al. 1988; Denmead et al. 1996). The goal of the following, highly simplified description is to illustrate that this convective mass exchange of CO_2 as inferred from the maximum power limit may in fact constrain the photosynthetic activity of terrestrial vegetation.

The drawdown of atmospheric CO_2 by photosynthesis is described by the mass balance of CO_2 of the atmospheric boundary layer. Noting the mass of CO_2 by $m_{CO_2,a}$ and the fluxes of photosynthesis and respiration by $J_{m,CO_2,ph}$ and $J_{m,CO_2,res}$ respectively, this budget is described by

$$\frac{dm_{CO_2,a}}{dt} = \rho_a \Delta z_{BL} \cdot \frac{n_{CO_2}}{n_a} \cdot \frac{d[CO_2]_a}{dt} = J_{m,CO_2,res} - J_{m,CO_2,ph} \qquad (10.30)$$

where Δz_{BL} is the height of the atmospheric boundary layer, n_{CO_2} and n_a are the molar mass of CO_2 and air, and $[CO_2]_a$ is the CO_2 concentration on a volume basis (e.g., expressed in ppmv). Note that CO_2 exchange with the free atmosphere is neglected in this formulation. This balance is illustrated in Fig. 10.6 for an average length of daylight of 12h with a photosynthetic rate that peaks at about 30 μmol CO_2 m^{-2} s^{-1} at solar noon (solid line in Fig. 10.6a). When averaged over the year, this photosynthetic rate corresponds to an annual carbon uptake by photosynthesis of about 3600 gC m^{-2} yr^{-1}, which is a typical value for the high rates of tropical rainforests (Beer et al. 2010). The release of CO_2 by respiration, which results from the metabolic activity of plants and consumers, is taken to be the average value of photosynthesis (dashed line in Fig. 10.6a), so that over the course of the day, the two fluxes balance and there is no net change in $m_{CO_2,a}$. This formulation of a steady state is equivalent to how it was formulated before in terms of heat storage changes and hydrologic fluxes.

Due to the temporal imbalance between $J_{m,CO_2,ph}$ and $J_{m,CO_2,res}$, there is a net exchange of CO_2 with the atmosphere at a rate $J_{m,CO_2,net} = J_{m,CO_2,res} - J_{m,CO_2,ph}$ (the dotted line in Fig. 10.6a). This net exchange results in variations in the carbon

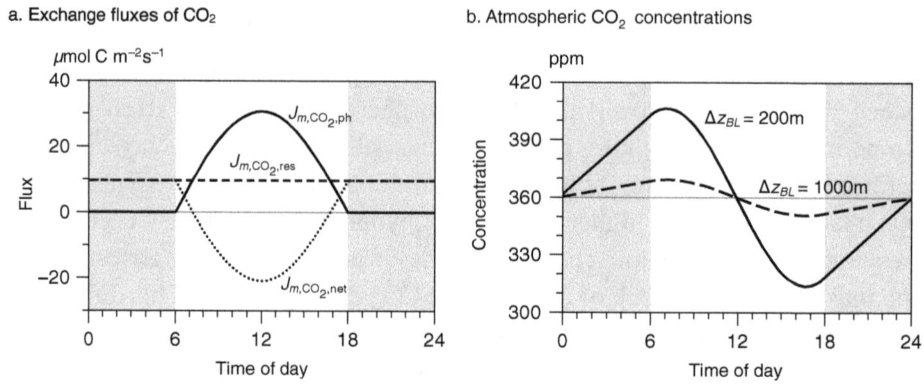

Figure 10.6 Diurnal variation of (a) photosynthetic CO_2 uptake, $J_{m,CO_2,ph}$, and release by respiration, $J_{m,CO_2,res}$, using values typical of a tropical rainforest. The net uptake of CO_2 by vegetation during the day, $J_{m,CO_2,net} = J_{m,CO_2,res} - J_{m,CO_2,ph}$, requires a depletion of CO_2 in the near-surface atmosphere. This variation is shown in (b) for two different, fixed heights of the convective boundary layer of the lower atmosphere, Δz_{BL}. Note that observed variations in CO_2 show a different variation because Δz_{BL} grows during the day.

stock in the atmospheric boundary layer, as shown in Fig. 10.6b, with the stock being reduced during the day when $J_{m,CO_2,net} < 0$. This is shown for two different heights of the boundary layer of 200 and 1000 m. As a boundary layer of 1000 m thickness contains more mass and thus a greater carbon stock, the variations in the CO_2 concentration due to surface exchange is smaller than for the case of the shallower layer of 200 m depth. Diurnal variations in the CO_2 concentration in the order of tens of ppm are typical in boundary layers (Wofsy et al. 1988; Denmead et al. 1996). Note, however, that because the boundary layer height is not static but typically increases during the day, the observed diurnal course of CO_2 concentrations look quite different from the idealized cases shown in Fig. 10.6b.

The net exchange of CO_2 requires the exchange of mass between vegetation and the atmospheric boundary layer. This mass exchange is described by the convective mass flux, which at the limit of maximum power and in the simple representation used here is given by eqn. 10.29. We can thus use this rate to infer the exchange rate of CO_2 between the vegetated surface and the atmosphere. To do so, we combine the flux J_m with the difference in CO_2 concentration between the boundary layer and the concentration at which CO_2 diffuses to the photosynthesizing tissues within leaves, $[CO_2]_v$,

$$J_{m,CO_2,ph} = J_m \cdot \frac{n_{CO_2}}{n_a} \cdot ([CO_2]_a - [CO_2]_v) + J_{m,CO_2,res} \tag{10.31}$$

This is a highly simplified expression, as the gas exchange of CO_2 between the canopy of the vegetative cover and the atmosphere is more complex. It involves the diffusive exchange of air through the stomata, small openings in the leaves that act as valves for the gas exchange between the interior air space of the leaves and the atmosphere. Stomata regulate this gas exchange of CO_2 jointly with the loss of water vapor from the interior air space of leaves. The loss of water results in plant transpiration and constitutes a large proportion of the evaporative flux. Stomata play a major role in shaping the gas exchange of plants with their environment to support their photosynthetic activity (Wong et al. 1979; Farquhar and Sharkey 1982; Ehleringer and Cerling 1995). As they link carbon uptake to water loss, photosynthetic activity and evaporation are strongly linked at the ecosystem level (Rosenzweig 1968; Law et al. 2002).

To get back to a simple description of the CO_2 gas exchange in relation to maximum power, we use the observation that the CO_2 concentration in leaves is about 70% of the concentration of ambient air (Ehleringer and Cerling 1995), so that $[CO_2]_v \approx 0.7 \cdot [CO_2]_a$. With this approximation and taking the average over the daytime period when $R_{s,s} > 0$, yields an expression for the mean photosynthetic uptake of CO_2 that is based on the convective mass exchange:

$$J_{m,CO_2,ph,avg} = 2 \cdot J_m \cdot \frac{n_{CO2}}{n_a} \cdot ([CO_2]_a - [CO_2]_v) = 0.6 \cdot J_m \cdot \frac{n_{CO_2}}{n_a} \cdot [CO_2]_a \tag{10.32}$$

where the factor of 2 originates from the steady-state condition in which $J_{m,CO_2,res} = J_{m,CO_2,ph,avg}$ when integrated over the whole day with $J_{m,CO_2,ph,avg} > 0$ being constrained to the daylight period, which is about half the whole day. Note that if there was no diurnal variation in solar radiation, there would be no need for net changes in CO_2 storage within the atmosphere. This exchange with the atmosphere, as formulated by eqn. 10.32, represents a formulation for photosynthetic activity that is based on a limiting flux of reactants, and represents the mass flux limitation for biotic activity that was already alluded to in Section 9.6.

When we combine this mass exchange-based rate of photosynthesis with expressions for J_m and H_{opt} from earlier (eqns. 10.29 and 10.16 for the case in which water does not limit evaporation), this yields an expression of the form

$$J_{m,CO_2,ph,avg} = \frac{3}{2} \cdot \frac{\gamma}{s+\gamma} \cdot \frac{n_{CO_2}}{n_a} \cdot \frac{0.3 \cdot [CO_2]_a}{c_p(T_s - T_a)} \cdot R_{s,s} = \epsilon_{lue} \cdot R_{s,s} \tag{10.33}$$

where in the latter expression, the factors were lumped into ϵ_{lue}. The motivation for doing so is to express the rate of photosynthesis in terms of the absorbed solar radiation. This is commonly done and referred to as a light-limited rate of photosynthesis (e.g., Monsi and Saeki 1953; Monteith 1972; Field et al. 1998), with the conversion efficiency, ϵ_{lue}, being referred to as the light use efficiency. The value of this parameter is well established by observations and is within the range of 17–81 · 10^{-3} μmol CO_2 μmol PAR^{-1} (Monteith 1977; Ruimy et al. 1995), with radiation expressed in terms of the photosynthetic photon flux density with a conversion factor of 1 W m^{-2} $\hat{=}$ 2.23 μmol PAR^{-1} (where PAR stands for photosynthetically active radiation). Using global mean values of $R_{s,s}$ = 165 W m^{-2}, $T_s = 288$ K, and $[CO_2]_a = 360$ ppm yields a value of $\epsilon_{\text{lue}} \approx 29 \cdot 10^{-3}$ μmol CO_2 μmol PAR^{-1} in eqn. 10.33, which is within the observed range. The expression given by eqn. 10.33 can also be compared to the rate of evaporation, which yields the water use efficiency, $\epsilon_{\text{wue}} = J_{m,CO_2,\text{ph,avg}}/E$, a parameter that is often used to characterize photosynthesis. Using the same expressions as earlier, this yields

$$\epsilon_{\text{wue}} = \frac{\gamma}{s} \cdot \frac{n_{CO_2}}{n_a} \cdot \frac{0.3\lambda \cdot [CO_2]_a}{c_p(T_s - T_a)} \tag{10.34}$$

Observations from ecosystem measurements estimate this efficiency to be about 3.2 g CO_2 (kg $H_2O)^{-1}$ with some variation between vegetation types (Law et al. 2002). Using mean conditions, the expression given by eqn. 10.34 yields a value of about $\epsilon_{\text{wue}} = 7.5$ g CO_2 (kg $H_2O)^{-1}$, which is somewhat too high compared to the observed range. This bias may be attributed to the global mean surface temperature being lower than the mean temperature at which most of the photosynthetic activity takes place. The photosynthetic uptake of CO_2 that is inferred here from the mass exchange between the surface and the atmosphere associated with the maximum power limit nevertheless yields estimates that match observed magnitudes rather well, suggesting that the mass exchange of CO_2 indeed acts as a constraint to terrestrial photosynthesis.

The transport-limited rate of CO_2 uptake can also be related back to the efficiency by which photosynthesis converts radiative energy into chemical energy. This efficiency is obtained by using the light use efficiency in combination with the chemical energy of about 479 kJ stored in a mole of glucose. It yields an efficiency of about 2%, which falls within the range of observed maximum efficiencies found in well-fertilized crops of 1–3% (Monteith 1972, 1978; Zhu et al. 2008; Amthor 2010). The efficiency is much lower than the 12% maximum efficiency derived from the thermodynamics of radiative energy conversion, as described in Section 9.6. The maximum efficiency of 2% inferred here relates to the thermodynamically constrained rate of mass exchange of CO_2 with the atmosphere, which in turn is driven by the solar radiative heating. The transport limitation imposed by the maximum power limit of dry convection thus may provide an alternative explanation for the low efficiency found in terrestrial photosynthesis.

The uptake of CO_2 by vegetation is certainly more complex than described here, as it includes effects of, for instance, water limitation, stomatal conductance and non-uniform transport characteristics within the boundary layer. Nevertheless, this order of magnitude estimate suggests that it is quite reasonable to interpret the dominant control on terrestrial photosynthesis in terms of the limit imposed by the physical mass exchange of CO_2 between the atmosphere and the vegetative cover. This interpretation is also consistent with the well-established CO_2 fertilization effect by which higher concentrations of CO_2 in the air can enhance photosynthetic rates, although the magnitude of this effect also depends on environmental conditions (Amthor 1995). When we accept this interpretation of terrestrial photosynthesis being primarily limited by CO_2 exchange, then the higher rates of photosynthetic activity on land compared to ocean can be explained by the more favorable conditions for convective mass exchange. The effects of vegetation on the physical functioning of the land–atmosphere system can also be understood in terms of an altered mass exchange, as described in the next section.

10.7 Thermodynamic consequences of vegetated land

We next look at the effects of vegetation on land surface functioning from the perspective of thermodynamic limits. The presence of vegetation strongly affects the physical properties and functioning of the land surface. Compared to bare ground, a vegetated surface is generally darker with a lower albedo (cf. Table 6.2), it absorbs solar radiation within a canopy with a low heat capacity, it has a greater ability to access soil water by its root system, it can control the rate of transpiration by the stomata, and through the heterogeneity in its canopy has a higher aerodynamic roughness. These physical differences affect the surface energy and water balance and its partitioning, the diurnal growth of the convective boundary layer, and ultimately the formation of convective clouds.

These effects then feed back to the conditions in which photosynthesis takes place, resulting in feedbacks between vegetation activity and surface–atmosphere exchanges. This notion of strong coupling is relatively well established (Betts et al. 1996) and shapes the observed conditions of the surface–atmosphere system on land. Such interactions are impressively demonstrated in observations in Western Australia, where human land use introduced a clear difference in vegetative cover that result in an almost as stark difference in cloud cover (Lyons 2002; Nair et al. 2011), as shown in Fig. 10.7. Clouds were found to preferentially form over native vegetation during the dry season, which has been attributed to the more intense radiative heating of the surface and a greater growth of the convective boundary layer, which then resulted in water vapor reaching its saturation. This example illustrates the close linkage between the surface energy balance, the convective boundary layer and clouds, and how this linkage is affected by vegetation.

The effects of vegetation on the surface–atmosphere system are commonly characterized by feedbacks that focus on surface temperature, water vapor deficit, the convective boundary layer (Betts et al. 1996; Raupach 1998), and climatic variables

Figure 10.7 An aerial photograph of Western Australia during the dry season, with native vegetation on the left and agriculture on the right. Note how convective clouds form over the native vegetation, but not over agriculture. Reproduced from Lyons (2002) with kind permission from Springer Science and Business Media.

at larger scales (Brovkin 2002; Bonan 2008). In this section, an alternative view of vegetation–atmosphere interactions is described that focuses specifically on the thermodynamic limit of dry convection and link these to the intensity of convective mass exchange, as shown in Fig. 10.8. Most effects in the figure are well established, yet the description here deviates in that the feedback on vegetation is focused on the role of convective mass exchange to maintain photosynthetic activity.

The vertical chain of linkages shown in the center of the figure summarizes the previous sections of this chapter: solar radiation is absorbed at the surface, heats it, creates buoyancy, and generates the kinetic energy associated with dry convection. The maximum power limit sets the constraint on the kinetic energy generated by dry convection and sets the magnitude of convective mass exchange. This mass exchange supplies CO_2 to the vegetated surface where it maintains CO_2 uptake by photosynthesis.

Terrestrial vegetation affects this chain in several ways. Three of the major effects are shown in Fig. 10.8 and result in feedbacks on photosynthetic activity. The first effect of the vegetative cover (labeled "loop A") is that it typically is associated with a lower surface albedo, which enhances the absorption of solar radiation at the surface. As this absorption is the main driver for surface–atmosphere exchange, this enhancement results in greater convective mass exchange, thus providing a positive feedback on photosynthetic activity.

The second effect deals with atmospheric heat storage ("loop B"). As the canopy of vegetation absorbs solar radiation rather than the Earth's surface, it absorbs it at a surface with a very low heat capacity. This causes the absorbing surface to heat up fast, thus generating buoyancy fast as well. The enhanced buoyancy generation then enhances the heat storage changes in the atmosphere while reducing the changes in the ground. It is consistent with a generally reduced ground heat flux found

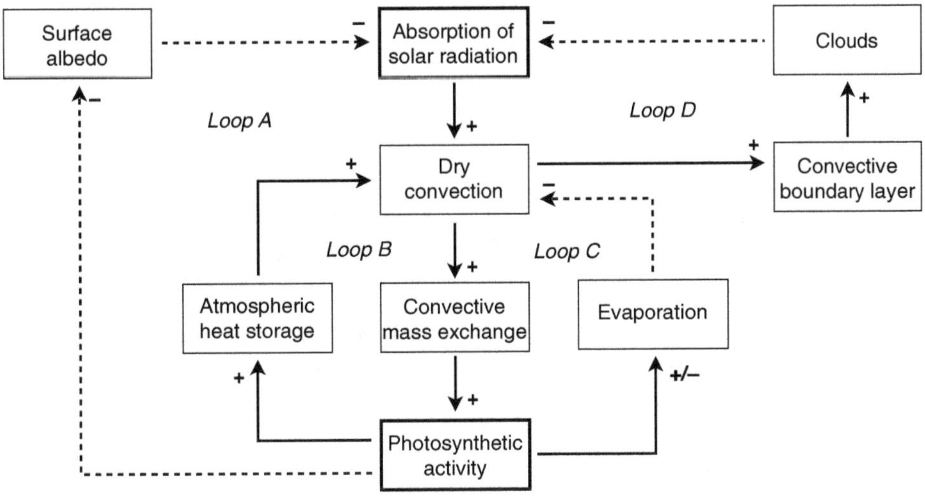

Figure 10.8 An overview of vegetation–atmosphere interactions in terms of the main linkages between absorption of solar radiation, convective mass exchange, and photosynthetic activity.

in forests, and with enhanced turbulent heat fluxes above forests. Rotenberg and Yakir (2010; 2011) demonstrated this effect with observations in a forest plantation in a desert in Israel and referred to this enhancement as the "convector effect" of forest canopies. As we have seen in Section 10.3, a greater buffering of the diurnal radiative imbalance by the atmosphere alters the maximum power limit to enhance convective exchange consistent with the convector effect. This enhancement of convective exchange would then also result in a positive feedback on photosynthetic activity.

The third effect deals with the inevitable water loss associated with CO_2 gas exchange during photosynthesis, by which vegetation shapes the evaporative flux from the surface into the atmosphere ("loop C"). This effect deals with two major components: on the one hand, vegetation enhances water availability by accessing soil water by the root system. The access to soil water storage allows the vegetative cover to maintain gas exchange for a longer time in dry episodes and this would enhance the evaporation rate into the atmosphere. This effect has been observed, for instance, in the Amazon rainforest (Nepstad et al. 1994), simulated by model simulations (Milly 1994; Kleidon and Heimann 1998), and found to be of substantial magnitude. On the other hand, water loss is regulated by stomata, in terms of their response to environmental conditions (Farquhar and Sharkey 1982) and in terms of their numbers on the leaves, which have been observed to vary considerably within Earth's history (Woodward 1987; Beerling and Royer 2002). Through stomatal regulation, the vegetative cover can reduce the transpiration rate

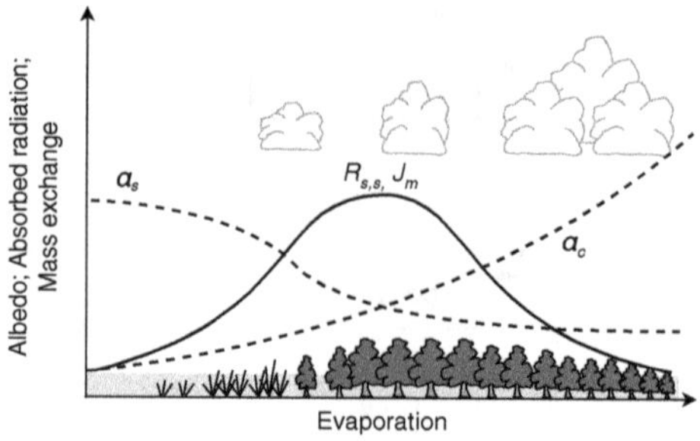

Figure 10.9 Illustration of a thermodynamic limit of maximized mass exchange J_m, due to a maximum in solar absorption at the surface $R_{s,s}$, in the surface–atmosphere system resulting from the contrasting effects of surface (α_s) and cloud (α_c) albedo associated with different values of evaporation. Modified after Kleidon and Fraedrich (2005). Reproduced with kind permission from Springer Science and Business Media.

to below the potential evaporation rate. The implication of stomatal functioning from the thermodynamic view is then that the enhanced sensible heat flux (as described in Section 10.4) acts to enhance dry convection and convective mass exchange, which could form a positive feedback on photosynthetic activity.

When these effects are taken together, one can imagine that the combination does not lead to some arbitrary outcome, but results in another form of thermodynamic limit in which the absorption of solar radiation at the surface is maximized (Kleidon and Fraedrich 2005). This form of optimality is illustrated in Fig. 10.9 and involves the effects of convective clouds within the boundary layer. The modulation of the surface energy balance by the effects described here affects the development of the convective boundary layer ("loop D" in Fig. 10.8). Greater absorption of solar radiation at the surface results in a greater variation in diurnal heat storage change (cf. eqn. 10.20). As this change takes place predominantly in the atmosphere, this would tend to result in a greater diurnal boundary layer development. A greater development is modulated by the partitioning between the sensible and latent heat flux, which affect the growth and the moistening of the boundary layer. Both of these effects can result in cloud formation, as is illustrated by the case study in Western Australia in Fig. 10.7, or more generally, by climate model simulations (e.g., Kleidon 2004b; de Arellano et al. 2014; Rochetin et al. 2014). The rate of evaporation from the surface, as modulated by and related to vegetation activity, has two major effects, which are shown schematically in Fig. 10.9. The first effect is that vegetation provides a darker absorbing surface, so that greater vegetation activity results in a lower surface albedo and thus in greater absorption of solar radiation. This effect is shown by the dashed line marked α_s in Fig. 10.9. The second effect is more indirect as it involves a greater rate of evaporation, which would result in greater cloud cover, as indicated by the dashed line marked with α_c. This effect likely has quite some uncertainty attached to it and can depend on other factors. Yet, in general, the combination of a greater diurnal boundary layer development with enhanced surface absorption and a greater evaporation rate should result in greater condensation within the atmosphere, thus enhancing cloud formation. The combination of the two effects should then result in an extremum state for the absorption of solar radiation at the surface at some intermediate rate of surface evaporation. This extremum state could potentially be reached by vegetation, as it modulates the absorption of solar radiation and the rate of evaporation. It would feed back to photosynthetic activity by a maximized rate of convective mass exchange, J_m, which to first order varies with $R_{s,s}$, thus minimizing the limitation imposed by mass exchange on photosynthesis. Sensitivity simulations with a climate model showed that these effects on cloud cover appear plausible (Kleidon 2004b) and can reproduce the sensitivity of stomatal density to CO_2 (Kleidon 2007). The thermodynamic limit to photosynthetic activity on land thus appears to be shaped not just by the ecophysiology of the vegetative cover or surface conditions, but ultimately by the emergent state of the whole vegetation–land surface–atmosphere system that results from the strong interactions between vegetation, the surface, and the atmosphere.

To summarize, terrestrial vegetation affects the conditions in the land surface–atmosphere system by several means and thereby the factors that shape the maximum power limit for surface–atmosphere exchange. These effects would then alter the conditions for photosynthesis, thus resulting in interactions between vegetation activity and the physical exchange fluxes of the land surface–atmosphere system. At the scale of the whole system, these interactions can result in a thermodynamic limit for photosynthetic activity in which the absorption of solar radiation at the surface is maximized, as shown in Fig. 10.9 and which maximizes the gas exchange between the surface and the atmosphere. While such a state appears somewhat speculative, it nevertheless reflects that the thermodynamic limit of biotic activity is likely to be set by the state of the *whole* system, a state that is strongly shaped by interactions with the physical environment. This is similar to the trade-off that results in the maximum power limit of convection, with the direct application of thermodynamics resulting in the Carnot limit, and the indirect effects on the temperature difference as a consequence of energy conservation resulting in the maximum power limit. For vegetation–land surface–atmosphere interactions, the effect on cloud cover may not be immediately obvious, but in the end, this effect is required from steady-state considerations of the water balance. Vegetation–atmosphere interactions viewed from a thermodynamic perspective can thus be seen as the means to shape the conditions of the whole land surface–atmosphere system to a state that is most conducive for photosynthetic activity by maximizing the rate of mass exchange.

10.8 Land in the Earth system

In this chapter, we focused on the thermodynamic setting on land, the aspects that distinguish land from ocean from a perspective of thermodynamic limits, and how these differences allow for the high levels of photosynthetic activity on land. We started by evaluating the thermodynamic limit to surface–atmosphere exchange driven by surface heating and dry convection, evaluated the effects of water availability, and linked the mass exchange associated with this limit to the carbon exchange associated with photosynthetic activity by terrestrial vegetation. Using global mean conditions, we estimated that the maximum power from dry convection is in the order of about 4 W m^{-2}, which is notably higher than the maximum power limit associated with large-scale motion of about 2 W m^{-2}, yet less than the $\approx$7 W m^{-2} that can maximally be generated by moist convection. We should recognize, however, that these forms of generating motion do not act independently but rather in concert, with their contributions varying in time. The global estimates provide a sense of the relative importance for these processes in generating surface–atmosphere exchange, with dry convective exchange likely playing a dominant role on land.

We then dealt with aspects specific to land that alter the maximum power limit for dry convection. The first aspect dealt with the diurnal cycle and how the surface–atmosphere system accommodates the imbalance in solar radiative forcing between day and night. We showed how the accommodation by heat storage changes below the surface or within the atmosphere affected the maximum power limit and resulted in a substantially different outcome, with the former resulting in constant heat fluxes throughout day and night while the latter showed a pronounced diurnal cycle in the heat fluxes. This outcome is broadly consistent with observed fluxes of the surface energy balance of ocean and land.

The second aspect dealt with the limiting effects of water availability on land and with the temporal imbalance between evaporation and condensation. When the latent heat flux is reduced due to the conditions at the land surface, the maximum power limit yields a compensating increase in the sensible heat flux. From energy balance considerations it was shown that when evaporation occurs without simultaneous condensation in the atmosphere, it leads to the well-known effect of evaporative cooling at the surface. The interpretation provided here is somewhat different from the common interpretation as it includes the treatment of the whole surface–atmosphere system.

We then linked the convective mass exchange of CO_2 with photosynthetic activity, arguing that it is this flux that limits photosynthesis rather than the availability of light. As mass exchange is driven by the absorption of solar radiation, this would nevertheless result in an apparent correlation between photosynthesis and absorbed solar radiation. This interpretation yields consistent estimates of photosynthetic rates in relation to absorbed radiation and evaporated water and it can explain the low, observed radiation use efficiency of photosynthesis in natural conditions.

When the major effects of vegetation on the land surface are then evaluated from this thermodynamic perspective, it results in a consistent picture in which vegetation effects act to enhance convective exchange. When linked to the limitation imposed by CO_2 exchange, this combination provides a positive feedback on photosynthetic activity, a feedback that is ultimately constrained by the formation of clouds in the boundary layer thus forming a thermodynamic limit to photosynthetic activity that is shaped by the whole surface–atmosphere system. The thermodynamic description of the vegetation–land surface–atmosphere system thus provides a novel and consistent view that is quite different from how these effects are commonly being interpreted, yet its outcome is broadly consistent with observations.

Taken together, these aspects would suggest that the high levels of photosynthetic activity on land are supported by a thermodynamic setting that allows for the greatest mass exchange between the vegetated surface and the atmosphere. The magnitude of this mass exchange is not only set by physical factors, but likely substantially enhanced by the biotic effects that vegetation has on the physical functioning of the land surface. It would thus seem that the aspects discussed here allow for critical, positive feedbacks on land by which biotic activity alters the physical conditions in such a way that they enhance mass exchange and the conditions for biotic activity (as shown in Fig. 10.8), thus resulting in the high levels of biotic activity on land. Note that while this outcome is to some extent speculative, particularly regarding the link between mass exchange and photosynthesis, the actual linkages between processes are relatively well established, so that this dominant feedback would nevertheless seem plausible. We will get back to this kind of feedback and biotic activity at the planetary scale when we deal with planetary evolution at the end of the book.

Although we evaluated land surface–atmosphere exchange using thermodynamics, the description in this chapter deviates quite substantially from previous work in which thermodynamics, and particularly entropy production, was used to evaluate biotic effects on the system (Ulanowicz and Hannon 1987; Schneider and Kay 1994b; Kleidon 2004a; Kleidon and Fraedrich 2005; Holdaway et al. 2010; Brunsell et al. 2011). These studies evaluated the surface energy balance from the perspective that life enhances entropy production to the largest possible extent, thereby maximizing the rate of entropy production. Using observations, this hypothesis is typically confirmed in these studies because vegetated surfaces have a lower albedo and thus absorb more solar radiation than non-vegetated surfaces, and typically at a colder temperature. While we did not focus specifically on entropy production in this chapter, what is described here is not in contradiction with these studies.

When we hypothesized about the potential maximization of absorbed solar radiation at the surface, as shown in Fig. 10.9, this would also result in a maximization of entropy production as a greater rate of surface heating necessarily results in greater entropy exchange and entropy production within the system. Yet the lines of reasoning that led to Fig. 10.9 involved a quite specific thermodynamic limit of convective motion, linked this to the limitation imposed by mass exchange for photosynthesis and cloud cover effects, and thus describes the processes involved in shaping the thermodynamic limit with more specific details. What this points out is that a simple evaluation of surface conditions with respect to its entropy production is likely to be too unspecific to evaluate the dominant factors that shape thermodynamic limits in the surface–atmosphere system.

The chapter highlighted the importance of looking at the land surface–atmosphere system as a coupled system in which the thermodynamic limit of maximum power acts to constrain the rate of mass exchange, consistent with the notion developed in the previous chapters. It provided an example at a more general level that highlights how thermodynamic limits and their manifestations become more complex when temporal variations are being considered, so that the highly simplified way in which these were treated here – or in the previous chapters – merely form a starting point. There are thus several aspects, such as the effects of moist convection or the coupling with the large-scale atmospheric circulation, that can refine the formulations presented here. Yet, this simplified formulation nevertheless provides a simple, thermodynamically consistent description of the functioning of the land surface and its interaction with vegetation that appears to provide adequate first-order estimates of the associated magnitudes.

At the end of this chapter, we should note that nowadays a substantial fraction of the land surface is no longer in a natural state, but strongly affected by human land use. The thermodynamic formulation of human activity is the scope of the next chapter, in which we will also get back to the effects of land use changes and their thermodynamic characterization.

11

Human activity

11.1 Human activity as a thermodynamic process

In the last chapter before the summary, we extend the thermodynamic description to human activity to formulate the last link represented in Fig. 1.5. This application of thermodynamics to human activity is rather important, as human activity increasingly shapes the functioning of the Earth system. As we will see in this chapter, thermodynamics can provide a generalized view of human activity, its limits as well as its impacts within the Earth system.

The impacts of human activity are noticeable at the planetary scale and are reflected in, for instance, the increase in atmospheric greenhouse gases such as CO_2 resulting in global climate change, stratospheric ozone depletion, the expansion of croplands and pastures, with about 40% of the land surface being placed under human use (Foley et al. 2005), and the alterations of the global cycles of nitrogen and phosphorus. The impact of human activity has reached such an extent that it has been suggested that the present day can be referred to as the new geologic era of the "Anthropocene" (Crutzen 2002). As the size of the human population as well as its energy consumption is likely to grow in the future, the effects of human activity on the Earth system are likely to increase as well. A description of the thermodynamics of the Earth system would be incomplete without a characterization of human activity and its effect on the Earth system.

Yet, the human role is often described as if it were separate from the functioning of the Earth system. There are certain aspects that are taken from the Earth to sustain human activity, such as food production or fossil fuels, as shown in Fig. 11.1a. Human activity then has impacts on the Earth system, for instance, in terms of land cover changes associated with food production, or global climate change due to the emission of CO_2 resulting from the combustion of fossil fuels. The purpose of this chapter is to describe human activity as a dissipative process that is embedded within the functioning of the Earth system, just as it was done for the other processes in the previous chapters. While this description may seem to be far fetched,

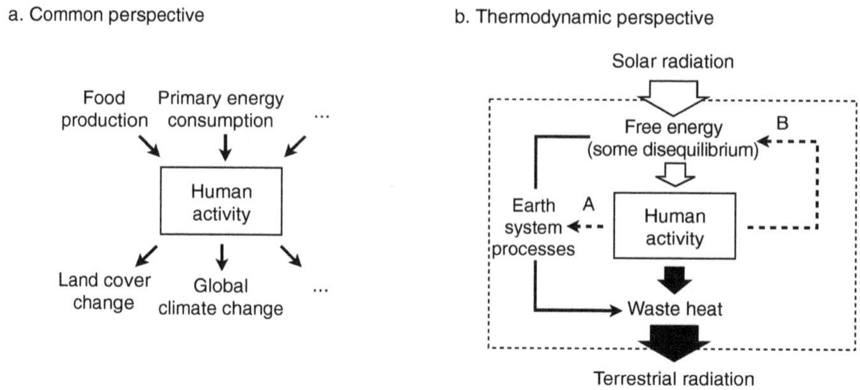

Figure 11.1 Illustration of human activity in (a) the common perspective and (b) the thermodynamic perspective described here. The dashed lines labeled A and B in (b) refer to the impacts of human activity.

it is important to note that at the very core of human activity is the consumption of energy, or better, the dissipation of free energy, and this free energy is drawn from the Earth system. This is illustrated by Fig. 11.1b, which shows human activity as a dissipative process similar to how it was introduced for the living cell and the whole Earth in Fig. 1.2 or described with respect to different life forms in Fig. 9.5.

As any other dissipative process, human activity is sustained by the use of free energy. This energy is taken up from the Earth system in the form of carbohydrates as food, and dissipated by the human metabolism as waste products and heat. In addition, humans use energy for heating, generation of mechanical work with machinery, production of fertilizers, generation of electricity, and other aspects of human activity (Smil 1999). This use of energy is generally referred to as primary energy consumption, and is not directly linked to the maintenance of the human metabolic activity. It also involves the appropriation of free energy from the Earth system, at present, mostly in the form of fossil fuels, which is then converted and dissipated by human use. Human activity in this chapter refers to these two forms of dissipation, the direct dissipation associated with maintaining metabolic activity and the indirect dissipation by consumption of primary energy which is referred to here as "externalized activity." This definition of human activity forms the basis for its thermodynamic characterization, for evaluating its limits as well as the impacts on the Earth system in the following.

At first sight, it would seem that the rates of energy consumption associated with human activity are very small when compared to other Earth system processes and particularly to the incident flux of solar radiation. This is, however, not the case, when we use generation and dissipation rates that drive the dynamics as a basis for comparision, rather than the magnitude of radiative or heat fluxes. To illustrate this point, consider the chemical energy dissipated by the metabolic activity of the world human population. The basal metabolic rate of a human body is about 100 W, which corresponds to a diet of 2000 kcal per day. This rate can, however, reach up to 1200 W when the human body is physically active (Campbell and Norman 1998), with sustainable levels of about 700 W (Hammond and Diamond 1997). When multiplied by a population of about $7 \cdot 10^9$ people, this yields a global rate of $0.7–4.9 \cdot 10^{12}$ W of chemical energy dissipation due to human metabolic activity. To place this number in proportion, recall that the power involved in sustaining the oceanic circulation is in the range of $2–7 \cdot 10^{12}$ W (cf. Section 7.4). The primary energy consumption is even greater at about $533 \cdot 10^{18}$ J in 2014 or $17 \cdot 10^{12}$ W (BP 2014). Note the important difference of this rate of free energy consumption to a heating rate. In terms of heating, the primary energy consumption is minute with a global mean rate of less than 0.1 W m^{-2}. This rate is much less than the mean absorption of solar radiation of 240 W m^{-2} and the increase in radiative forcing of about 1.5 W m^{-2} due to the enhanced concentration of greenhouse gases caused by the consumption of fossil fuels (Forster et al. 2007). This comparison illustrates that when human activity is viewed in terms of energy conversion rates rather than heating rates, present levels of human activity represent a thermodynamic process of considerable magnitude, substantiating the view promoted by Crutzen (2002) that humans shape a planetary force, resulting in the new era of the Anthropocene.

With this relevance of human activity as a planetary, thermodynamic process being established, we can then relate the sources of energy for human activity to the processes of the previous chapters and their respective limits. The energy used to sustain the human metabolism relates to food production, which uses the products of photosynthesis, mostly from land. Food production is then ultimately limited by the factors that limit photosynthetic activity, reduced by the losses associated with the further conversion into food. Primary energy is also taken from the Earth system. The forms of energy being used can broadly be characterized into forms that deplete a stock of free energy, such as fossil fuels (for which the depletion rate is much greater than the generation rate, Dukes (2003)), fission or fusion, or forms that are sustainably generated within the Earth system, which represent the different forms of renewable energy. As the previous chapters describe the thermodynamic limits by which different forms of energy are generated within the Earth system, these then set the limits by which the primary energy needs of human activity can be met with renewable energy. While these are only two factors that are associated with human activity, these should nevertheless give us a perspective on the magnitude of the two major limitations of human activity at the planetary scale.

The impacts of human activity on the system can also be described in this thermodynamic view. The consumption of photosynthetic products as well as the combustion of fossil fuels both deplete the chemical disequilibrium between carbohydrates and atmospheric oxygen that was established by photosynthetic life, at present in the case of food production, or at some time in the distant past in the case of fossil fuels. Human activity thus depletes the chemical disequilibrium of the Earth system, just as the living cell in general and chemotrophic or heterotrophic organisms in particular (cf. Fig. 9.5). The consequences of this depletion are then reflected in the change of the land surface that is referred to as land cover change, or in the increase in CO_2 concentration in the atmosphere in the case of fossil fuel combustion. The effects can be broadly categorized into two types, shown by the arrows labeled A and B in Fig. 11.1. Type A describes a reallocation effect: human energy consumption depletes a stock of free energy at the expense of a natural process which would deplete this form of energy. An example for this effect is the agricultural expansion into naturally forested regions, as the products of photosynthesis are consumed by human activity rather than by natural ecosystem processes. The result of this change is that an increase in human activity is associated with a reduction in the activity of the natural biota. Type B describes the consequences of human energy consumption on the rate by which free energy is generated within the Earth system. This type of effect is more complex and can act in both directions, to reduce or enhance this generation rate, and we defer examples of these effects to later in this chapter.

This categorization of human effects can then be used to look at the expected evolution of human activity in the future. To do so, we concentrate on looking at human activity as a thermodynamic process taking place within the Earth system. Human activity is evaluated in the same way as the Earth system processes in the previous chapters in terms of thermodynamic limits and in terms of the feedbacks that shape the limit and the evolutionary dynamics towards it. In doing so, we can then infer how a sustainable future of human activity may look like, and what consequences this entails for the Earth system as a whole. We leave out many aspects that this entails, particularly regarding socioeconomic dynamics. These have been looked at using thermodynamics and energy for a long time, noting that socioeconomic activity and energy consumption are intimately linked (Ostwald 1909; Lotka 1921; Ayres and Kneese 1969; Georgescu-Roegen 1971; Odum 1973; Ayres and Nair 1984; Cleveland et al. 1984; Daly 1992; Cleveland and Ruth 1997; Hall et al. 2001; Garrett 2009; Liao et al. 2012). It is interesting to note that conventional economic theories generally view energy consumption as a consequence of economic activity rather than as a relevant constraint. Alternatively, in the field of ecological economics, energy consumption is rather seen as the actual driver of the economic process (Ayres and Nair 1984; Daly 1992; Hall et al. 2001; Ayres et al. 2003). The previous chapters that dealt with natural processes support this latter view, so that here we view human activity as a thermodynamic process and socioeconomic dynamics as the outcome and manifestation of human energy consumption. By placing human activity in the context of the Earth system and its limits, we can then relate these estimates to well-known concepts such as the human carrying capacity (e.g., Cohen 1995) or the limits to growth (Meadows et al. 1972) that deal with the limits to human activity at the planetary scale.

This approach to formulating human activity as a dissipative process within the Earth system sets the scope for this chapter. We first look at human activity in terms of its metabolic energy needs and how this relates to the environment shaped by the Earth system. We then describe the two aspects of human dissipative activity in terms of the consumption of food and primary energy. These two forms of dissipative activity are referred to as food acquisition and primary energy acquisition here to emphasize that these forms of energy are taken from the Earth system and converted to meet the human energy need. We look at these acquisitions in greater detail in terms of the factors that shape their limits by drawing links to the previous chapters. We then evaluate the consequences of human activity on the Earth system, formulate the feedbacks in greater detail, and place these in the evolutionary context of a sustainable future. The chapter closes with a brief summary in which human activity is placed back in the context of the whole Earth system.

11.2 Energy and human activity

We start the description of human activity with the general requirement for chemical energy to maintain metabolic activity of any form of life. This requirement follows well-established, allometric relationships that relate metabolic energy consumption to body mass. Such relationships extend to the mass of the different organs of an organism, and in particular, to the ratio of brain to body mass (Jerison 1955, 1985). These relationships help us to understand which aspects of the human species are different to the rest of the biota from a viewpoint of energy and shed some light on how the conditions of the Earth system favored the evolution of such a species, and intelligent life in general. From the perspective of these energetic constraints, we can then understand how the additional consumption of primary energy overcomes some of these constraints and results in greater levels of human activity.

The best-known and well-established allometric relationship is Kleiber's law (Kleiber 1932), which describes the metabolic energy consumption of organisms as a function of their body mass. It states that the metabolic activity at rest, or the basal metabolic rate, of a living organism increases with its body mass m. This rate J_{bmr}, is inferred from measurements by the consumption of oxygen by an organism and is expressed by

$$J_{bmr} = \alpha \cdot m^{\beta} \quad (11.1)$$

where α is a species-specific parameter and the exponent has a relatively uniform value of $\beta = 3/4$ across species. The value of β can be understood and predicted as the outcome of optimized fractal branching networks within organisms, such as cardiovascular systems in mammals, that minimize their frictional dissipation (West et al. 1997), as was already described in Section 5.7. By distributing mass and energy, these networks sustain the circulatory requirements of multicellular forms of life. This interpretation of Kleiber's law tells us that a greater organism can distribute mass and energy more effectively within the body, so that it requires proportionally less energy to maintain a greater body mass. Note that Kleiber's law does not explain body size variations with temperature that are known as Bergmann's rule in biogeography (e.g., Ashton et al. 2000).

When organisms are physically active by moving, foraging, climbing, or by performing other forms of physical work, it has been found that their maximum sustained metabolic rates are constrained to be less than seven times the basal metabolic rate (Hammond and Diamond 1997). While this observation is empirical, it nevertheless suggests a physical constraint on the maximum rate by which organisms are metabolically active and can perform physical work in their environment. This constraint would also apply to human activity.

The value of the parameter α varies greatly across species. Table 11.1 shows a number of values for different species that are taken from Zotin (1984). It shows that more primitive organisms, such as sponges and snails, have much lower energetic

Table 11.1 *Values of the species-specific parameter α that characterizes the intensity of the basal metabolic rate. Based on estimates provided by Zotin (1984).*

Type or class	Value (10^{-3} W $g^{-3/4}$)
Single-celled animals (Protozoa)	0.098
Sponges (Porifera)	0.140
Jellyfish, corals (Coelenterata)	0.192
Worms (Oligochaeta)	0.512
Crabs, shrimps (Crustacea)	0.727
Snails, mussels (Mollusca)	0.895
Amphibia	1.45
Reptiles	2.46
Insects	3.15
Mammals (lowest)	6.7
Mammals (insectovores)	20.8
Mammals (rodents)	24.2
Birds	26.1
Mammals (carnivores)	29.2
Mammals (primates)	36.5
Songbirds	43.2

requirements than mammals and birds of the same body mass. The high rates for mammals and birds are linked to the maintenance of homeostatic temperatures within the body (Allman 1999). As temperature affects the rates of chemical reactions, the regulation of body temperature allows an organism to much better control the metabolic reactions within the body and decouples them from environmental influences. Yet, it comes at the price of much greater requirement for food for the same body mass. Note that primates, to which humans belong, show the second highest value of the parameter α in the list. This implies that humans represent a species that has one of the highest requirement for metabolic energy in relation to their body mass.

The mass of internal organs generally follow allometric relationships such as Kleiber's law as well (Gould 1966). When we specifically focus on the mass of the brain in relation to body mass, then deviations from such allometric relationships are of interest as these represent conditions in which the brain mass is greater than expected from the allometric relation (Count 1947; Jerison 1955; Martin 1981). It has, for instance, been found that primates have a brain that is about 2.3 times greater than the mass of non-primates and that fruit eaters have greater brains than leaf eaters for the same body mass (Allman 1999). Such deviations are characterized by so-called encephalization (Jerison 1977, 1985), which characterizes such deviations in the brain-to-body mass ratio. Humans have the highest encephalization among mammals (Martin 1981). Brain tissue is metabolically expensive tissue– it has a mass-specific metabolic rate that is about nine times the average metabolic rate in humans (Aiello and Wheeler 1995). Because the energetic costs of greater brains do not seem to substantially affect the overall allometric scaling of the basal metabolic rate in organisms, it has been suggested that greater brains are offset by a smaller digestive system, but that this requires a greater quality in the diet (Aiello and Wheeler 1995). The greater quality in the diet is in turn linked to a larger share of animal-derived food. As this food is ultimately derived from photosynthesis and its conversions by food webs, the ability to maintain a high quality diet is thus linked to the overall activity of the biosphere.

Another energetic aspect related to large brains is that brain tissue is particularly sensitive to heat. In this context, it has been suggested that the evolution of bipedalism during human evolution in the past three million years facilitated the evolution towards larger brains in humans as the upright posture of the body allows for a greater ability to lose heat (Falk 1990; Wheeler 1991a,b). The role of heat loss in an organism can be illustrated by placing the metabolic heat generation in the order of 100 W or more in humans in the context of the fluxes that heat and cool the Earth's surface. As we have seen in the previous chapters, the radiative and convective fluxes that heat and cool the surface are also in the order of 100 W m^{-2}, so that the additional heating by metabolic activity imposes a substantial contribution to the energy balance of an organism at the surface. The ability of a human body to lose heat to the environment thus depends on the conditions of the Earth system, and has been suggested to potentially form an additional constraint to the evolution of large brains in humans (Schwartzman and Middendorf 2000; Kleidon 2009a).

To summarize these metabolic aspects associated with human activity, the metabolic activity of organisms including the human species appears to be strongly constrained, as reflected by the widespread presence of allometric scaling laws and by the finding that the sustained metabolic activity of organisms is less than seven times the basal metabolic rate. These act as constraints that limit the ability of organisms in general and the human species in particular to perform physical work within the Earth system. Physical work is in turn associated with the ability to acquire food resources from the environment. The development of a disproportionally large brain in the human species likely comes at a large energetic expense requiring a higher quality diet, but it increases the capacity to process information (Jerison 1985) which affects foraging behavior (Foley and Lee 1991). In other words, a larger brain helps to acquire food of higher quality, possibly with less effort, but at the same time, it comes at an energetic expense. The evolution of agriculture and the extension to a human domination of Earth's ecosystems of the present day (Vitousek et al. 1997) can be seen as an extension of this altered foraging behavior that was enabled by large brains. It results in an enhanced acquisition of food from the products of photosynthesis, thus allowing for a greater level of human activity.

At the end of this section, we want to place human activity and its dynamics into a more abstract, yet somewhat speculative picture of how it interacts with the thermodynamics of the Earth system at large. This picture is shown in Fig. 11.2. Human metabolic activity is placed at the center of the diagram, with a certain fraction of it representing the activity of the brain. This activity is sustained by the acquisition of food of a certain quantity and quality, which is derived from photosynthesis taking place within the Earth system. The acquisition of food requires foraging, which represents the physical work involved in walking, hunting, and gathering. This physical work, shown by path A in Fig. 11.2, is sustained by metabolic activity and is constrained to be less than seven times the basal metabolic rate. This path is common to all living organisms. The success of food acquisition in terms of quantity and quality can be enhanced by the information processing of the human brain. This is shown by path B in the diagram. Because of the unusually large brain-to-body mass ratio in humans, it would seem that this path represents a competitive advantage that human species have compared to other species. The brain activity also allows humans to develop and use tools, and, more importantly, utilize other forms of energy from the Earth system. This utilization of other forms of energy, reflected in the consumption of primary energy, allows for additional work to be

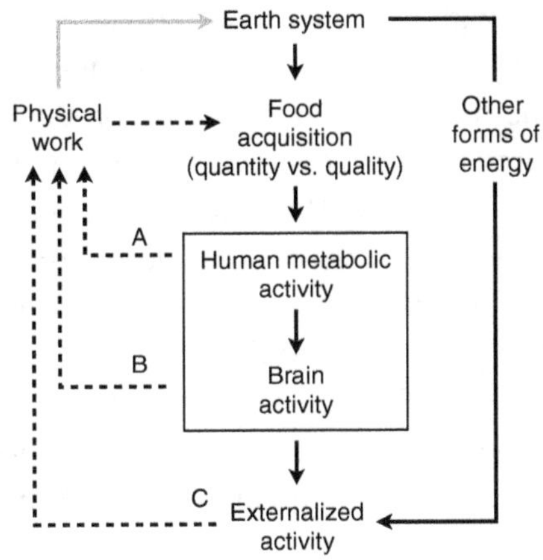

Figure 11.2 Schematic diagram to show how human metabolic activity is maintained by acquisition of food from the Earth system. Physical work is required for this acquisition, directly by the work involved in foraging (path A), and indirectly by brain activity (path B) which can make the acquisition more effective. Primary energy consumption by humans, which uses other forms of energy generated by the Earth system, is shown here as "externalized activity." This form of activity further enhances the quantity of food acquisition (path C), e.g., by mechanized agriculture, but requires other forms of energy to be sustained.

performed in the acquisition of food (path C), for instance in agriculture. It would seem that this latter aspect allows humans to overcome the allometric constraints of the human body to perform physical work and process information by developing technology. It also seems that the ability of the human species to utilize other forms of energy and thereby "externalize" metabolic activity makes the human species unique within the Earth's biosphere. This externalized activity has also been referred to as "industrial metabolism" (Ayres 1989), "socio-economic metabolism" (Fischer-Kowalski and Haberl 1998), and the metabolism of the technosphere (Haff 2013), but here we refer to it in more general terms as "externalized activity."

While not all of human activity is represented by this diagram, it nevertheless accounts for the factors that sustains it at a basic physical level and that directly links to the thermodynamic description. The following two sections deal with the acquisition of food from the biosphere to provide the energy to sustain metabolic activity and the acquisition of primary energy to maintain externalized activity. This is then followed by the effects that human activity has on the Earth system, as shown by the grey arrow in Fig. 11.2.

11.3 Limits to food acquisition

The acquisition of food, the direct form of energy acquisition used to sustain human metabolic activity, is related to the appropriation of a fraction of the chemical free energy generated by photosynthesis. This requirement for human activity links its limits on acquiring food to the limits that constrain photosynthesis and how much of it can be used to provide food. The limit on food production links two concepts established in the literature: the first concept is the human appropriation of net primary productivity (Vitousek et al. 1986; Rojstaczer et al. 2001; Haberl et al. 2007, 2014), which quantifies the extent to which terrestrial productivity is used by humans and provides food resources, among other uses. The second concept is the human carrying capacity (e.g., Cohen 1995; Franck et al. 2011), which has long been used to derive estimates for the maximum population size of humans that can be maintained sustainably on Earth. In this section, these two concepts are related to the thermodynamic description of human activity of the previous section and used to estimate the limit on food production within the Earth system as well as the factors that shape and alter this limit.

The upper limit by which the terrestrial biosphere can provide chemical energy as food to human use is set by the limits to photosynthesis. In Section 9.6, this limit was estimated to be about 12% of the incoming solar radiation if the conversion of solar radiation is the factor that limits photosynthesis, although observed efficiencies are less than 3% (Monteith 1978; Amthor 2010). The chemical free energy generated out of this less than 3% of the absorbed solar radiation by the photosynthetic conversion process is used by plants to grow and maintain the plant tissues that allow them to absorb light and obtain water, carbon dioxide, and nutrients from the environment. About half of the generated chemical energy is consumed by plant respiration, that is, by the metabolic activity that sustains the maintenance and growth of plants. The remaining half results in the production of biomass, which is referred to as the net primary productivity. The net primary productivity sets the starting point for the chemical free energy that can be appropriated to human use.

As with most of the previous energy conversions, there is only a certain fraction of the net primary productivity that can be appropriated to human use in a sustainable way. Such a maximum in appropriation can be demonstrated by the following, simple considerations, which are based on the numerical simulation results of Kleidon (2006). We consider a region with a certain biomass of vegetative cover, m_{bio}, and which has a certain maximum level of net primary productivity, $J_{m,\text{CO}_2,\text{npp,max}}$, that is shaped by environmental limitations. The actual level of net primary productivity, $J_{m,\text{CO}_2,\text{npp}}$, is expressed as an increasing function of biomass, m_{bio}, as a greater biomass allows extraction of more resources from the environment, and which saturates at the maximum. This relationship is written as

$$J_{m,\text{CO}_2,\text{npp}} = \frac{m_{\text{bio}}}{m_{\text{crit}} + m_{\text{bio}}} \cdot J_{m,\text{CO}_2,\text{npp,max}} \tag{11.2}$$

where the parameter m_{crit} describes how fast the productivity saturates with biomass. The biomass is in turn described by a mass balance, which balances the biomass growth by $J_{m,CO_2,npp}$ with litter production and subsequent heterotrophic respiration, or, more generally, biomass turnover, and the rate by which humans extract biomass from the vegetative cover, $J_{m,CO_2,human}$:

$$\frac{dm_{bio}}{dt} = J_{m,CO_2,npp} - \frac{m_{bio}}{\tau_{bio}} - J_{m,CO_2,human} \tag{11.3}$$

where the rate of biomass turnover is expressed in a simple relationship in terms of the biomass and a residence time, τ_{bio}. In steady state, $dm_{bio}/dt = 0$, in which case, the human appropriation can be expressed by

$$J_{m,CO_2,human} = J_{m,CO_2,npp} - m_{bio}/\tau_{bio} \tag{11.4}$$

Note that by considering a steady state of the mass balance, we do not deal explicitly with the specific events of harvesting, which can put most of the standing biomass to human use. The steady state described here refers to the mean state over several harvesting cycles which require periods of regrowth. Human appropriation as described here refers to the combined effect of the intensity and frequency of harvesting, but we do not distinguish between these aspects in this representation.

In the case of no human appropriation, the natural rate of biomass turnover balances the net primary productivity and yields a steady-state biomass of $m_{bio,nat} = J_{m,CO_2,npp} \cdot \tau_{bio}$. In this case, all biomass formed by plants is consumed by natural, heterotrophic activity. In the presence of human appropriation, the biomass is reduced to a value in steady state of $m_{bio} = (J_{m,CO_2,npp} - J_{m,CO_2,human}) \cdot \tau_{bio}$. Because this reduced biomass would then result in a lower net primary productivity, the rate of human appropriation has a maximum value, as shown in Fig. 11.3. The particular value of the maximum possible rate of appropriation as well as the value for the biomass at which this rate is maximized depends on the two parameters of the model, the residence time of the biomass τ_{bio}, and the critical biomass m_{crit}. A greater residence time lowers the turnover of biomass, thus shifting the maximum to a higher value at a higher biomass. A greater value of the critical biomass results in lower productivity for a given biomass and a lower maximum for the human appropriation. While the particular value of how much of the net primary productivity can be put to human use depends on these parameters that encapsulate complex relationships between biomass, productivity, and turnover, this simple model nevertheless illustrates that not all, but only a fraction of the net primary productivity can be appropriated by humans.

While empirical relationships have been used to illustrate this maximum in human appropriation, the underlying mechanism is nevertheless similar to the purely physical processes described in the previous chapters. This example is

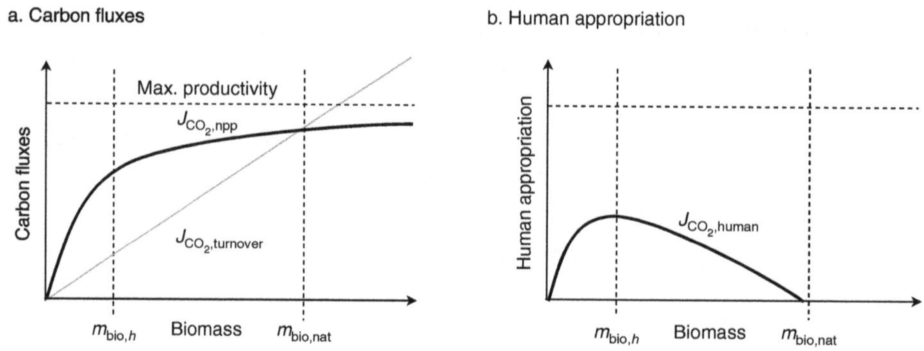

Figure 11.3 Variation of (a) net primary productivity, $J_{m,CO_2,npp}$, and biomass turnover, $J_{m,CO_2,turnover}$ and (b) human appropriation, $J_{m,CO_2,human}$, with biomass in steady state. The dashed horizontal line marks the maximum productivity permitted by environmental constraints. The dashed vertical line at a high biomass value marks the natural steady state in which $J_{m,CO_2,npp} = J_{m,CO_2,turnover}$, while the vertical line at low biomass marks the state at which the human appropriation, $J_{m,CO_2,human} = J_{m,CO_2,npp} - J_{m,CO_2,turnover}$, is at a maximum value. The biomass in the natural state and at the maximum of human appropriation are marked by $m_{bio,nat}$ and $m_{bio,h}$.

particularly similar to the example described in Section 4.6 on the maximum power that can be derived from kinetic energy. In that example, frictional dissipation and conversion of kinetic energy into other forms act as two competing processes to consume kinetic energy. The maximum conversion rate resulted from the trade-off between a greater momentum flux associated with the conversion and a lower wind speed. Here, biomass represents the stock of free energy, and biomass turnover as well as human appropriation both deplete biomass. This results in a similar trade-off: a greater conversion flux associated with human appropriation results in a lower biomass, and thereby in a reduced generation of biomass.

This simple estimate of a maximum in appropriation does not account for the extent to which the appropriated biomass can be used to yield edible, high quality food in the form of sugars, fats, and proteins. This restriction is described by the harvest index, which expresses the mass of the harvested product in relation to the total biomass of the plants. The upper limit of this index is estimated to be up to 40% for current crops that were grown in optimum conditions (Wheeler et al. 2008).

When these losses from solar radiation to chemical energy generation by photosynthesis to food production are combined, this results in a very low efficiency of converting sunlight to food. Given that the efficiency of photosynthesis is observed to be less than 3%, biomass production is about 50%, and the maximum in human appropriation is probably also in the order of 50% of biomass production (Kleidon (2006) found this to be about 33%) of which only 40% can be used as food, this results in an overall maximum conversion of 0.3% of solar radiation into food. Applied to the mean absorption of solar radiation of 165 W m^{-2}, this translates into a minimum requirement of at least 200 m^2 of productivity to sustain the basal metabolic rate of a single human. When this number is scaled up to a population of about $7 \cdot 10^9$ humans, this translates into a minimum requirement of $1.4 \cdot 10^{12}$ m^2 of productive land to sustain the metabolic energy needs of the human population. This estimate translates into a minimum of at least 2% of the productive land surface. While this estimate is rather crude, it nevertheless illustrates that the metabolic energy requirements of humans translate into a sizable requirement for productive land.

Observed human land use was estimated to be about 33% in 2000 (Ramankutty et al. 2008) and thus notably higher than the 2% minimum estimate. About 2/3 of the 33% of human land use relates to pastures and raising livestock such as cattle and sheep (cf. Table 11.2). As this form of food production involves an additional trophic level, that is, the appropriated productivity sustains livestock of which only a fraction results in food production, this is associated with a lower efficiency in converting solar radiation to food. The productivity of this land fraction corresponds to the human appropriation of net primary productivity from terrestrial ecosystems (Vitousek et al. 1986; Rojstaczer et al. 2001; Imhoff et al. 2004; Haberl et al. 2007, 2014). The associated flux was estimated in these studies to range from 10 to 55% of the terrestrial productivity, although it is important to note that different methodologies were used with some estimates also including the indirect effects of human activity on the terrestrial productivity of natural ecosystems and that these estimates also include the appropriation of wood, which is used as building material

Table 11.2 *Extent of human land use in different biomes in terms of cropland and pastures. Based on estimates given by Ramankutty et al. (2008).*

Biome	Area 10^6 km^2	Cropland 10^6 km^2	Pasture 10^6 km^2	Human use %
Tropical evergreen forests	16.77	1.81	1.48	19.6
Tropical deciduous forests	5.86	1.58	1.43	51.4
Temperate forests	9.58	2.45	1.42	40.4
Boreal forests	8.20	0.13	0.15	3.4
Mixed forests	14.96	1.16	0.71	12.5
Savanna	19.18	3.02	6.49	49.6
Grassland	14.29	2.74	7.25	69.9
Shrublands	17.93	1.94	7.02	50.0
Deserts	15.24	0.13	1.22	8.9
Tundra	7.01	0.04	0.92	13.7
Rock/ice	1.21	0.00	0.02	1.7
All land	130.23	15.00	28.11	33.1

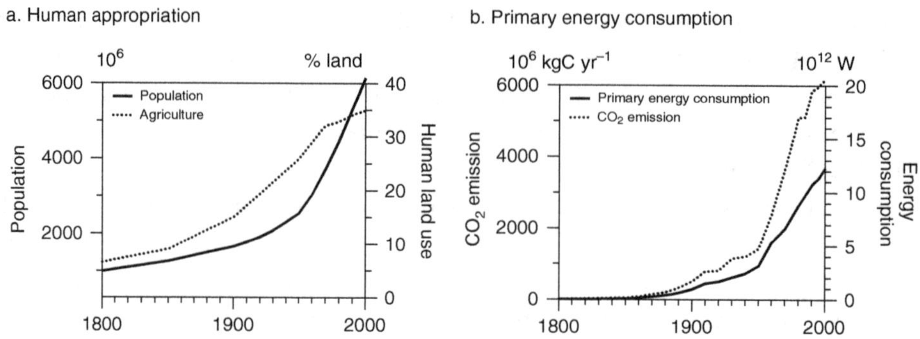

Figure 11.4 Trends in human activity over the last 200 years reflected in terms of (a) population size (solid) and human use of land cover (dotted) and (b) primary energy consumption (solid) and associated CO_2 emissions (dotted). Data are taken from the HYDE 3.1 data base (Klein Goldewijk et al. 2010, 2011).

and fuel, but not as food. Haberl et al. (2007) estimate the appropriated productivity in terms of human harvest to be about $8.2 \cdot 10^{12}$ kg C yr^{-1}, which represents about 12% of the terrestrial productivity. The authors further estimate that about 50% of the appropriation is associated with cropping and about 28% with grazing, so that $6.4 \cdot 10^{12}$ kg C yr^{-1} is associated with food production, or about 10% of the terrestrial productivity. These estimates deal with conditions around the year 2000. The historical development of land use and population size over the last 200 years is shown in Fig. 11.4a and illustrates clearly how related land cover change and human population size are to each other.

We can extend these trends further into the future and ask what size of a human population can be sustained by the productivity of the Earth's biosphere, a concept referred to as the human carrying capacity. Such estimates have been done for a long time, with different methods (Cohen 1995), and exhibit a remarkable range from less than a billion to above 1000 billion humans, with a median range of 7.7–12 billion humans. The approach to estimate the human carrying capacity that is closest to a thermodynamic characterization relates the food requirement of a human to food production to estimate a maximum population size. This measure directly relates to our thermodynamic approach as it relates the human metabolic energy need to the maximum chemical free energy that can be appropriated from the productivity of the biosphere. By relating these two terms, an energetically-based estimate for the human carrying capacity can be derived, although it should be noted that other factors, such as the availability of freshwater or other resources, are not accounted for as potential limitations. This focus on energy as the ultimate limiting factor may be justified because other resources, such as water and nutrients, can in principle be generated by technological means using energy, for instance, by desalination of seawater or industrial nitrogen fixation. Using this approach with the current extent of croplands and crop modeling, Franck et al. (2011) estimated a carrying capacity of about 11.4 billion humans for present-day conditions.

There are several ways by which the carrying capacity could be enhanced. Crop yields could quite substantially be improved by better management or breeding. In rain-fed regions, crops typically produce only 50% of their maximum yield (Lobell et al. 2009), so that a better water and nutrient management could potentially increase crop yields by 50% or more (Mueller et al. 2012). Through breeding, a greater fraction of photosynthesis could be allocated to harvested biomass, or photosynthesis could reach the thermodynamic limit of radiation conversion of 12% by genetic engineering or other means, or even higher with artificial photosynthesis (Barber 2009) (although the low efficiency of photosynthesis in natural environments may be explained by a CO_2 transport limitation, as in Section 10.6). Such improvement in crop yields allows for a greater carrying capacity without an expansion of croplands into natural lands.

When such agricultural expansion is considered, the human carrying capacity could in principle be substantially larger. Franck et al. (2011) estimated this capacity to be 282 billion people when croplands are extended to all productive lands. We could further envision a potential extension of agriculture into currently unproductive lands by irrigation. While naturally such an expansion would likely be unsustainable in steady state because of the inherent climatological limitations of freshwater availability, the inclusion of technological means such as seawater desalination in this expansion could nevertheless be sustainable with an external input of energy. For this expansion to be sustainable, the energetic gain from this expansion would, at a minimum, need to more than outweigh the energetic costs associated with the technology. This energetic evaluation is known as the energy return on investment (EROI) (Pimentel et al. 1973; Murphy and Hall 2011a,b), a concept that generally describes if a certain activity is going to be energetically sustainable. Two climate modeling studies (Ornstein et al. 2009; Bowring et al. 2014) showed that such an agricultural expansion, or “desert greening,” could be energetically sustainable with desalinated water because the energetic gains by enhanced photosynthesis would more than compensate for the energetic costs of desalination. A further means to expand croplands and food production could be “floating farms” in the oceans. Such expansions could substantially increase the carrying capacity, but they would require primary energy to sustain the expansions beyond natural limits. What these considerations illustrate is that when viewed from a thermodynamic perspective, the current population size is well below the carrying capacity of the Earth, although a development towards this capacity would likely change the functioning of the surface and thus of the Earth system rather considerably and would require non-trivial technological means (e.g., Marchetti 1979) as well as external energy input to make this sustainable.

The limits to food acquisition to sustain human activity are thus difficult to quantify because technological means provide new possibilities to enhance and extend food production in the Earth system beyond the natural level. An aspect that is not quite apparent in these considerations is that even current levels of crop yields by industrialized agriculture demand a considerable input of additional energy associated with machinery, transportation, and fertilizer production. Pimentel et al. (1973) estimated this input of primary energy to be about a third of the energy content of harvested corn. Also, large-scale desalination of seawater that would provide more freshwater to achieve greater productivity comes at an energetic cost (Elimelech and Phillip 2011). As long as these energetic costs can be met by the acquisition of primary energy, the use of such technological means does not make this an unsustainable option to enhance food production, but it requires a look at the sustainability of primary energy use.

11.4 Limits to primary energy acquisition

Human activity consumes a substantial rate of primary energy that is associated with manufacturing, fertilizer production, transport, heat and electricity generation, and other forms of activity that were referred to earlier as externalized activity (Fig. 11.2). The primary energy to sustain these activities are currently drawn from the Earth system mostly in the form of fossil fuels, geologic deposits of carbohydrates in the form of coal, oil, and natural gas that were once produced by photosynthesis. These deposits are currently extracted at a much greater rate than at which they were formed during Earth's history, thus making the current consumption of primary energy unsustainable (Ciamician 1912; Hubbert 1981; Dukes 2003) even though the proven oil reserves of today are likely to last for fifty years or longer (BP 2014). Based on the statistics of commercially traded fuels and renewable energy, the average rate of primary energy consumption during 2014 was estimated to be about $17 \cdot 10^{12}$ W (BP 2014), with renewable energy contributing a share of about 2%. Note that the consumption of primary energy is substantially higher than the metabolic energy needs, with an average of about 2400 W per person compared to the 100 W associated with the basal metabolic rate of a human. This consumption of energy has increased substantially over the last 200 years with the growth in the human population and industrial activity as shown in Fig. 11.4b. One should also note that primary energy describes mostly the energy content of fuels that are combusted and to some extent converted into other form, e.g., electricity or chemical or material conversions. The efficiency associated with these conversions has substantially increased in the last century (Smil 2000). The thermal efficiency of power plants, for instance, has increased from about 5% in 1900 to 55% in 2000. During this time period, the energetic cost of producing ammonia used in the production of fertilizers declined from about 100 GJ per ton NH_3 down to 30 GJ. This combination of an enhanced rate of primary energy consumption and an increased efficiency of its conversion implies that the work performed by externalized human activity has increased disproportionally more over the last century than the rate of primary energy consumption.

This section focuses on the extent to which externalized human activity can be driven by forms of renewable energy, that is, energy that is sustainably generated by Earth system processes. To do so, we draw on the estimates of the previous chapters in which these generation rates were estimated. Specifically, we relate the limits to radiative energy conversion from Chapter 6 to limits of solar renewable energy and the limits of kinetic energy generation from Chapter 7 to wind energy. We then evaluate the extent to which these can be further converted to human use. The focus here is specifically on those forms of energy that are renewable, that is, generated by Earth system processes, and not on forms of energy such as fission or fusion which are associated with the depletion of a stock.

We start with an overview of the energy fluxes and natural generation rates of the Earth system, which are summarized in Fig. 11.5. The starting points are the

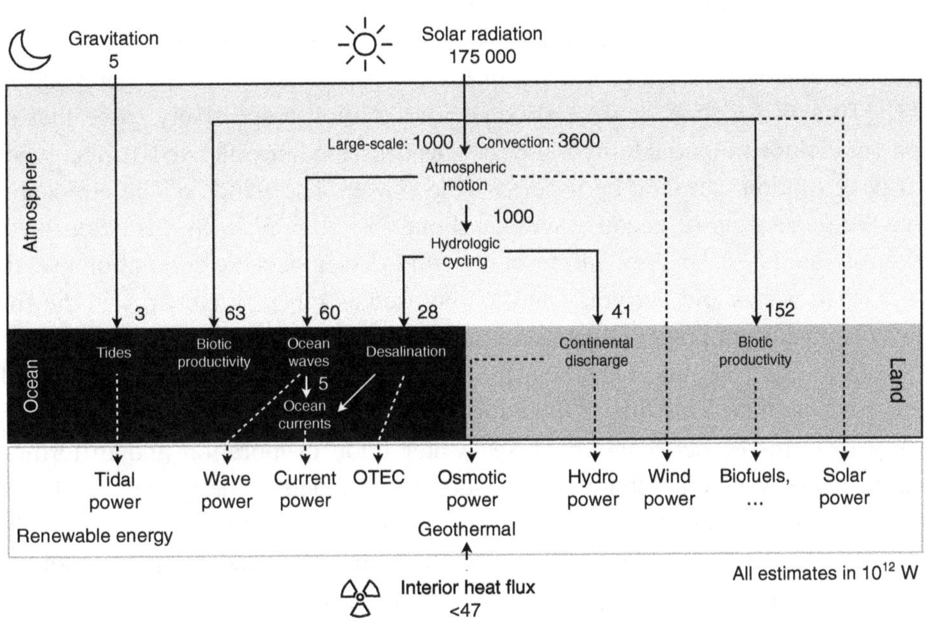

Figure 11.5 Overview of potential sources of renewable energy with estimates given in numbers. Estimates are given in units of 10^{12} W as global averages, based on different sources and taken from previous chapters. The dashed arrows show which forms of energy can be used by different types of renewable energy that are shown at the bottom of the figure.

three types of forcing of the Earth system: (i) the solar radiative forcing with a total radiative energy flux of about $175000 \cdot 10^{12}$ W (a flux of 1370 W m^{-2} over the cross section of the Earth, Section 6.3), (ii) the geologic forcing represented by the interior heat flux with about $47 \cdot 10^{12}$ W (section 7.7, from Davies and Davies (2010)), and (iii) the gravitational or tidal forcing of other planetary objects, primarily the Moon and the Sun, which result from the depletion of their rotational energies and contribute a total of about $5 \cdot 10^{12}$ W to the Earth system, of which about $3 \cdot 10^{12}$ W are reflected in ocean tides (Kagan and Sündermann 1996). These types of forcing are subsequently converted into different types of energy, and are linked to different types of renewable energy technology, as shown in the bottom of the figure.

From the figure it is immediately clear that solar radiation is by far the largest forcing of the Earth system and has thus the greatest potential as a source of renewable energy. It can directly be converted into solar power. It is used by photosynthesis and converted into chemical free energy, with generation rates of $63 \cdot 10^{12}$ and $152 \cdot 10^{12}$ W for marine and terrestrial productivity respectively, as derived in Section 9.6. Terrestrial productivity is used for food production, but biomass can also be used as a source of renewable energy, e.g., in forms of biofuels (Field et al. 2007). Most of the absorbed solar radiation, however, results in differential heating, which then generates large-scale and convective motion. These conversions were estimated in Chapters 7 and 8 to be in the order of 2 and 7 W m^{-2} or 1000 and $3600 \cdot 10^{12}$ W for large-scale and convective motion respectively (note that this does not include potential temporal effects as described in Chapter 10). The kinetic energy of motion can then be used as wind power. The winds of the atmosphere drive the generation of ocean waves of about $63 \cdot 10^{12}$ W, with a fraction of this power of about $5 \cdot 10^{12}$ W resulting in the wind-driven oceanic circulation (Section 7.4). Ocean waves and currents can then be used as renewable energy in the form of wave and current power (Isaacs and Schmitt 1980).

A substantial fraction of the heat flux involved in convection is represented by the latent heat flux. This flux is linked to hydrologic cycling, which generates free energy in terms of desalination of sea water upon evaporation and in terms of potential energy when water vapor condenses in clouds. The power involved in the desalination of ocean water was estimated in Section 9.4 to be $28 \cdot 10^{12}$ W, and this power is associated to some extent with the maintenance of the thermohaline circulation of the ocean. The power involved in ocean water desalination is related to different forms of renewable energy (Isaacs and Schmitt 1980). The difference in osmotic pressure due to the difference in salinity between river water and sea water at the river mouth can be used as osmotic power. The difference in salinity in ocean water relates to the maintenance of cold temperatures in the deep ocean by the thermohaline circulation, which maintains a temperature difference with the surface ocean that can be exploited by a technology referred to as ocean thermal energy conversion (OTEC) (Zener 1973; Nihous 2006). When precipitation reaches the land surface, it has a certain potential energy that is converted to the kinetic energy of overland and river flow. In Section 8.8, this generation rate was broadly estimated to be about $41 \cdot 10^{12}$ W, which can then be used as hydropower. These first-order estimates set the magnitude of the different forms of energy within the Earth system that can be converted further by technology into renewable energy.

In the following, we consider the conversion limits for solar and wind power, two types of renewable energy that show the greatest availability in the Earth system, as shown by the estimates in Fig. 11.5.

Solar radiation can be used by at least two different types of technology, with their theoretical conversion limits described in Section 6.5: first, direct concentrated solar (DCS) technology concentrates solar radiation by mirrors to a small area to heat a fluid to a high temperature. The fluid with a high temperature is then used to drive a heat engine, as in a conventional power plant. This technology requires direct solar radiation, that is, solar radiation that is concentrated into a small solid angle and not scattered by clouds because it is only then that solar radiation can be concentrated to a small volume to obtain temperatures that can go potentially as high as the emission temperature of the Sun. The conversion limit is very high, about 93%, so that almost all of the direct solar radiation at the Earth's surface can be converted into another form, e.g., electricity. Using satellite-derived estimates of the direct solar radiation at the Earth's surface (line A in Table 11.3), Kleidon et al. (2016) estimated the theoretical potential to be $41400 \cdot 10^{12}$ W globally (line D in Table 11.3).

The second form of solar energy technology is photovoltaics, which uses the short wavelengths of solar radiation for conversion into electric energy. This

Table 11.3 *Global estimates of incident solar radiation at the Earth's surface and limits for converting solar radiation and derived forms of energy at the Earth's surface into other forms usable to humans. All estimates are given in units of* 10^{12} *W. Based on estimates given by Kleidon et al. (2016) using the NASA Surface meteorology and Solar Energy (NASA-SSE) data set version 6.0.*

Description	Global	Land	Ocean
A: Direct solar radiation (surface)	44500	12500	32100
B: Diffuse solar radiation (surface)	34500	9900	24600
C: Total solar radiation (surface A + B)	79000	22400	56700
D: Direct concentrated solar (93% of A)	41400	11600	29800
E: Photovoltaic (73% of C)	57700	16300	41400
F: Biomass (1.5% of C)	1200	300	900
G: Wind power (<0.2% of C)	133	38	95

technology does not depend on direct solar radiation, so that all of the incident solar radiation at the surface can be converted, yet at a lower maximum efficiency of about 73%. Because of the greater radiative flux that can be used (line C in Table 11.3), the conversion limit for photovoltaics is higher with a theoretical potential of $57700 \cdot 10^{12}$ W globally (line E in Table 11.3). Even when these estimates are reduced to land area and to current efficiencies of industrial photovoltaic cells of about 20%, these estimates translate into potentials that far exceed the human demand for primary energy. Note that in comparison, the potential of photosynthesis to yield renewable energy is substantially lower due to the low efficiency of photosynthesis and further losses in generating biomass (line F in Table 11.3).

The next limit we evaluate is the maximum level by which kinetic energy of the large-scale atmospheric circulation can be converted into electric energy by wind farms. To start, it is important to recognize that not all winds can be converted, but only a fraction, because otherwise no heat would be transported with the flow, which would prevent the generation of kinetic energy. There is thus an upper limit to how much can be converted (Miller et al. 2011a), which was already described in general terms in Section 4.6.

The starting point are the two estimates given in Fig. 11.5 with respect to motion. While the power associated with large-scale motion is the dominant form associated with horizontal flow that is used by wind turbines, the power associated with convection can also be used as an energy source, a technology that is commonly referred to as solar chimneys, or solar updraft towers (Haaf et al. 1983; Schlaich et al. 2005; Zhou et al. 2010). Such towers concentrate the vertical convective flux to generate power from the updraft. In sunny, dry areas with a high sensible heat flux, this technology could potentially appropriate a substantial fraction of the power associated with dry convection. The maximum efficiency for the generation of convective motion, as derived in the previous chapters, is nevertheless in the order of 10%, so that the potential of this form of energy generation is necessarily much lower than for the direct use of solar energy. In the following, however, we focus on the more common technology of wind turbines that use the horizontal momentum of the large-scale atmospheric circulation.

To describe this limit of wind power from the large-scale flow, we use a setup as illustrated in Fig. 11.6a, following the studies of Miller et al. (2011a) and Gans et al. (2012). The large-scale circulation supplies the near-surface atmosphere with horizontal momentum by a vertical flux to the surface, which was referred to as F_{acc} in Section 4.6. In a natural steady state in the absence of wind farms, this momentum flux from the free atmosphere is balanced by turbulent friction F_{fric}. The frictional force is typically expressed as a quadratic function of wind speed (cf. eqn. 7.8 in Section 7.3), so that this balance of forces shapes the wind speeds near

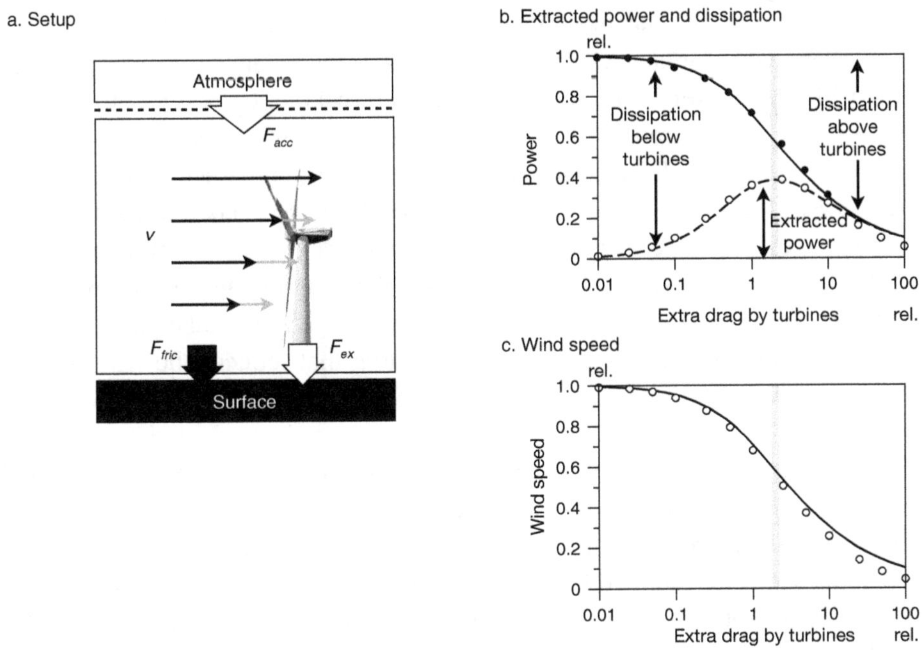

Figure 11.6 The limit to wind power illustrated (a) in a conceptual diagram describing the momentum balance that describes the resulting wind speed v consisting of an accelerating momentum flux F_{acc}, a friction force F_{fric}, and a force exerted by wind turbines F_{ex}, (b) in terms of frictional dissipation and the maximum power that can be extracted from the system, and (c) the reduction of wind speed. This description is identical to Fig. 4.7 and applied to limits of wind power. The vertical grey line marks the value at which the extracted power is maximized. The circles in (b) and (c) show the corresponding values over land from a set of climate model simulations (after Miller et al. 2011a).

the surface. When wind turbines are present, another force F_{ex}, associated with the turbines enters this balance, which exert a drag on the flow that spins the turbine blades and drives the generators that convert the mechanical energy of the wind into electric energy. Since the friction force depends on the wind speed, we can use this balance of forces in steady state to infer the wind speed (cf. eqn. 4.36). As the rate of energy removal by the turbines is given by the product $F_{ex} \cdot v$, we can then infer how this balances of forces, the wind speed, and the partitioning between frictional dissipation and kinetic energy removal by the turbines changes with different values of the drag F_{ex}. As this drag represents the size or number of turbines, we can then identify the limit of how much of the kinetic energy of the system can be converted into wind power at large spatial scales.

This sensitivity is shown in Fig. 11.6b and c for the partitioning of the kinetic energy influx into frictional dissipation below and above the turbines and the kinetic energy extracted by the turbines. The more momentum the wind turbines remove from the flow with a greater value of F_{ex}, the more the balance of forces is shifted, resulting in a lower velocity and less dissipation of kinetic energy below the turbines. Because of the trade-off between greater momentum extraction and wind speed, it was shown in Section 4.6 that kinetic energy is extracted at the greatest rate when $F_{ex,opt} = 2/3 \cdot F_{acc}$ (cf. eqn. 4.38), resulting in a maximum extraction of about 38% of the kinetic energy that would naturally be dissipated (cf. eqn. 4.40 and Fig. 11.6b). Fig. 11.6c shows how the greater the drag exerted by the wind farm reduces the wind speed and thus shifts turbulent dissipation to take place above the wind farm. In the extreme case of substantial drag that reduces the wind speed to near zero, the wind turbines act as if they move the resting "surface" at the level of the wind turbines, so that most of the kinetic energy is dissipated by turbulence above the wind farm. Also shown in Fig. 11.6b and c are results from climate model simulations of large-scale wind power that simulate fluid dynamics in greater detail (Miller et al. 2011a) yet substantiate the interpretation by the simple model.

It can further be determined from continuity considerations originally performed by Betz (1920) that a third of this energy is dissipated by wake turbulence behind the turbines, so that only 2/3 of the extracted kinetic energy is converted to useful work (Corten 2001). This yields an overall upper limit of 26% for the conversion into electric energy. With a downward kinetic energy flux of about 1 W m^{-2} into the boundary layer, this yields a maximum potential of wind power from large-scale motion of about 0.26 W m^{-2}, or $133 \cdot 10^{12}$ W when integrated globally. As large-scale motion is caused by differences in absorbed solar radiation, it can be expressed as a fraction of the mean absorbed solar radiation and is added as line G in Table 11.3. While this estimate is used here to illustrate the magnitude of the limit, climate model simulations that explicitly include the effect of large wind farms on the atmospheric flow largely agree with these estimates (Keith et al. 2004; Miller et al. 2011a; Jacobsen and Archer 2012), as well as with the described mechanism by which wind speeds are substantially reduced at the limit to about 58% of the value in the absence of wind farms (see circles in Fig. 11.6b and c). Similar kinds of limitations by the effects on the flow have been described by energy conversion from tidal currents in the ocean (Garrett and Cummins 2007) and with respect to OTEC energy conversion (Nihous 2006).

While the estimates performed here establish order of magnitudes, the approach that we use is unique as it represents an approach based on thermodynamic limits and relate these to the forcing and resulting interactions in an Earth system context. Estimates of the magnitude of renewable energy within the Earth system have been done for a remarkably long time, going back to estimates by Thomson (1881) and Heinrich Hertz in 1885 (as described in Mulligan and Hertz 1997) to the present-day (Hubbert 1981; Rogner et al. 2000; Rosen and Scott 2003; Herrmann 2006;

IPCC 2011). These estimates are typically performed from bottom up and are based on observations. This focus on observations, rather than on the mechanisms and limits by which the associated forms of energy are generated, can lead to gross misrepresentations of renewable energy potentials that are thermodynamically impossible to accomplish. For instance, the high wind speeds in the jet streams of the atmosphere have been referred to as "wind in some locations contains at least ten times the energy of sunlight" (Vance 2009), with the erroneous conclusion that the associated wind power potential is vast. Climate model simulations in this regard show that the high wind speeds of jet streams represent near geostrophic flow (cf. Section 7.6), so that these speeds result from the near absence of friction, and not from a strong force (Miller et al. 2011b). The potential of wind power from jet streams have thus been estimated to be rather small in the order of $5 \cdot 10^{12}$ W, which is contrary to studies that claim these geostrophic winds to have 100 times more power than the current primary energy demand (Archer and Caldeira 2009), but that do not account for the interaction of turbines with the flow.

Likewise, the theoretical potential for geothermal renewable energy was described by the heat content of the top 10 km of the Earth's crust (Goldstein et al. 2011). This potential provides the incorrect impression that the heat content of the crust could be drawn down to 0 K to generate energy, which is clearly impossible. While the Earth's interior stores a vast amount of thermal energy (see Table 2.2), this size of heat storage does not give any indication of the maximum sustainable rate by which thermal energy can be extracted from the interior. The bottleneck that keeps the heat loss from the interior so low is the thermal conductivity of the solid Earth. Yet, on the other hand, if this conductivity was greater, the Earth would have already cooled down much more throughout its history and lost a greater fraction of its internal energy. Hence, the sustainable rate of geothermal energy has to be less than the geothermal heat flux indicated in Fig. 11.5.

In summary, the overview of the different energy conversion rates within the Earth system given here demonstrates the strength of the thermodynamic approach, because it allows a clear formulation of physical limits, but it also needs to account for the interaction between the conversion process and the Earth system. Using this view, it is apparent that the human demand for primary energy is of similar magnitude to many Earth system processes. It demonstrates that by far the greatest source of sustainable energy is solar energy, as it is potentially available at a rate that substantially exceeds all other sources of renewable energy because it uses a flux of low entropy radiation that has not been converted into another form. This should come as no surprise, as most other forms shown in Fig. 11.5 originate indirectly from conversions of solar radiation, yet at inevitable losses every time energy is converted into another form. What thermodynamics tells us is that the limits to sustainable use of these indirect forms of energy need to be substantially smaller than the incident flux of solar radiation into the Earth system.

11.5 Thermodynamic consequences of human activity

We have now dealt with the magnitudes and limits of human energy acquisition associated with maintaining metabolic and externalized activity. As a next step, the effects of this energy acquisition is evaluated from a thermodynamic perspective, getting back to the two effects A and B shown in Fig. 11.1. In doing so, we focus on the impacts of human activity that are associated with energy use, which does not encompass all impacts, but likely the largest share of these. The main focus in evaluating the impacts is on the thermodynamic state, which can then be related to changes in the more common metrics of climate change, such as changes in surface temperature.

The effects of human activity are described in terms of the general template shown in Fig. 11.7a. In thermodynamic terms, the consumption of free energy represents a dissipative process, which is driven by the depletion of a certain form of thermodynamic disequilibrium. We can thus broadly label thermodynamic disequilibrium as the cause or driver of the dissipative process. The consequence, or effect, of this process is then to reduce the state of thermodynamic disequilibrium within the system. This sequence of thermodynamic disequilibrium to dissipative

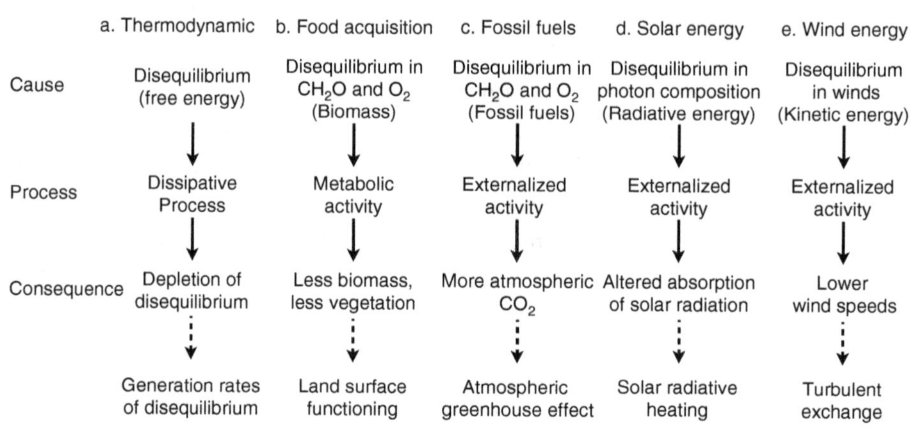

Figure 11.7 Thermodynamic characterization of human impacts. Panel (a) shows the general thermodynamic description of a state of thermodynamic disequilibrium that is associated with a certain form of free energy, the dissipative process that this disequilibrium drives, and the consequence of the process. Panels (b) to (e) apply this thermodynamic characterization to food acquisition and consumption of primary energy in terms of fossil fuels, solar, and wind energy associated with human activity.

process to consequence is then embedded in the Earth system, in which particular processes act to generate the state of thermodynamic disequilibrium from the planetary forcing. Ultimately, whether it evolves to a greater magnitude depends on the feedbacks that connect the dissipative process with the rate by which disequilibrium is being generated.

The forms of energy consumption by human activity associated with metabolic activity and what was referred to as externalized activity follow the same thermodynamic sequence, as illustrated in Fig. 11.7b–e. Food acquisition by human activity is associated with the consumption of biomass from the environment at a certain rate. Biomass represents a form of chemical free energy and chemical disequilibrium with respect to the carbohydrates stored in biomass and the molecular oxygen in the atmosphere. When biomass is appropriated by human activity at a certain rate, which represents a fraction of the net primary productivity, it depletes this form of disequilibrium, reflected in less biomass. As human appropriation competes with the natural decomposition of biomass by heterotrophs, it inevitably results in a lower dissipative activity of the heterotrophic part of the terrestrial biosphere. Furthermore, the appropriation of biotic productivity is associated with a different state and functioning of the land surface, as was already illustrated by the sharp contrast in land use in Western Australia shown in Fig. 10.1. The manifestations of the impacts on the land surface on typical climatic variables such as surface temperature are, however, not quite so straightforward. Typically, croplands are associated with a higher albedo, so that less solar radiation is being absorbed, but the ability of the surface to evaporate is also reduced. While the first effect of less absorption should result in a cooling of the surface, the reduced ability to evaporate should result in a warming. The overall effect on surface temperature can thus result in both, a cooling or a warming, depending on the magnitude of both effects. As absorption of solar radiation is reduced, the overall effect is nevertheless that the land surface–atmosphere system is less dissipative with reduced convective heat fluxes. Furthermore, model estimates indicate that the conditions induced by land cover change result in climatic conditions that are less favorable for photosynthesis when the enhancing effects of irrigation and fertilizer use are not considered (Kleidon 2006; Smith et al. 2014).

The effect of fossil fuel consumption is shown in Fig. 11.7c. The chemical disequilibrium associated with fossil fuels also relates to carbohydrates and atmospheric oxygen as in the case of food acquisition, with the difference that the carbohydrates resulted from past productivity and were separated from the atmosphere in geologic formations. The major effect of depleting this disequilibrium is the enhanced concentration of atmospheric CO_2. As CO_2 is a major greenhouse gas, it adds to the optical depth in the radiative transfer of terrestrial radiation and results in the well-known effect of global warming. In this sense, an enhanced concentration of atmospheric CO_2 represent a state of the Earth system that is closer to geochemical equilibrium, although the consequences of this disequilibrium are more closely related to the role of CO_2 as a greenhouse gas.

The use of renewable energy as a form of primary energy is also associated with the depletion of states of thermodynamic disequilibrium. Solar energy technology (Fig. 11.7d) uses the disequilibrium in solar radiation either with respect to its high concentration into a narrow solid angle in case of direct concentrated solar conversion or in terms of its highly energetic photons in case of photovoltaic energy conversion. Both technologies use the state of radiative disequilibrium of solar radiation that is naturally dissipated when solar radiation is absorbed and emitted at much lower temperatures into a broad solid angle (as shown in Fig. 6.1). The effect of this technology is that the absorption of solar radiation takes place at the converter, so that it affects the way that solar radiative heating occurs at the surface.

In case of wind energy (Fig. 11.7e), wind farms generate energy out of the disequilibrium represented by the difference in wind speed between the lower atmosphere and the Earth's surface. The effect of this depletion is a reduction of wind speeds, as demonstrated by the example in the previous section. The lower wind speeds then reduce the turbulent exchange between the surface and the atmosphere, with is likely to affect the surface energy balance.

These examples illustrate that human activity inevitably has impacts on the Earth system because the energy that is required to sustain its activity is inevitably associated with the depletion of certain forms of thermodynamic disequilibrium. That impacts on the Earth system are inevitable are simply a consequence of the sheer magnitude of human energy consumption. Yet, the impact also depends on which disequilibrium is being depleted. If we consider a hypothetical scenario in which the current primary energy demand of $17 \cdot 10^{12}$ W is derived entirely from renewable energy, then this would require about half of the maximum potential for wind energy on land, but only about 0.1% of the maximum potential of solar energy on land. The relative impact on the process from which the energy is drawn from is thus much greater when wind energy is used in comparison to solar energy. What this means is that a lower number of steps of conversion of the planetary forcing into a particular form of renewable energy results in a greater potential as an energy source and in lower impacts of its use.

There is a critical aspect that sets solar energy use apart from the use of any other form of renewable energy. In the natural state, most of the solar radiation is absorbed and directly converted into heat, resulting in substantial entropy production during the absorption process at the low prevailing temperatures of the Earth (cf. Fig. 6.8). What this means is that most of the low entropy in solar radiation is not used to generate free energy by natural processes and is merely dissipated upon absorption. When solar radiation is used by solar energy technology, this dissipative loss is reduced as a fraction of the absorbed solar radiation is converted into free energy of another form. This can result in an overall enhancement of free energy generation within the Earth system. With other forms of renewable energy, this is quite different. As the estimate of the maximum power limit for kinetic energy indicates that the atmosphere operates near its thermodynamic limit, any use of wind energy by human technology cannot increase kinetic energy generation within the atmosphere unless the radiative forcing is altered. Greater wind energy use thus leaves the magnitude of kinetic energy generation unaffected or even lowers the generation rate while reducing the share that is naturally dissipated by turbulence. This comparison illustrates that there are two different impacts that human activity can have on the overall thermodynamic functioning of the Earth system: it can enhance or lower the overall magnitude of free energy generation within the system. Expressed differently, human activity represents a sink of the free energy generated within the Earth system, but with the use of solar-based technology, human activity could also act as a source of free energy generation within the Earth system.

These contrasting effects of solar energy to other renewable energy on the Earth system are important when we consider potential futures of human activity that represent steady states in terms of free energy generation rather than transient states in which stocks are depleted as in the case of fossil fuel use. These are schematically shown in Fig. 11.8, following Kleidon (2012). This figure visualizes the overall free energy generation within the Earth system by its different processes as a "pie," with the size given by the overall free energy generation of the planet indicated by G_{planet}. Human activity is driven by consuming a piece of this pie, indicated by the fraction G_{human} in the figure. As the human population and its energy use is likely to grow in the future, the share of the pie appropriated by humans is likely to increase. This increase can be accommodated in two different scenarios, shown as cases A and B in Fig. 11.8.

In what is shown as case A in Fig. 11.8, the increase in human energy consumption is accomplished by an increase in the share, but the effects on the planetary free energy generation are such that it shrinks the size of the pie. The increase in the human share as well as the shrinking of the overall size results in a reduction of free energy generation available to drive natural processes including the biosphere. This represents an adverse effect of human activity on the Earth system in thermodynamic terms, in which the increase $\Delta G_{human} > 0$ results in a decrease in $\Delta G_{planet} < 0$. It results in the overall Earth system becoming less dissipative and being represented by a thermodynamic state that is closer to thermodynamic equilibrium. Examples for this scenario are the human appropriation of net primary production and the use of wind energy. Both examples appropriate free energy from the natural process, which at sufficiently large scale are likely to reduce, and not increase, the generation rate of free energy by natural processes.

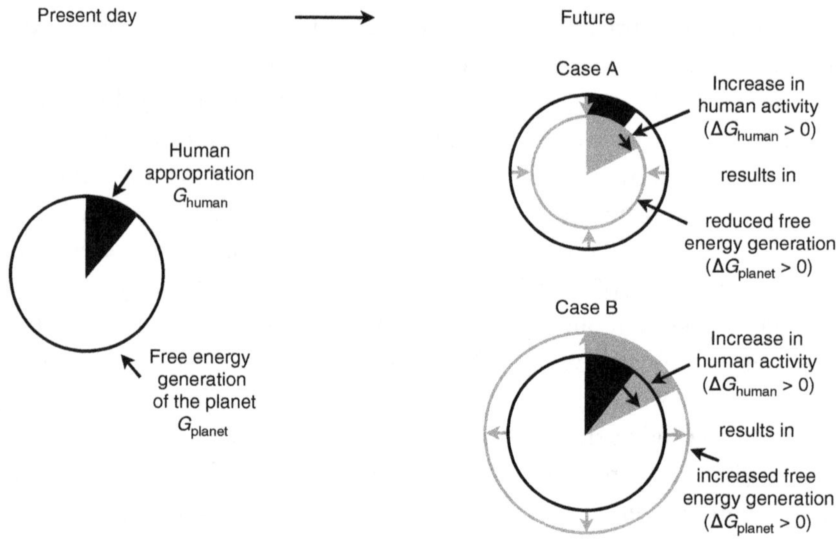

Figure 11.8 Two potential cases of human impact on the Earth system associated with a future increase in human activity. The current free energy generation by the Earth system G_{planet}, is shown by the size of the pie on the left, with the share of human energy consumption G_{human}, shown as the black segment (not drawn to scale). When the demand for food and primary energy increases in the future, this can result in the two different cases A and B shown by the grey pies on the right. Adapted from Kleidon (2012).

Case B in Fig. 11.8 is different as the increase in the human share is associated with a greater overall size of the pie, so that the increase $\Delta G_{human} > 0$ results in $\Delta G_{planet} > 0$. Examples for this scenario are the expansion of agriculture into currently non-productive areas, such as deserts, and photovoltaics. The expansion of agriculture into non-productive areas with irrigation would likely increase the overall productivity of the terrestrial biosphere, while photovoltaics is able to generate more free energy with the absorption of solar radiation than natural processes. Both of these examples require a certain investment of energy associated with the technology that these involve. An expansion of agriculture into naturally unproductive areas requires, for instance, energy for irrigation and fertilization, while photovoltaics requires energy in the manufacturing of the solar panels. A critical success criterion for these examples is that more energy is gained than invested, which was introduced earlier as the energy return on energy invested (EROI) and which is used in the evaluation of sustainability (Pimentel et al. 1973; Murphy and Hall 2011a,b). Formulated in terms of this concept, case B in Fig. 11.8 represents a scenario in which the energy return due to an enhanced human activity does not just increase energy generation for human activity, but for the whole planetary system.

These two scenarios can be linked back to the general dynamics of a thermodynamic process as described in Chapter 5. In Chapter 5, the dynamics of a process to its thermodynamic limit was described in general terms. It was argued that thermodynamic processes evolve to their limit, thus generating free energy at their maximum possible rate, although this evolution to the maximum is non-trivial and likely involves the formation of structures and the organization of flows. When applied to human activity as a thermodynamic process, we could also expect it to evolve towards its thermodynamic limit. Case B of Fig. 11.8 would represent an evolutionary pathway towards a greater level of free energy generation and thus towards this limit. The evolutionary pathway as well as the limit are, however, also non-trivial as these would require the technological means to follow the path described by case B. Drawing energy from fossil fuels as in the present day, however, is unsustainable as it draws primary energy from the depletion of a stock.

Case B links to a thermodynamic view of how human activity can develop sustainably within the Earth system. This thermodynamic view of sustainability (see also, e.g., Dincer and Rosen 2005) links closely to the common definition of sustainability, which refers to a development that "meets the needs of the present without compromising the ability of future generations to meet their own needs" (World Commission on Environment and Development 1987). The scenario represented by case B represents such a development as it does not compromise the ability of future generations to sustainably appropriate energy from the Earth system to human use.

This evaluation of human impact certainly does not encompass all aspects of human activity and neither does it deal with all aspects associated with energy use. Yet, it provides a fundamental perspective of impacts that are associated with human energy use, a central aspect of human activity. As energy is generated from forms of thermodynamic disequilibrium within the Earth system, human energy use is generally associated with depleting thermodynamic disequilibrium associated with different processes within the Earth system, and the impacts can be understood in these terms, as illustrated by the examples. The thermodynamic view thus allows us to identify a general direction of impacts of human activity on the Earth system and link these to general effects associated with dissipative processes on the thermodynamic state of the system.

11.6 Human activity in the Earth system

In this chapter, we viewed human activity as a thermodynamic process that is maintained by the dissipation of free energy. In doing so, the many socioeconomic dimensions of human societies are not considered as we reduced the focus on the essential energy needs associated with human activity. This reduction to a

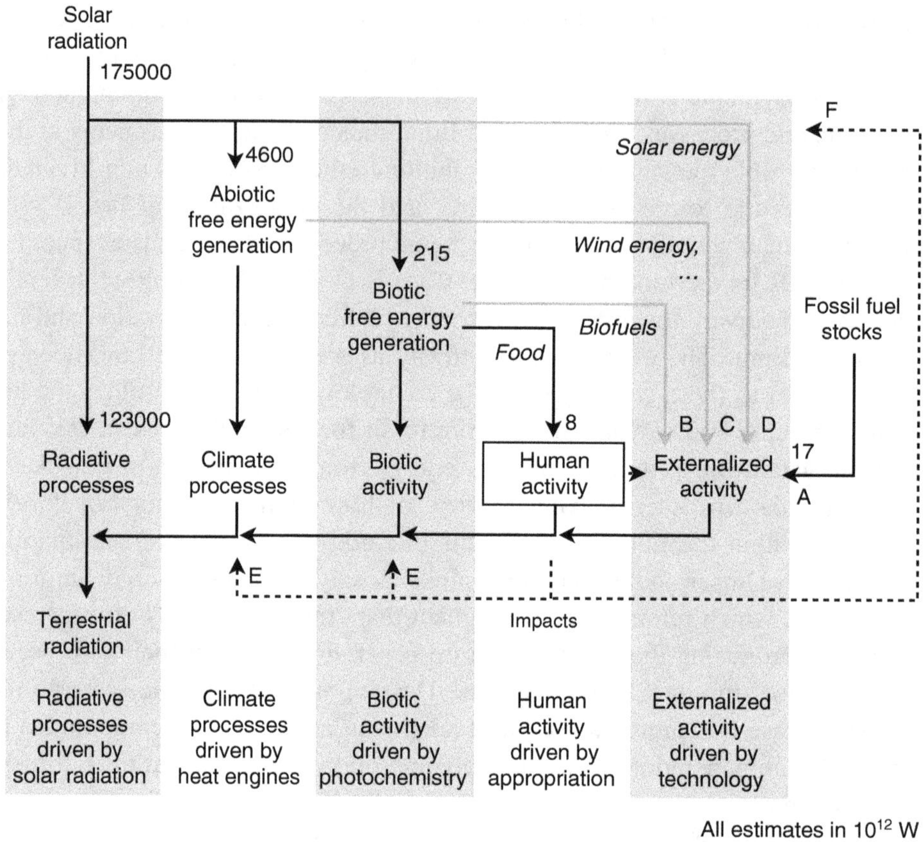

Figure 11.9 Summary of human activity in the Earth system in terms of its appropriation of energy in relation to other Earth system processes. The numbers describe the magnitude of radiative fluxes or free energy generation rates in 10^{12} W.

thermodynamic process has the advantage that it allows us to treat and analyze human activity in the same way as any other physical process described in the previous chapters. We can then compare human activity to other Earth system processes using the same thermodynamic aspects in terms of its magnitude, its connections to other Earth system processes, the factors that limits its strength as well as the impacts that it has on the Earth system.

This view of human activity in the context of the Earth system is summarized in Fig. 11.9. The need to maintain the metabolic activity of the human body is directly associated with energy consumption. This energy is taken up in chemical form in the form of food, which is appropriated from the chemical free energy generated by photosynthesis. With about $8 \cdot 10^{12}$ W, this appropriation of chemical energy from the biosphere already represents a sizable share of the overall biotic free energy generation of about $215 \cdot 10^{12}$ W (land and ocean combined).

A larger share of human energy consumption is associated with what we referred to as externalized human activity. It refers to the energy consumption and physical work associated with socioeconomic activity and which is at present associated mostly with the consumption of fossil fuels. This externalized consumption represents about $17 \cdot 10^{12}$ W and is derived from the depletion of fossil fuel stocks (arrow A in Fig. 11.9). It allows humans to perform more physical work in the Earth system compared to what the human body – or any other species – could perform, thus likely to result in a positive feedback on the levels of human activity. Taken together, the human population currently consumes a total of about $25 \cdot 10^{12}$ W, which is a considerable rate when compared to the rates by which processes within the Earth system generate and dissipate free energy.

In the figure, the abiotic rate is summarized by the estimate of $4600 \cdot 10^{12}$ W of abiotic free energy generation by heat engines that are associated with the generation of convective and large-scale motion, but it should be noted that this is by far the greatest term, with many processes operating at much lower power, as for instance can be seen by the more detailed estimates shown in Fig. 11.5. By this comparison of magnitudes, it becomes evident that human activity represents a planetary thermodynamic process of substantial magnitude, which substantiates the planetary-scale impacts of human activity and Crutzen's (2002) notion of a new geologic era of the Anthropocene on thermodynamic grounds.

The planetary-scale impact of human activity associated with this energy consumption can also be interpreted in thermodynamic terms. As any form of free energy consumption relates to a depletion of thermodynamic disequilibrium, many aspects of human impact can be understood as such a depletion effect, which then relate further to changes in state variables such as surface temperature, biomass, wind speeds, and the like. The consumption of fossil fuels represents the depletion of the chemical disequilibrium between carbohydrates and atmospheric oxygen that was generated by photosynthesis in the Earth's past. Agricultural food production represents the depletion of chemical free energy associated with biomass generated by present-day photosynthesis. When fossil fuels are increasingly replaced by renewable forms of energy, these relate to a depletion of disequilibrium as well, as shown by arrows B, C, and D in Fig. 11.5.

These impacts affect the Earth system in two ways: first, when free energy of the Earth system is consumed by human activity, it is not dissipated by a physical process or by biotic activity, thus reducing the dissipative activities of these processes (arrow E in Fig. 11.5). Biomass appropriated by humans cannot be consumed within natural food webs and decomposers, and the kinetic energy of the wind appropriated by wind turbines is not dissipated by near-surface turbulence. The second path, shown by arrow F in Fig. 11.5, deals with the impacts of human activity on the generation rates. In this respect, it was shown that the use of solar energy as a renewable energy source stands out from other sources of energy because it allows for a substantially greater free energy generation out of solar radiation than what is possible by heat engines or photosynthesis. It is this type of effect associated with human activity that can enhance the free energy generation of the planet and that would seem to be consistent with the notion of a sustainable future. Such a future would seem to be able to overcome the natural limitations imposed by the Earth system far beyond the levels before humans became the dominant species, but this would require specific forms of technology and primary energy to be sustainable.

This description of human activity is rather general and does not deal with all aspects and impacts. It nevertheless serves as a basis to describe human activity as a thermodynamic process, to understand its impacts, and to compare it to other thermodynamic processes within the Earth system.

This chapter completes the thermodynamic description of Earth system processes from the planetary drivers to human activity and associated impacts. As a next, final step, we will place these processes together into a holistic thermodynamic view of the Earth system as well as its dynamics and evolution, and relate it to the topics raised in the introduction of this book.

12

The thermodynamic Earth system

12.1 Thermodynamic foundations of the whole Earth system

We have now dealt with the basics of thermodynamics and its application to a range of dominant Earth system processes, from the radiative forcing of the planet to human activity. At the end of this book, the goal is to synthesize these thermodynamic components to a comprehensive picture of how the Earth system functions as a whole and how its dynamics is a reflection of the second law, and to illustrate what insights this picture may provide for the questions raised in the motivation for the book. This chapter closes with a perspective of the possible future directions.

The main motivation for formulating the Earth system in thermodynamic terms is that thermodynamics is so general that it is applicable to all Earth system processes. It thus provides a unifying basis for describing the directions, connections, and interactions of processes so that we can get an understanding of how the whole system functions and evolves. While for thermal energy and heat this thermodynamic formulation is common and straightforward, other processes such as radiation or motion are less commonly formulated in thermodynamic terms. Yet, when this is done with the use of conjugate variables, Earth system processes can be dealt with in the same units of energy, thus making them comparable, and the dynamics of these processes can be formulated in terms of energy conversions. Here, thermodynamics provides another critical component for a unifying basis. The second law of thermodynamics formulates the overall direction of these conversions and imposes a condition that results in thermodynamic conversion limits. These thermodynamic limits act as relevant constraints to the dynamics of Earth system processes. The first and second law of thermodynamics thus provide the bare minimum of essential physics to consistently formulate Earth system processes and their interactions.

Equally important in establishing a thermodynamic foundation of the Earth system is to place thermodynamics in the context of the Earth system. This was illustrated in the introduction by Fig. 1.5, which shows how the planetary forcing creates gradients that are further converted to motion, hydrologic and geochemical cycling, biotic and human activity and that result in the dynamics of the Earth system. This hierarchical view of Earth system processes allows us to separate the drivers from the driven processes and apply thermodynamic limits to these conversions. Yet, it is equally important to recognize the strong effects that the driven processes have on the drivers, resulting in strong interactions. These interactions with the driving gradient affect the thermodynamic conversion limit, resulting in the maximum power limit as the outcome of this strong interaction. While the Carnot limit is well established as a thermodynamic limit, the interaction of the resulting processes with the driving gradient by the constraints imposed by the energy balances are typically not considered in the formulation of thermodynamic limits. The previous chapters showed several examples highlighting that this interaction is highly relevant. It results in the thermodynamic limit of maximum power, which predicts magnitudes of processes and energy conversions that are largely consistent with observations. This limit is closely associated with the proposed principle of maximum entropy production (MEP, Section 4.7), except that the more detailed formulation and interpretation of the maximum power limit of a heat engine is easier to interpret. The success of the maximum power limit is not trivial, as it implies that natural processes evolve to and operate near their thermodynamic limit. The dynamics that are behind this evolution were attributed to two dynamic feedbacks, a fast, positive feedback by which a flux enhances the power, and a slow, negative feedback that depletes the driving gradient. The combination of thermodynamics, feedbacks, limits and a hierarchical view of the Earth system thus provides a basis for establishing the magnitude of its processes and a description of its first order interactions that shape the functioning and evolution of the whole system. It yields a picture of the Earth system in which processes act to accelerate the dynamics in the direction of the second law, thus making it a highly dissipative system, and probably making it a system that is as dissipative as possible.

This thermodynamic foundation can be applied to several, general questions that were raised in the introduction and illustrate which insights can be learned from this foundation. The first question deals with the conditions for life that shape a habitable environment. As a starting point, we can use Schrödinger's (1944) general view of the living cell as a thermodynamic system far from equilibrium. Any cell is embedded in its environment, and a vital component of the environment is to supply the raw materials at a sufficiently high rate to sustain the cell's metabolism. In doing so, the ability of the living cell to "make a living" is linked not just to its ability to harvest free energy from its environment, but equally so to the ability of the living organism to exchange the raw materials of its metabolism with its environment. This ability to exchange mass is a characteristic of the environment and is constrained by the maximum power limit. We encountered this restriction when we described chemotrophic lifeforms near hydrothermal vents as the sites at which life probably first emerged and showed that it yields a reasonable prediction for photosynthetic rates of the present day. A habitable environment could thus be seen as one that is able to exchange mass between the living organisms and their environments. As this ability to exchange mass is thermodynamically constrained and relates to the planetary forcing, this thermodynamic view of the Earth system can provide a novel perspective on the habitability of planetary environments that goes beyond geochemistry and that we will extend upon later on.

The second question deals with the consequences of life for the planet, and whether these yield general effects, evolutionary directions, and optimal conditions at the planetary scale. This question is addressed by the Gaia hypothesis of Lovelock (Lovelock 1972a,b; Lovelock and Margulis 1974), which states that the Earth system is regulated to an optimum state by and for the biosphere. We can evaluate this question also from the thermodynamic foundation described here, which in the end can result in a similar outcome as proposed by the Gaia hypothesis. As any dissipative process, life depletes a form of disequilibrium to sustain its metabolic activity, so that biotic activity has the general effect of depleting a state of thermodynamic disequilibrium in the Earth system. By far the greatest disequilibrium is represented by the spectral composition of solar radiation at the surface of the Earth, which is depleted by photosynthetic activity. Yet, thermodynamic considerations suggest that it may not be light that limits photosynthetic activity as the observed maximum efficiency of light use by photosynthesis is well below the theoretical limit of the radiative conversion. In the previous chapters, it was proposed that the rate of CO_2 mass exchange may play a substantial role in limiting the photosynthetic activity on land, while it is well established that in the ocean, it is the exchange of nutrients that limits marine productivity. This would imply that the transport and exchange of mass could potentially form a relevant constraint in maintaining biotic activity, possibly more important than the availability of light. As biotic activity alters the atmospheric composition and the intensity of the greenhouse effect, it affects the radiative forcing and thus the physical constraints that shape the limit to convective mass exchange. One can thus imagine positive environmental feedbacks on biospheric activity, similar to what Lenton (1998) referred to as environmental feedbacks to growth. The outcome of this feedback could result in a maximized photosynthetic activity, not in terms of light absorption, but rather in terms of the mass exchange that can be accomplished by turbulent–convective exchange between the biosphere and its immediate environment of the surface ocean and the near-surface atmosphere. As this mass exchange relates to the intensity of solar radiative heating, one can further imagine that the biotic feedbacks on the Earth system act to maximize not just the photosynthetic activity of the biosphere, but also the mass exchange by maximizing the absorption of solar radiation at the planetary scale to the extent possible. Such a state would then represent a maximization for and by the biosphere and thus share the main features that were formulated by the Gaia hypothesis. Yet, it would not be postulated, but formulated on a more general thermodynamic basis as a hypothesis that processes in complex systems evolve to and maintain states at their thermodynamic limit. Such a general outcome of a thermodynamic system applied to the whole Earth would then provide a basis to interpret events and the evolutionary history of the biosphere and the Earth system as we can relate these trends to an evolutionary direction towards thermodynamic limits and how these limits can be shifted to higher levels.

The third question raised in the introduction of the book addresses human activity in an Earth system context, asks about the limits to human activity and what thermodynamics can tell us about a sustainable future. In the book, we dealt with this question in two ways. At the abstract level, the need for food and primary energy associated with human activity is used to condense human activity to a thermodynamic, dissipative process that is driven by consuming free energy from the Earth's environment. While this abstraction may seem rather reductionist, it serves the purpose of making human activity comparable to other thermodynamic processes of the Earth system in terms of how this energy is generated, what the limits of its acquisition are, the associated impacts with this acquisition, and the resulting interactions between human activity and the Earth system. We can then use this description and speculate about its future evolution and what this would imply for the Earth system. At the practical level, the thermodynamic description of the Earth system in combination with natural conversion limits of different forms of energy yields a consistent estimate of the different potentials for food production, for renewable energy, and how much of these can sustainably be used.

The following sections summarize the thermodynamic foundation of the Earth system and use these three questions to synthesize the main points and estimates of the book.

12.2 The thermodynamic Earth system

The thermodynamic description of the whole Earth system is based on common thermodynamic concepts and common descriptions of Earth system processes. This book brings these two aspects together to a holistic perspective of the Earth system and applies thermodynamics from the planetary forcing to the process under consideration. While the importance of interactions is well established in Earth system science and the dissipative nature of non-equilibrium systems is well recognized in thermodynamics, the combination of these two aspects results in bringing interactions to thermodynamics, and non-equilibrium thermodynamics to Earth system science. It adds insights to the common formulations in thermodynamics, in which systems are typically described with fixed and prescribed boundary conditions in which the state of the system does not interact with the conditions at and fluxes across the boundary. It further adds thermodynamics to the common view of the dynamics of Earth system processes, which are typically not being viewed as the outcome of a non-equilibrium dissipative system that in every step and in its totality follows the second law and are subject to thermodynamic limits. This combination of non-equilibrium thermodynamics and Earth system science establishes the foundation of the thermodynamic Earth system. It provides a perspective of the functioning of the Earth system that is grounded on general physical principles and the second law of thermodynamics in particular.

A brief summary of the general aspects of thermodynamics start with the first and second law and the characterization of the system state in terms of how far it is maintained away from thermodynamic equilibrium. While the first law describes the conservation of energy when energy is converted from one form into another, the second law describes the direction in which the dynamics of internal processes within a system take place towards establishing thermodynamic equilibrium. Dissipative systems are maintained in a state of disequilibrium by the fluxes at the system boundary, which are associated with a net entropy exchange with the surroundings of a system.

With this general description of thermodynamic systems, the combination of the first and second law for an energy conversion process results in the well-established Carnot limit of a heat engine. We derived the Carnot limit here directly from the first and second law in steady state (Section 4.2). This derivation is much simpler and more general than the typical derivation in textbooks that specifically use the Carnot cycle to derive the limit. When the work done by a heat engine remains within the system and is dissipated there, as in the case of the atmosphere, the Carnot limit can be extended to account for this dissipative heating. It was straightforward to include this dissipative heating term to obtain the Carnot limit of a dissipative heat engine (Section 4.4) that has been used in atmospheric science (e.g., Renno and Ingersoll 1996; Bister and Emanuel 1998). It is important to note that the Carnot limit formulates the limit to energy conversion in a general way without making specific assumptions about how the conversion process takes place. The Carnot limit thus conveniently allows us to establish an energy conversion limit from the conditions at the boundary of the system without the need to explicitly consider the details of how the conversion takes place inside the system.

When the Carnot limit is applied to Earth system processes such as atmospheric motion, then the interaction of this process with the boundary conditions of the heat engine need to be accounted for. This can be done in a steady-state setting of the energy balances. The combination of the Carnot limit with the energy balances results in a fundamental trade-off by which a greater heat flux through the heat engine is accompanied with a reduced driving temperature difference, thus resulting in the maximum power limit (Section 4.5). Such interactions and their effect on thermodynamic limits that take place at a somewhat broader description of the system are typically not described in thermodynamics. These effects on the boundary conditions is typical for Earth system processes. In the case of convection, for instance, this interaction of convective heat fluxes with their driving temperature difference is rather strong, which results in the maximum power limit. Furthermore, in the application of this limit to the diurnal cycle, it was shown that temporal variations and storage effects alter this limit (Section 10.3). Thermodynamic limits in the Earth system are thus shaped not just by the first and second law, but also by interactions with the boundary conditions and aspects of variability.

The maximum power limit does not only apply to thermal energy conversions, but also to further conversions into other forms of energy as well (see also Odum and Pinkerton 1955). For instance, we dealt with a maximum power limit to wind energy conversion into renewable energy which shared the same features, with the momentum balance acting as a relevant constraint. The resulting dynamics are then a consequence of the second law, aim to reduce the free energy within the system, and result in a reduced state of thermodynamic disequilibrium. For instance, in the case of atmospheric convection, this results in a substantial cooling of the surface of the Earth, and thereby reduces the disequilibrium represented by the temperature difference between the surface and the radiative temperature of the atmosphere. It is within the Earth system context that work, dynamics, and depletion of disequilibrium are all connected. The resulting dynamics act to accelerate the depletion of the disequilibrium in the direction imposed by the second law that in total result in a more dissipative thermodynamic system at the planetary scale.

The most direct application of thermodynamic limits to the Earth system deals with the conversion of heating to motion. This sets the limit to atmospheric and interior motion that can be generated from the planetary forcing, as described in Sections 6.6, 7.4, and 7.7 and with the magnitudes summarized in Fig. 12.1. The maximum power limit acts to constrain the generation of kinetic energy (shown by the thermal conversion efficiency in Fig. 12.1), the associated heat flux as well as the temperature difference, as these three aspects are linked by the constraints imposed by the maximum power limit and the associated energy balances. As the resulting heat fluxes and temperature differences compare well with observed magnitudes, this agreement suggests that planetary motion on Earth operates near its thermodynamic limit. It results in a maximum generation of about $4600 \cdot 10^{12}$ W of kinetic energy, with most of it being associated with dry and moist convection and the large-scale circulation contributing a maximum of about $1000 \cdot 10^{12}$ W to this estimate.

This thermodynamic limit of maximum power has relevance to planetary processes beyond motion as it constrains the intensity of mass exchange (Section 7.5). As the heat transported by motion is constrained by thermodynamics, it also constrains the rates by which other characteristics are being exchanged, such as

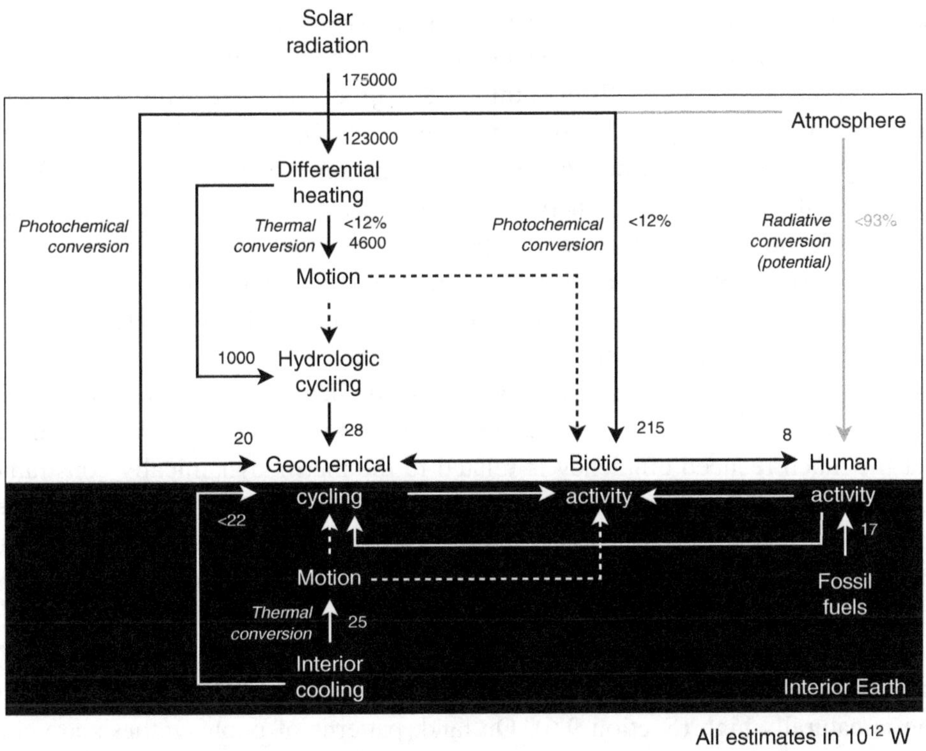

Figure 12.1 Estimates of planetary drivers and energy conversion limits for different Earth system processes with the connections that were outlined in Fig. 1.5 in the introduction. Black arrows indicate flows of energy, while dashed arrows indicate the effects of mass exchange limitations. All estimates are given in units of 10^{12} W and are described in greater detail in the respective chapters.

water vapor, carbon dioxide, and other compounds. This constraint is indicated in Fig. 12.1 by the dashed arrows. The thermodynamic constraint of maximum power thus translates into a constraint on the rate of material cycling within the system. This link was used to describe the limit to hydrologic cycling in Section 8.5 and to geochemical cycling in Section 9.5. The implied limit on hydrologic cycling results in a magnitude consistent with observations, and suggests a magnitude of potential energy generation of about $1000 \cdot 10^{12}$ W. It also provides different expressions for the sensitivities to surface warming of the saturation state, of evaporation and precipitation that relate to the sensitivity of the associated latent heat flux, and aspects such as thunderstorms or hurricanes that relate to the associated moist power (Section 8.6) for which different sensitivities were previously reported.

Geochemical cycling is driven by the chemical free energy generation through a variety of conversion processes (Chapter 9). Atmospheric photochemistry, particularly the photodissociation of molecular oxygen, directly converts solar radiation into chemical energy. Chemical free energy is also generated by hydrologic cycling, directly by the desalination of seawater and indirectly by lightning-based chemistry related to moist convection. These atmospheric processes were estimated to generate chemical free energy in the order of about 20 and $28 \cdot 10^{12}$ W. The chemical free energy generation by interior processes was constrained by the interior heat flux. If reduced by the generation of kinetic energy of interior motion derived by maximum power in Section 7.7, this leads to an upper limit of about $22 \cdot 10^{12}$ W.

In contrast, using estimates of present-day photosynthetic activity, it was shown that biotic activity generates substantially more chemical free energy than abiotic processes. Photosynthesis, as the main energy generation process for biotic activity, is also a photochemical process, yet its actual conversion efficiency is well below the theoretical 12% maximum conversion efficiency. We speculated that the much reduced efficiency is related to the thermodynamically constrained rate of mass exchange, indicated by the dashed arrow in Fig. 12.1, because it sets a limit to the rate by which living organisms can exchange resources with their environment to sustain their metabolisms. That this mass exchange acts as a relevant constraint is consistent with the current view that the supply of nutrients constrains and shapes observed patterns of marine productivity and with the theory that life first emerged in hydrothermal vents at the ocean floor where this mass exchange rate is naturally high (Section 9.6). On land, patterns of photosynthesis are commonly described in terms of water and light limitations, yet it is well established in the literature that photosynthesis operates well below the conversion limit for solar radiation, which may imply that light may not be the limiting resource. The thermodynamic limitation represented in mass exchange can explain this apparent discrepancy, because it suggests that solar radiation does not act as a limiting energy source for terrestrial photosynthesis, but notably as a limiting radiative heating source to drive mass exchange between the surface and the atmosphere (Section 10.6).

Last, but not least, human activity draws upon the chemical energy generated by photosynthesis to feed its metabolic activity and fossil fuels to sustain what we have referred to here as externalized activity. In Chapter 11, we discussed that food acquisition accounts for about $8 \cdot 10^{12}$ W while primary energy consumption represents about $17 \cdot 10^{12}$ W, which are notable magnitudes when compared to the other estimates in Fig. 12.1. This substantial magnitude of human activity emphasizes its role as a planetary process with planetary scale impacts on the Earth system.

With this, we established the thermodynamic connections from the planetary drivers to human activity, as outlined in Fig. 1.5 with estimates given in Fig. 12.1.

It is important to note that while this view is described in a top-down fashion, from the planetary forcing to human activity, interactions play a critical role in establishing the state of the system as reflected in the estimates shown in Fig. 12.1. This critical role can be seen in the basic trade-off that sets the maximum power limit by which the flux depletes the driving gradient. While for the maximum power limit, the feedback to the driving temperature gradient is direct and relatively fast, the effects of the cascade of Earth system processes that are driven by planetary motions are more subtle, indirect, and can act on much longer time scales. These effects, no matter how subtle or slow, feed back to the planetary drivers, thus resulting in a thermodynamic system in which the emergent state is strongly determined by planetary interactions. These interactions involve the interactions with hydrologic cycling. The radiative characteristics of the different states of water play a key role as they strongly shape the radiative environment, in terms of the water vapor content in the atmosphere, which is the dominant greenhouse gas of the Earth's atmosphere, and by the presence of clouds, snow, and ice, which affect the albedo. These effects strongly affect the solar and terrestrial radiative fluxes in the Earth system and thereby feed back to the planetary radiative forcing (cf. Section 8.7). Geochemical cycling affects the chemical composition of the atmosphere and crust, which in turn affect the respective radiative and material properties and the magnitude of the greenhouse effect (Section 9.8). Biotic activity alters geochemical cycling and the surface energy balance on land (Section 10.7). The energy needed to sustain human activity and which is acquired from the Earth system depletes states of thermodynamic disequilibrium (Section 11.5), culminating into the human impact on the planetary system that is reflected in the current trends of rising atmospheric CO_2 concentrations.

This yields a thermodynamic Earth system in which processes operate in an overall thermodynamic direction described by the second law and are subjected to thermodynamic limits, yet interactions also play an important role as they feed back to the planetary drivers and shape the limits. Overall, these dynamics and interactions yield a system that is as dissipative as possible, thus accelerating the advance in the direction of the second law to the extent possible. It yields a perspective of the Earth system in which energy conversions and transport constraints are the bottlenecks of the system, yet that these constraints can be overcome, for instance, by internal cycling of nutrients within ecosystems (as described in Section 9.7) or by technological means associated with human activity. It thus provides quite a different view on how the Earth system works, in which direction it evolves to, and which aspects constitute the bottlenecks of the system from a view that is based on common climatological variables, such as surface temperature and wind speeds.

12.3 Habitable planets and life

This view of the thermodynamic Earth system links environmental conditions to biotic activity that goes beyond simple thermodynamic requirements for life as formulated by Boltzmann (1886) and Schrödinger (1944). While these authors recognize the role of low entropy sources to sustain metabolic activity, the additional

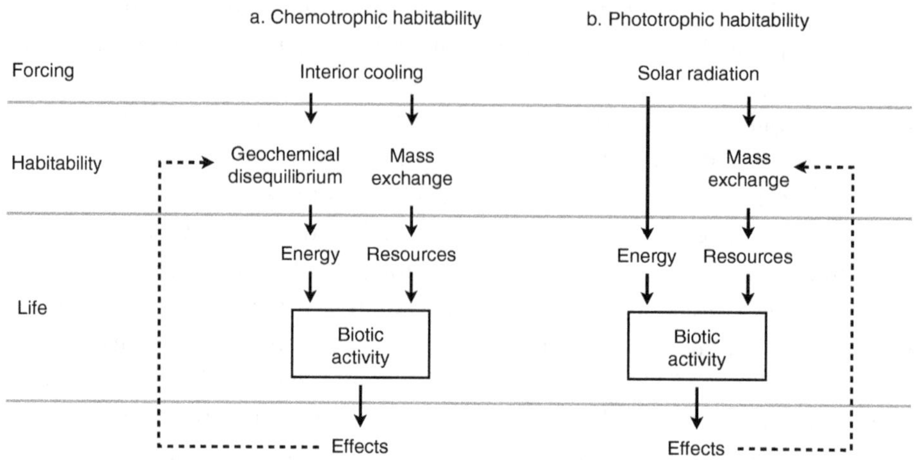

Figure 12.2 Illustration of two forms of habitability related to (a) geologic and (b) climatic factors of the Earth system applicable to chemotrophic and phototrophic life.

constraint that emerges from this thermodynamic view is related to the ability of the planetary environment to sustain mass exchange between biotic activity and its surroundings. This additional thermodynamic constraint is illustrated in Fig. 12.2 for the forcing associated with the cooling of the Earth's interior, which relates to the habitability of chemotrophic life, and that associated with the solar radiative forcing, which relates to the habitability of phototrophic life.

As any life form, or in fact, any chemical reaction, requires material exchange of its reactants and products to be sustained in a steady state of thermodynamic disequilibrium, it would appear that this ability of the planetary environment to exchange mass of different chemical composition is a key factor in shaping planetary habitability. A more habitable environment would then be characterized by an environment that has a greater ability to exchange mass between the biota and its environment, thereby enabling a greater metabolic activity. From this requirement, we should then be able to infer what a more habitable environment should look like in terms of the planetary characteristics in combination with its forcing. Furthermore, this view can then inform us about the conditions for the emergence of life and of the potential thermodynamic signatures of inhabited planets.

Planetary habitability is commonly defined in terms of the conditions that allow for the presence of liquid water (Huang 1959), although other factors such as surface temperature also play a role (Lineweaver and Chopra 2012; Seager 2013). Liquid water can only exist for a certain temperature and pressure range, with the temperature of a planetary surface being related to the solar radiative forcing. As the solar radiative forcing decreases with increasing distance of a planet to its star, there is a certain range of distances in which the conditions at the surface of the planet are neither too hot nor too cold to allow for liquid water, which is referred to as the habitable zone.

To evaluate this condition, internal factors of the planetary atmosphere need to be accounted for, particularly regarding the greenhouse effect on surface temperature related to the concentrations of water vapor and carbon dioxide (Kasting et al. 1993). This definition of planetary habitability is illustrated here in terms of the planets of the solar system in Table 12.1. This table lists the planets as well as the Earth's moon in the sequence of increasing distance from the Sun as well as its planetary properties and conditions, with the distance of the planet to the Sun, d_{planet}, its incident solar radiative flux $R_{s,\text{in}}$, the planetary albedo α_{planet}, the mean absorption of solar radiation R_s, the observed surface temperature T_s, the computed radiative temperature T_r from R_s, and the maximum conversion limit from radiative to free energy $G_{\text{rad,max}}$, as well as the limit from radiative heating to kinetic energy by convection $G_{c,\text{max}}$, and large-scale motion, $G_{\text{ls,max}}$. The habitable zone typically stretches from around the orbit of Venus to Mars, noting that the surface conditions on Venus and Mars could be considerably changed by their respective greenhouse effects.

The thermodynamic view adds a more dynamic perspective to this definition of habitability. While liquid water is a very suitable medium to dissolve and transport chemical compounds, the view here adds the dynamic constraint that water – or any chemical compound in any form – also needs to be able to physically move within the planetary environment. This is similar to the concept of power supply and demand which was used by Shock and Holland (2007) to define habitability. The starting point to assess the suitability of planetary environments to generate motion is the potential to generate free energy from the solar radiative forcing. The upper bound of this limit, $G_{\text{rad,max}}$, is described by the direct conversion of solar radiative energy into free energy, as given by eqn. 6.28, an aspect that relates to the rate of planetary entropy production and that is easily estimated from absorbed solar radiation and the radiative temperature of the planet (Aoki 1983).

While this rate sets the upper potential, the relevant limit related to mass exchange of climate processes is much lower and described by the maximum power limit associated with the conversion of radiative heating into physical work by convection and large-scale motion, $G_{c,\text{max}}$ and $G_{\text{ls,max}}$ (cf. Sections 6.6 and 7.4). These two limits are derived in simplified form and illustrated for the solar system in Table 12.1. The limit to convective motion is derived from the Carnot limit applied to half of the absorbed solar radiation and the temperature difference between the surface and the radiative temperature. Note that this is a simplification, because it does not explicitly include the effect of the atmospheric greenhouse and it does not use the solar radiation absorbed at the surface but rather the total (on Venus, for

Table 12.1 *Planets of the solar system and their climatic and thermodynamic characteristics in terms of their distance to the Sun d_{planet}, their incident solar radiation $R_{s,in}$, planetary albedo α_{planet}, absorbed solar radiation R_s, observed surface temperature T_s, radiative temperature T_r, maximum total power from radiative conversion $G_{rad,max}$, and maximum power for convective and large-scale motion, $G_{c,max}$, and $G_{ls,max}$.*
Data of the first five columns taken from NASA planetary fact sheet, http://nssdc.gsfc.nasa.gov/planetary/factsheet/.

Planet	d_{planet} (10^9 m)	$R_{s,in}$ (W m^{-2})	α_{planet} (frac.)	R_s (W m^{-2})	T_s (K)	T_r (K)	$G_{rad,max}$ (W m^{-2})	$G_{c,max}$ (W m^{-2})	$G_{ls,max}$ (W m^{-2})
Mercury	57.9	9126.6	0.14	1957.6	440.2	431.1	1762.3	20.2	9.79
Venus	108.2	2613.9	0.90	65.4	737.2	184.3	62.3	24.5	0.33
Earth	149.6	1367.6	0.31	237.3	288.2	254.3	223.3	13.9	1.19
Moon	149.6	1367.6	0.12	300.9	253.2	269.9	282.1	*	1.50
Mars	227.9	589.3	0.17	122.3	208.2	215.5	116.2	*	0.61
Jupiter	778.6	50.5	0.52	6.1	163.2	101.7	5.9	1.1	0.03
Saturn	1433.5	14.9	0.47	2.0	133.2	76.8	1.9	0.4	0.01
Uranus	2872.5	3.7	0.5	0.5	78.2	53.2	0.5	0.07	0.00
Neptune	4444.5	1.5	0.4	0.2	73.2	44.8	0.2	0.04	0.00
Pluto	5870.0	0.9	0.6	0.1	48.2	35.4	0.1	0.01	0.00

*: No mechanical power can be derived because $T_s < T_{rad}$.

instance, most solar radiation is absorbed in the atmosphere). The limit to large-scale motion is derived as in Section 7.4, using 40% of the absorbed solar radiation to approximate the insolation difference between tropical and extratropical regions. This estimate is also a simplification, as, for instance, Venus rotates very slowly so that the dominant horizontal radiative gradient is the difference between the day and night hemispheres. In this case, heat transport is directed towards depleting this radiative difference, for which a thermodynamic limit can also be established (Lunine and Lorenz 2002; Lorenz 2010).

The proxies for the limits of generating motion and mass exchange given in Table 12.1 nevertheless provide insightful, first-order approximations. Mercury has the highest limit due to its proximity to the Sun and its low albedo, yet because it has no atmosphere, this potential to generate motion and drive mass exchange cannot be realized. To achieve this limit, it requires a fluid that is radiatively active so that a radiative temperature difference can be generated between the absorption at the surface and the emission from the atmosphere, or between greater absorption in tropical regions and greater emission in extratropical regions. The lack of an atmosphere on Moon also implies that this potential cannot be achieved. As for the outer planets, the radiative forcing is so low that little motion is generated and dissipated, thus representing a low ability to exchange mass. Hence, even if the surface temperatures were sufficiently high to be suitable for liquid water, the ability of the outer planets to generate free energy from the Sun is strongly limited, which is simply attributable to the large distance from the Sun. This only leaves Venus and Mars as planets with climate systems that have a comparable ability to exchange mass as the Earth system.

It is important to recognize that the ability of a planet to exchange mass does not relate to the wind speeds of the atmosphere. Observations of planetary jet streams show that the outer planets have jet stream velocities that are much greater than on Earth despite a lower radiative forcing, an observation referred to as jet stream paradox (Catling 2005). One should note, however, that jet streams represent near geostrophic flow (cf. Section 7.6), which has greater wind speeds the less dissipative the flow (cf. eqn. 7.29), so that greater geostrophic wind speeds do not seem to be in contradiction with a less dissipative planetary system. It would thus seem that a greater ability for mass exchange should be associated with more dissipative atmospheric flow, which in turn should be linked to weaker jet streams.

These aspects illustrate that the thermodynamic limits on generating motion can be related to the habitability of a planet by providing the means to sustain mass exchange. While mass exchange is insufficient to explain life, it nevertheless constitutes a necessary condition that refines the definition of a habitable environment that can support the origin of life and maintain a high level of biotic activity. In the context of the evolution of life on Earth, this condition is consistent with the current understanding that life first emerged as chemotrophic life in hydrothermal vents at the sea floor (Russell and Hall 1997, see also Section 9.6). These sites exhibit a naturally large rate of mass exchange so that they are likely to have sufficiently high fluxes of chemical free energy to be utilized. As hydrothermal vents relate to mid-oceanic ridges, which require sea floor spreading, plate tectonics may be an important requirement for the origin of life and in shaping habitability (Spohn 2012). Yet, chemotrophic life depletes chemical disequilibrium in the environment, as indicated by the dashed arrow in Fig. 12.2a, and has little means to alter mass exchange. It would thus seem to be constrained to a low dissipative activity.

For phototrophic life, this mass exchange is generated mostly by winds at the ocean surface and by surface heating due to the absorption of solar radiation on land. This intensity of mass exchange is not a fixed property of the system, but is affected by the strength of the greenhouse effect, hydrologic cycling, and directly by vegetation functioning on land (cf. Chapter 10.7). As biotic activity affects the atmospheric composition and the strength of the greenhouse effect, the effects of biotic activity can potentially result in positive feedbacks by which phototrophic life enhances mass exchange (as shown by the dashed arrow in Fig. 12.2b), thereby possibly making planetary conditions much more habitable.

With greater levels of biotic activity, life has inevitably a greater effect on its environment. The thermodynamic implication of this effect can then in principle be used to remotely identify life on planets, as proposed by Lovelock (1965). However, as already discussed at the end of Chapter 9, the dominant effect of chemotrophic (and heterotrophic) life is to deplete chemical disequilibrium in the environment. The effect of this form of life would then be impossible to distinguish from an abiotic process that depletes the disequilibrium as well, so that chemotrophic life appears unlikely to be detectable from the thermodynamic state of the planetary environment. In case of phototrophic life, the direct planetary imprint is likely to be a greater magnitude of chemical disequilibrium as solar radiation was used to generate chemical free energy. The major form of disequilibrium generated by photosynthesis is represented by the oxic atmosphere and the reduced carbon compounds in biomass and other organic carbon stocks. This chemical disequilibrium between atmospheric oxygen and reduced carbon compounds at the surface could, in principle, be detectable from space, thus potentially forming a signature of a habitable planet, even though the magnitude of disequilibrium is not indicative of how active the biosphere would be (cf. Section 9.9). It would thus seem that habitability is intimately related to the thermodynamics of a planetary system. Yet, the description provided here goes beyond a state of chemical disequilibrium within the atmosphere and concerns the ability of a planetary system to enhance mass exchange by internal processes as a basic requirement to sustain chemical and metabolic reactions at high dissipative rates.

12.4 Planetary evolution, optimization, and regulation

The thermodynamic view of the Earth system can be applied to its past evolutionary history and related to a possible evolution towards an optimal state at the planetary scale. We described qualitatively in Chapter 5 how a thermodynamic process would likely evolve to its limit by its dynamics. The description was based on two dynamic feedbacks that act upon the power associated with the process, with a fast positive feedback of the heat flux on the power and a slow negative feedback associated with the depletion of the gradient by the heat flux (see Loop A in Fig. 12.3). This description was focused on the dynamics and the thermodynamic limit of a single process. The goal in the following is to extend this line of reasoning to the thermodynamic Earth system so that we can formulate what we can expect from the thermodynamic evolution of the whole Earth system, which characteristic behavior we can expect from its emergent steady state and which thermodynamic role life is likely to play in this system. This perspective follows up on Lotka's seminal work (Lotka 1922a,b) on maximum power and biological evolution by natural selection and extends it to maximum power of the whole Earth system.

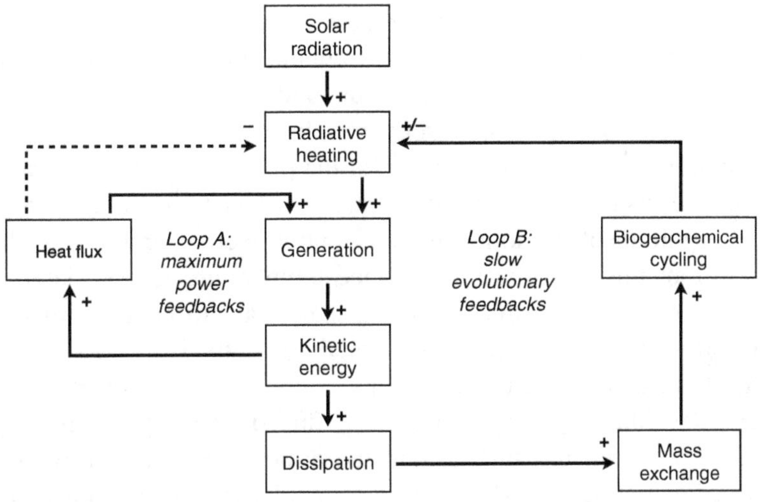

Figure 12.3 A combined view of the feedbacks associated with the maximum power limit on generating motion and the linkages with the biogeochemical cycling of water, carbon, and other compounds. The evolution towards the maximum power limit likely involves two relatively fast feedbacks (Loop A), shown on the left. When mass exchange acts as a limitation on biogeochemical cycling, then this process can result in a slow feedback on the radiative forcing (Loop B) that would result in thermodynamically constrained evolutionary trends directed towards a radiative forcing that would permit more power and mass exchange.

The extension of feedbacks to the thermodynamic Earth system requires the integration of processes that act across a range of time scales. The dynamics of convection towards the maximum power limit likely operate on short, sub-daily time scales, as the dynamics and the response of the driving temperature gradient react comparatively fast. This can, in principle, be further quantified by the characteristic time scale set by the ratio of kinetic energy to power, as in Section 5.2. The associated mass exchange then acts as a limitation to other processes, such as the hydrologic cycle, and potentially biotic activity and geochemical transformations at the surface. These processes take place at successively longer time scales of weeks, years, and thousand of years, so that these processes take longer to reach a steady state with the radiative forcing and mass exchange. Yet, these processes may also alter the radiative forcing, in terms of the solar radiative heating at the surface, for instance, by changes in ice or cloud cover, in vegetation, or by changes in the atmospheric composition that alter the greenhouse effect. Such changes in the radiative forcing could then result in slow feedbacks and evolutionary trends in the Earth system as they alter the thermodynamic limit on mass exchange, as shown by loop B in Fig. 12.3. These slow trends would then be thermodynamically constrained because mass exchange would act as the bottleneck for these processes. The trends would be thermodynamic in that these would successively result in greater power associated with the processes, from the power involved in generating motion to the power involved in biogeochemical cycling, and likely result in states of greater thermodynamic disequilibrium and a more dissipative Earth system.

To illustrate that such feedbacks are plausible and relevant and may result in an optimum state for mass exchange at the planetary scale, we evaluate how mass exchange reacts to surface temperature. As surface temperature is strongly affected by the magnitude of the atmospheric greenhouse effect and thus by the atmospheric composition, this sensitivity to surface temperature is associated with different intensities of biogeochemical cycling that relate to changes in the atmospheric composition. This sensitivity is illustrated qualitatively in Fig. 12.4.

The starting point for the sensitivity is the expression of mass exchange at the surface, described by the sensible heat flux divided by the driving temperature difference between the surface and the atmosphere, ΔT, as in Section 10.5. In a warmer greenhouse world with a given solar radiative forcing, the temperature difference ΔT would be enlarged, as sketched by the dashed line in Fig. 12.4, thus requiring less convective exchange to accomplish the convective heat fluxes. Such a decrease in convective heat flux with surface warming has been inferred from the maximum power limit in Section 8.6, described earlier (Betts and Ridgway 1989), and found in climate model studies of global warming (Bony et al. 2013). In the presence of an evaporating surface, the greater value of the slope of the saturation vapor pressure, s, with a warmer surface temperature further reduces the sensible heat flux and thus the associated mass exchange (Kleidon and Renner 2013a).

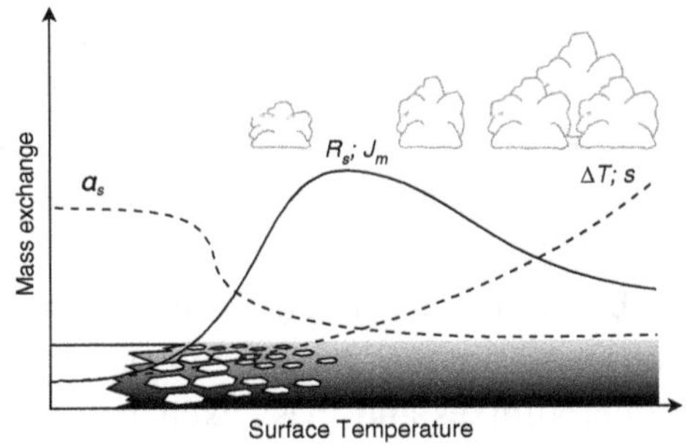

Figure 12.4 Illustration of a plausible, thermodynamically optimal state of maximum mass exchange J_m, between the surface and the atmosphere at the planetary scale. The state would result from the contrasting effects of surface temperature on the extent of ice and snow, affecting surface albedo α_s, the surface absorption of solar radiation R_s, the temperature difference between the surface and the atmosphere ΔT, and the slope of the saturation vapor pressure curve s. Reproduced after Kleidon and Fraedrich (2005) with kind permission from Springer Science and Business Media.

In addition, the absorption of solar radiation at the surface should decrease, as a higher concentration of water vapor would absorb more solar radiation within the atmosphere. These factors would suggest that mass exchange decreases in magnitude with a warmer surface temperature due to a number of factors.

On the other hand, the colder the surface temperature, the more likely the presence of ice and snow. This presence would increase the surface albedo, shown by the dashed line labeled α_s in Fig. 12.4, reducing the surface solar radiative forcing and thus mass exchange. Even though other relevant effects, such as seasonal variations, horizontal transport by the large-scale atmospheric circulation, are not accounted for in this simplified picture, the basic physical mechanisms described here suggest that we could expect a maximum rate of mass exchange on a planet with a cool temperature that is above the freezing, but not too warm. This state likely represents a state with greatest absorption of solar radiation at the surface, the greatest rate of conversion into other forms of energy, and the strongest biogeochemical cycling. It would thus seem that this state of the Earth system is a state that is thermodynamically optimal with a maximum rate of dissipative activity at the planetary scale.

When we link the coupled dynamics of the maximum power state of mass exchange and biogeochemical cycling shown in Fig. 12.3 with the planetary optimum state shown in Fig. 12.4, we can derive a thermodynamic baseline scenario of how a planet with life should evolve towards the optimal state (Kleidon 2009b). Biotic activity would play a key role in this baseline, as it fuels and modulates biogeochemical transformations and cycling with the supply of chemical free energy generated by photosynthesis. Once life emerged, we would expect that its further evolution would be directed towards higher levels of biotic activity, as described by Lotka (1922a) as "... natural selection tends to make the energy flux through the system a maximum, so far as compatible with the constraints to which the system is subject." Lotka's work links the well-established notion of biological evolution by natural selection to the thermodynamic evolution of free energy generation and dissipation by the biosphere that is subjected to the constraints imposed by mass fluxes and mass exchange of the environment. This thermodynamic evolution would be associated with changes in geochemical cycling through food webs such that nutrient limitations would be alleviated by faster internal cycling within the biosphere, as described by the cycling ratios of Volk (1998) and demonstrated in Section 9.7. Biotic activity would thus reach a state of maximum power associated with its metabolisms that is given by the physical constraints set by the planetary environment, particularly regarding its ability to exchange mass. The effects of biotic activity could then alter geochemical cycling within the Earth system that would affect the radiative forcing and thereby the slow, positive feedback between biotic activity, geochemical cycling, and mass exchange that would tend to alleviate these constraints. The final outcome of this coupled evolution would be the thermodynamic optimal state shown in Fig. 12.4 in which mass exchange is maximized, thus maximizing the planetary habitability as formulated in the previous section.

This optimal state would be shaped entirely by the physical factors that determine the thermodynamic limit to exchange mass. It mirrors the suggested optimal state on land that related to the intensity of evapotranspiration in Section 10.7, with the difference being that the state described here would apply to the planetary scale and involve more generally the combination of solar and greenhouse forcing on mass exchange. It would then be reasonable to expect that the coupled system at this optimal state would react with negative feedbacks when perturbed, similar to how a system at a state of maximum power would react to perturbations, as described in Section 5.5. This outcome would then share great similarities to the behavior described by the Gaia hypothesis of Lovelock (1972a,b): the planetary thermodynamic optimal state would represent a homeostatic state that is regulated "for and by the biosphere," as it would represent a state that allows for maximized biotic activity by maximized mass exchange and that is shaped by the effects of biotic activity on the physical constraints on mass exchange. While such a strong interaction between life and Earth and its effect on the evolutionary past of the Earth system is well recognized and referred to as the coevolution of life and Earth (e.g., Lenton et al. 2004; Retallack 2007), this coevolutionary dynamics is typically not expressed as a thermodynamic evolution with a specific trajectory towards greater power, disequilibrium, and a thermodynamic limit at the planetary scale.

While this picture is clearly speculative and needs further substantiation, what we want to do in the following is to illustrate that this thermodynamic baseline scenario is nevertheless plausible and could serve as a first guess of what to expect from thermodynamics of the evolution of an inhabited planet. In this baseline, we would expect a general trend in which the biosphere and the Earth system generate more free energy, thus resulting in an evolution towards an increasingly more dissipative system and a greater level of thermodynamic disequilibrium up to the thermodynamic optimal state. We can then ask whether the evolutionary past of the Earth system followed this baseline. To illustrate this, the dominant reconstructed trends in atmospheric composition over Earth's history in relation to evolutionary changes in the biosphere are shown in the top part of Fig. 12.5 (based on Catling 2005; Holland 2006), while the bottom part sketches the hypothesized thermodynamic trend. The focus here is not on the precise magnitude of these trends or their timing, but rather whether these are indicative of a general thermodynamic trend towards a more dissipative Earth system due to biotic activity.

We should first note that the planetary drivers have changed over the history of the Earth. On early Earth, the solar forcing was about 70% of today's forcing, and the interior heat flux as well as the tidal forcing were much stronger (Stevenson

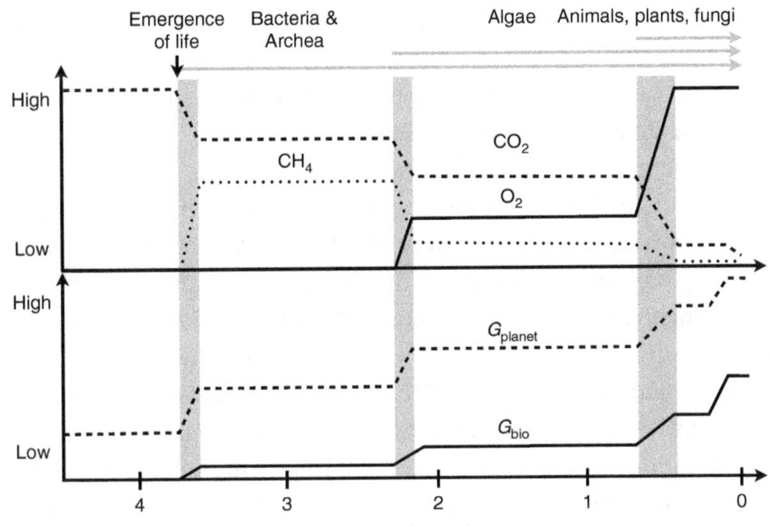

Figure 12.5 Qualitative illustration of (top) the reconstructed evolutionary history of the Earth's atmospheric composition in relation to life and (bottom) its speculated characterization in terms of thermodynamic trends associated with the rates of free energy generation by the biosphere, G_{bio}, and of the whole planet, G_{planet}. Trends are not drawn to scale and timings are approximations. Transition periods are marked by the grey areas. Time is given in billion of years before the present (b.p.).

et al. 1983; Kagan and Sündermann 1996). These differences would have resulted in a different planetary forcing of the early Earth. The much lower solar radiative forcing together with the absence of geologic indications for frozen water were used by Sagan and Mullen (1972) to suggest that the early Earth likely had a much stronger greenhouse effect than the present day to maintain temperatures that were well above the freezing point of water. This initial state of the atmospheric composition was subsequently transformed by geochemical weathering reactions (Walker et al. 1981) that resulted in distinct changes and transitions in the atmospheric composition, from an initially reducing environment to a final oxidizing state (Kasting 1993). This trend is reflected in the increasing concentration of atmospheric oxygen in Fig. 12.5, with transition periods marked in grey that coincide with the emergence of additional, and more complex forms of life.

From a thermodynamic perspective, this trend of increasing atmospheric oxygen relates to an increase in the geochemical disequilibrium of the Earth system. Note that this increase likely did not proceed as a uniform trend, but was probably characterized by some transition periods of substantial change that are marked by the grey areas in Fig. 12.5. Geologic reconstructions indicate that these transition periods may have been associated with episodes of large-scale glaciations that are referred to as "Snowball Earth" periods (e.g., Hyde et al. 2000). Such periods would have been less dissipative because of the much higher surface albedo of the glaciated surface. Yet, as these periods were transient, they probably simply reflect an overshoot on the cold side of the optimum showed in Fig. 12.4 on the path to a more dissipative state at the planetary scale.

After life emerged some 3.8 billion years ago, biotic productivity introduced the first geochemical changes in the environment. Its activity likely released some oxygen, which was consumed by the oxidation of the crust. The decomposition of organic material likely proceeded in an anoxic environment, resulting in the production of methane and elevated concentrations of this greenhouse gas in the atmosphere. During this stage, the chemical disequilibrium would then be reflected mostly by the reducing conditions of the atmosphere and the oxidized state of the crust. Once photosynthetic life emerged and became abundant around 2.2 billion years ago, conditions changed to an oxic environment with greater concentrations of atmospheric oxygen. These changes would have resulted in disequilibrium between the oxic atmosphere and the reduced biomass, and would have been enhanced from the previous stage. As the maintenance of a greater extent of chemical disequilibrium requires a greater generation rate, this enhanced disequilibrium state suggests a greater photosynthetic activity. A third major transition occurred about 0.5–0.8 billion years ago during the Cambrian explosion, characterized by cell differentiation, the emergence of multicellular life, and the emergence of vascular plants (e.g., Lenton et al. 2004). These evolutionary changes were accompanied by a further increase in atmospheric oxygen and a reduced level of carbon dioxide. As this would represent an even greater extent of chemical disequilibrium, it suggests that photosynthetic activity was further enhanced by these evolutionary changes.

Consider two further examples of transitions that occurred on land during the last 400 million years of planetary evolution that likely enhanced biotic activity without a marked planetary change in the atmospheric composition and thus may not have altered the chemical disequilibrium. These changes are marked symbolically by the most recent increase in G_{bio} in Fig. 12.5. The first example relates to the emergence and spread of flowering plants, or angiosperms, around 120 million years ago. Flowering plants are characterized by a hydraulic architecture that allows for a much greater photosynthetic capacity and greater rates of leaf gas exchange (Brodribb and Feild 2010; Feild et al. 2011), so that the spread of flowering plants likely resulted in greater terrestrial productivity and change in the terrestrial biosphere (Boyce et al. 2009). The second example concerns the evolution of grasses and the expansion of grasslands, which likely enabled the evolution of large mammals including hominoids during the last 35 million years (Retallack 2001; Bobe and Behrensmeyer 2004). As was described in Section 9.7, the presence of grazers can enhance nutrient cycling in ecosystems as decomposition of organic matter in humid digestive systems within mammals can proceed at a faster pace than in the abiotic, dry environment.

These examples are indicative of an evolution of the Earth system that successively allowed for greater biotic activity, resulting in greater chemical free energy generation, G_{bio}, and greater geochemical and physical changes in the planetary environment that are marked to some extent by a greater chemical disequilibrium and that should be related to greater power G_{planet}. This trend is consistent with the thermodynamic baseline scenario. What this trend does not inform us about are the changes in the physical dynamics of the Earth system and the extent to which these act to limit the necessary mass exchange associated with biotic activity, as described by the basic feedback shown in Fig. 12.3. Such constraining effects of mass exchange on biotic activity are currently not considered, so that this aspect would require further evaluation in the future. Our thermodynamic baseline scenario would predict that this constraining effect on mass exchange would play a major role, and would be associated with substantial changes in the physical climate system during the evolutionary past towards greater free energy generation, that is, larger values of G_{planet}, and a greater ability to transport and exchange mass between the biosphere and the planetary environment. It would seem that this thermodynamically constrained baseline scenario for the evolution of an inhabited planet towards greater dissipation and disequilibrium is, while speculative, at least a plausible scenario.

12.5 Human activity and a sustainable future

As a next step, we want to extend this thermodynamic baseline scenario into the future and ask what it would imply for human activity and the future evolution of the planet. This question would allow us to sketch out possible future scenarios of human activity that would be sustainable within the Earth system and understand the impacts that human activity has at a basic level. A critical component for this thermodynamic perspective of human activity is to formulate and view it as a dissipative process that operates within the Earth system, as it was described in Chapter 11. When we treat human activity in this abstracted thermodynamic formulation, its activity is measured in terms of an acquisition rate of free energy, G_{human}, and its impact can be measured in terms of its effects on states of disequilibrium within the Earth system. It allows us to apply the insights from purely physical processes and see how these advance to their thermodynamic limits to the evolutionary dynamics of the coupled human–Earth system and speculate what the likely outcome of this evolution would be in the near and in the distant future.

To link this thermodynamic view to the more common view of anthropogenic global change, we first note that much of the impacts of human activity on the Earth system are caused by food production and consumption of fossil fuels. Both aspects involve the appropriation of chemical free energy from the Earth system in terms of carbohydrates from biomass or geologic deposits. When this energy is consumed by human activity, it depletes the associated state of thermodynamic disequilibrium and leaves less free energy behind to drive natural processes. With this general interpretation, the effects of human activity in terms of global warming and land cover change were linked to a depletion of thermodynamic disequilibrium in Section 11.5. In this view, it would seem that human activity has mostly a "detrimental" effect on the Earth system as its effects bring the Earth system closer to thermodynamic equilibrium and reduce its ability to generate free energy.

A critical aspect that can result in the opposite effect is a more efficient use of solar radiation by human-made technology. A specific example for such technology is photovoltaics, which generates electric free energy from absorbed solar radiation. As described in Section 11.4, this technology has a much greater conversion efficiency than a heat engine that is driven by the solar radiative heating of the surface. This can then result in a large increase in free energy generation available to human use, as shown by the grey arrow in Fig. 12.1. It also depletes a form of thermodynamic disequilibrium that is represented by the photon composition of solar radiation, which is naturally dissipated when absorbed and converted into heat; but in contrast to abiotic absorption, photovoltaics generates free energy upon absorption.

This energy adds further options to human activity. For instance, it could be used for seawater desalination, thereby making more freshwater available, which could then sustain an agricultural expansion into currently unproductive areas where water is naturally scarce. Such an agricultural expansion could result in a sustainable expansion of the terrestrial biosphere. This additional energy could then also drive greater transformation and transport of mass beyond natural rates of geologic processes (Haff 2010). It would thus effectively add a new "loop C" to Fig. 12.3 that would describe effects of such enhanced free energy generation made available by human technology on mass exchange and biogeochemical cycling. This additional free energy generation can then effectively decouple human-driven processes from the limitations of mass exchange by natural processes within the Earth system. This decoupling is similar to how biotic activity decouples itself from the limits of geochemical cycling by a faster internal cycling of nutrients. It can then result in a more dissipative state of the Earth system that is beyond the level shown in Fig. 12.4 because technology allows for more efficient means to convert radiative forcing into free energy.

We can use this consideration and sketch two possible scenarios of the future evolution of the Earth system that mirror the two scenarios described at the end of Chapter 11 in Fig. 11.8 and which extend the past evolutionary trends sketched in Fig. 12.5. Both of the two scenarios consider an increase in human energy consumption due to future growth in population and lifestyle changes. The motivation for considering only scenarios of increased energy consumption is that so far we have considered thermodynamic processes that evolve to higher rates of power (and dissipation) up to their limit, so that it would appear that it is a basic, thermodynamic trend of dissipative processes to evolve to increased dissipative states, if allowed for by the constraints of the system, rather than to states of reduced energy consumption.

Scenario A in Fig. 12.6 describes an increase in human activity, G_{human}, yet it does not result in an enhanced capacity for natural processes to generate free energy

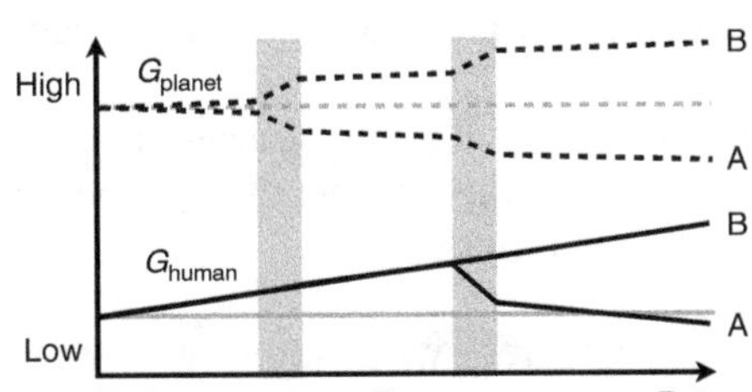

Figure 12.6 Two possible scenarios A and B for the future evolution of the Earth system with human activity, sketched in terms of the two rates of human energy consumption G_{human}, and planetary free energy generation G_{planet}. The grey lines indicate the present-day reference level of the two rates. The shaded areas refer to non-linear transitions with examples described in the text.

beyond the natural state. In this scenario, human activity draws its energy from natural processes, specifically biotic activity to meet metabolic needs and it uses, e.g., forms of renewable energy that are indirectly generated by solar radiation. As both forms of energy consumption deplete states of thermodynamic disequilibrium in the Earth system, both would eventually result in a reduced rate of free energy generation G_{planet}. Manifestations of this trend would, for instance, be an agricultural expansion into tropical rainforests or the use of wind energy as a renewable energy source. What is also illustrated in Fig. 12.6 is that this progression likely proceeds in a stepwise way, where environmental feedbacks can result in rapid changes once certain thresholds are passed, a concept known as tipping points (Lenton et al. 2008). One example for such a tipping point is the Amazonian rainforest, where a certain extent of land cover change may result in a regional reduction of rainfall which may then reduce the ability of the regional environment to sustain high rates of biotic productivity (Nepstad et al. 2008). Scenario A would thus represent a scenario that in the long run would likely result in a lower level of human activity that can be sustained by the Earth system with a lower ability of the planet to generate free energy.

Scenario B is also described by an increase in human activity so that it initially matches scenario A. A marked difference in this scenario is that the increase in human energy consumption is met by technological means that enhance free energy generation of the biota and the planet as a whole (grey arrow in Fig. 12.1). Such a trend would be manifested in, e.g., an agricultural expansion into currently unproductive regions with the use of technology, like the greening of deserts through irrigation, and the use of solar energy as renewable energy in areas where the absorption of solar radiation currently results only in a radiative heating of the surface. Scenario B would allow for an evolution of the planet towards a more dissipative state that reflects greater levels of energy conversions, higher levels of biotic activity, stronger biogeochemical cycles, and which can sustain greater levels of human activity. It is a scenario that Lotka (1922a) already described as follows:

> "that the influence of man, as the most successful species in the competitive struggle, seems to have been to accelerate the circulation of matter through the life cycle, both by enlarging the wheel and by causing it to spin faster."

This trend would likely also progress with stepwise changes due to tipping points, yet with the important difference that the resulting environmental feedbacks would facilitate the original change. An example for this type of feedback is the enhancement of rainfall in scenarios of desert greening that are maintained by large-scale irrigation (Ornstein et al. 2009; Bowring et al. 2014). Scenario B would thus represent a scenario that would sustain greater levels of human activity in the long-term future.

These two scenarios can be contrasted against well-known concepts that deal with the future of human activity in the Earth system, the "limits to growth" report by Meadows et al. (1972) and the "planetary boundaries" concept proposed by Rockström et al. (2009). Both of these concepts use the preindustrial conditions of the Earth system as a reference state that determines the limit to future levels of human activity. Although these concepts do not use a thermodynamic approach and do not deal with limits to energy conversions, in a sense they refer to a static, or reduced ability of the Earth system to sustain human activity. This is in contrast to the scenario B, which illustrates that it may be possible to enhance the ability of the Earth system to sustain greater levels of human activity by technological means, particularly by the more efficient use of solar radiation, a notion that has long been recognized (e.g., Ciamician 1912; Glaser 1968). Such a technology-based enhancement of the ability of the Earth system would be a non-trivial exercise as it involves environmental effects and consequences that would need to be carefully evaluated. One may also oppose human activity and technology having such a key role in shaping the future evolution of the Earth system. Yet, the author thinks that merely on thermodynamic grounds, scenario B may simply represent an unavoidable outcome of thermodynamic evolution at the planetary scale in which dissipative systems are driven to their limit and that includes the human dissipative system including the options made available by technology. This is, of course, a speculative interpretation, even though energy conversions are indisputable at the core of human activity and planetary functioning and energy conversions are indisputably constrained by thermodynamics.

We can also extrapolate scenario B into the more distant future and speculate about its outcome in terms of planetary evolution. It has been suggested, for instance, that to extend energy generation solar panels be deployed into space (Glaser 1968; Hoffert et al. 2002), which could result in energy generation rates that are in excess to what is available on Earth. Furthermore, Dyson (1960) suggested that an advanced civilization may use all of the solar free energy by forming a sphere around the star, which would enhance the energy potential even further. The technology that would be necessary to achieve this state is, of course, at present unimaginable. So even though these scenarios may sound more like science fiction, it is important to note that such scenarios would nevertheless in principle seem thermodynamically feasible and would follow a trajectory of enhancing entropy production in the universe (Chaisson 1998; Lineweaver and Egan 2008).

What these considerations suggest is that a thermodynamic perspective of human activity within the Earth system contributes a novel perspective to the formulation of scenarios of the future. From this perspective, it is evident that the rates of energy conversions associated with human activity are comparable to other Earth system processes, so that human activity forms a planetary process with unavoidable, thermodynamic imprints on the state of the system just as any other dissipative process. An important aspect of human activity is that it adds further options for energy conversions to the planet through human-made technology that are unavailable to the Earth system in the absence of human activity. This is particularly the case for the direct conversion of solar radiation into free energy, for instance, by photovoltaic cells, which is much more efficient in generating free energy than the natural heat engines driven by solar radiative heating. This additional means to generate free energy by human technology should allow for a sustainable growth of human activity into the future that is accompanied by greater rates of geochemical cycling and greater rates of energy conversions within the Earth system in general. In doing so, human activity would then follow a trend of what would be expected of a dissipative process as it evolves to its thermodynamic limit.

12.6 Perspectives

In summary, this book describes the thermodynamic foundations to define Earth system processes and their interactions that shape the functioning of the whole system. The second law of thermodynamics, which sets a fundamental direction, plays a particular role in this description. This direction may not always be straightforward to be noticed, as, for instance, in the case of geostrophic flow. Yet in the end, it is this direction of the second law that shapes the overall dynamics of the whole Earth system with its many different processes.

Several examples in this book showed that the limits that are associated with the second law yield estimates for Earth system processes consistent with observations, demonstrating that the second law acts as a highly relevant constraint, particularly for sustaining motion. This view further suggests that this constraint on motion results in transport limitations associated with mass exchange that couple motion to hydrological and geochemical transformations and cycling within the system. It shows how important interactions are in shaping limits in the coupled system – directly in the form of depleting gradients and the maximum power limit, and indirectly in the form of feedback loops involving geochemical cycling, biotic and human activity, and the planetary forcing. It further suggests that the resulting evolution of the system is driven by feedbacks that result in dynamics towards thermodynamic limits and greater dissipation. This thermodynamic foundation of the Earth system puts the second law into practical use of critical relevance that goes far beyond its sometimes mystical appearance. It thus provides a theoretical foundation for Earth system science that is rooted in the very basic laws of thermodynamics.

Yet, this book can only provide a starting point. It is far from being complete in the treatment of compartments of the Earth system, it would be interesting to explore its relation to other theoretical descriptions, and speculations have been made that would need to be substantiated further. The omissions concern, for instance, the highly relevant role of clouds, for which thermodynamics may set limits to moist convection that would help to constrain the resulting dynamics of cloud cover and the response to change. Ocean processes have been largely neglected here, as are many other compartments, such as ocean and atmospheric geochemistry, land surface hydrology, soil science, geomorphology, ecology, as well as socioeconomics. Hopefully, this book nevertheless provides a perspective and some references as starting points of how thermodynamics can help to formulate the dynamics of these compartments in terms of thermodynamic directions and limits, in terms of their linkages to other processes of the Earth system and in terms of transport limitations as potential bottlenecks.

On the theoretical side, there are concepts in complex system science, such as self-organized criticality, bifurcations, multiple steady states and tipping points, fractal networks (which were briefly mentioned in Chapter 5) and turbulence theory that have not been addressed here. It would likely be rather instructive to relate these theories to the non-equilibrium thermodynamic context of the Earth system in greater detail. A particular relevant aspect that was described in Chapter 5 regards the role of structures and patterns in Earth system processes and the extent to which these represent the means of a process to reach its thermodynamic limit. We referred to this as "maximization by structures" (Section 5.7). While this hypothesis seemed plausible, it requires a more in-depth evaluation. If confirmed, this hypothesis would provide a thermodynamic approach to aggregate the high spatiotemporal variability at small scales to a larger scale by using the associated limit, which would be very useful to better describe the effect of such structures in models of Earth system components without an explicit formulation of the dynamics at small scales. Future work on this theoretical side would make the description of dynamics that are involved in the thermodynamic maximization more quantitative than what has been described in Chapter 5.

The description provided in this book also addressed the proposed principle of maximum entropy production (MEP). While MEP has been suggested to be applicable to a range of Earth system processes, it is unspecific about which process should maximize its rate of entropy production as all of the various processes of the Earth system produce entropy. The thermodynamic description of the Earth system in this book provides a more specific picture of how thermodynamic limits apply to Earth system processes and thereby addresses criticisms that were raised regarding MEP (Volk and Pauluis 2010). An example for this more specific picture is given by the thermodynamics of hydrologic cycling, for which it was formulated here that it is the convective motion generated by the moist heat engine that limits the intensity of hydrologic cycling rather than the immediate availability of energy or the entropy production associated with the phase change. As these processes in the planetary context all link to absorbed solar radiation, the maximization of power by moist convection is likely to result in a similar outcome as the maximization of the associated entropy production. In this sense, the formulation provided here does not contradict previous applications of MEP to Earth system processes, but rather develops it further in a more refined description of the processes.

A couple of speculative implications have been made in this book that would need to be evaluated against observations. This relates particularly to the role of mass exchange limitations that have been suggested at several places and which have the potential to yield a different view of the relevant constraints for the associated processes. This includes the example of hydrologic cycling as described earlier, as well as the example of carbon exchange of terrestrial vegetation. Both formulations yield consistent estimates and outcomes, and place a focus on the critical exchange processes at the system boundary and feedbacks that alter the conditions at the system boundary. This critical exchange process is often expressed in simulation models by some empirical parameterizations and it is not clear whether these respond in the same way as what would be expected from the maximum power limit. The thermodynamic approach formulated here may provide a better way to parameterize this exchange, yet this would need to be evaluated in greater detail.

Yet, even at this point there are several indications that the thermodynamic foundations described in this book provide a compelling picture of the Earth system. It allows for plausible, first-order estimates of the magnitude of Earth system processes that are based on simple, yet physical formulations with a minimum of empirical parameters. It provides a theoretical basis that allows us to understand and hypothesize what the dominant limitations and bottlenecks are of these processes and how they react to change. It allows us to make elaborate guesses of how the climate system of hardly known exoplanets may look like. The derived estimates are then solely based on the laws of thermodynamics and do not rely on increasingly comprehensive, but also increasingly incomprehensible numerical simulation models of the Earth system. Such simple and robust predictions provide an additional means to the complex descriptions of the Earth system with numerical models to understand the Earth system and its response to change. They complement modeling approaches with estimates that are physically based and transparent in their derivation. This foundation should then help us to gain a better, and more profound understanding of the Earth system and how it interacts with change.

To close, the perspective of the Earth system that we gain from this thermodynamic foundation is well described by what Aristotle noted more than 2000 years ago, namely that "the whole is greater than the sum of its parts." The parts of the Earth system are easily identified as the different spheres of the Earth, the atmosphere, the oceans, ice sheets, land, the lithosphere, biosphere, and the human sphere. Earth system processes link these spheres to each other and to the planetary forcing. Thermodynamics provides the additional constraints on the direction and limits of these processes that then form the emergent dynamics of the thermodynamic Earth system.

Glossary

Activity is used here to describe the magnitude of dissipation of a thermodynamic process associated with a state far from equilibrium. The term is also used in chemistry as a thermodynamic quantity representing the effective concentration of a particular component in a solution or other system. 8, 325

Adiabatic describes a condition in which no heat is added to or removed from a system. 78

Affinity is a thermodynamic quantity in chemistry that describes the tendency of two compounds to combine with another. 35, 228

Albedo describes an optical property of an object that represents the proportion of incident radiation that is reflected. 124, 134

Anthropocene is a term to refer to the era in Earth's recent history that is strongly shaped by human activity. 291

Arrow of time is a term used to describe temporal dynamics as the consequences of the second law and the increase in entropy. 4

Available potential energy represents the fraction of potential energy within a system that can be converted into other forms of energy, typically kinetic energy. 156

Basal metabolic rate is the rate by which an organism at rest consumes chemical free energy by its metabolic reactions. 295

Biosphere is the region of the Earth occupied by all living organisms. 241

Boundary describes the delineation of a certain volume that defines a thermodynamic system. 21

Buoyancy is a force that is associated with the depletion of potential energy within a system. 154

Carnot limit is the thermodynamic limit of converting heat into work set by the combination of the first and second laws of thermodynamics. 74

Closed system is a thermodynamic system that exchanges only energy with its surroundings. 23

Complex system is a system in which processes are linked in a close or complicated way. 1

Condensation refers to the phase transition of a compound from the gaseous to the liquid state. 189

Conjugate variables is a set of two variables for which the product describes a form of energy. The pair is formed by one intensive and one extensive variable. 24

Conservation law is a physical principle that formulates that the total value of a physical quantity or parameter (such as energy, mass, linear, or angular momentum, charge) remains constant in a system which is not subject to external influence. 22

Convection is a form of heat transfer accomplished by a moving fluid that is caused by the tendency of hotter and therefore less dense material to rise, and colder, denser material to sink under the influence of gravity, which consequently results in transfer of heat. 157, 190

Coriolis force is an apparent force that results from a mass moving in a rotating system. The Coriolis force acts perpendicular to the direction of motion and to the axis of rotation. On Earth, the effect tends to deflect moving objects to the right in the northern hemisphere and to the left in the southern and is important in the formation of cyclonic weather systems. 157

Degrees of freedom describes a number of independently variable factors affecting the range of states in which a system may exist. 109

Diffuse radiation is radiation that is spread over a wide solid angle. 133

Diffusion is a form of heat transfer accomplished by the random movement of molecules. 95

Direct radiation is radiation that is concentrated into a narrow solid angle, such as solar radiation. 133

Disequilibrium is defined in thermodynamics as a state that deviates from the state of thermodynamic equilibrium. Since equilibrium is defined by the maximum in entropy, disequilibrium is characterized by an entropy below its maximum value. 4, 56

Dissipation is the depletion of free energy through its conversion into heat. 50

Dissipative structure is a spatiotemporal organization of a process that is associated with dissipation and that shows a reproducible statistical behavior. 7, 101

Dissipative system is a thermodynamic system in which irreversible processes result in an increase of entropy within the system. 8

Dynamics refers here to the fluxes that deplete disequilibrium within a system. 63

Efficiency is the ratio describing the conversion of an energy flux into power. 75

Energy is the property of matter and radiation which has the capacity to perform work, such as causing macroscopic motion. 19

Entropy is a thermodynamic quantity representing the unavailability of a system's energy for conversion into work. It can be viewed as the dispersal of energy within a system at the microscopic level. There are forms of entropy related to radiation, the chemical configuration of electrons in atoms and molecules, and to rotational, vibrational and translational modes in molecules. 19

Entropy production is the rate by which a process increases the entropy of a system. 54

Equilibrium in thermodynamics is defined as the state of maximum entropy. 3, 46, 56

Evaporation refers to the phase transition of a compound from its liquid to its gaseous state. 189

Evolutionary dynamics are described here as dynamics that extend beyond energy and mass balances. They are characterized by a change in free energy generation, dissipation, and the thermodynamic state of the system. 99

Exchange flux describes the transport of a physical property such as energy, mass, or momentum across the boundary of a thermodynamic system. 21

Exergy is a term commonly used in the engineering literature to refer to free energy. 61

Extensive variable is a thermodynamic variable that depends on the size of the system. 24

Externalized activity is used here to describe a dissipative process caused by humans, but is not directly associated with the human metabolism. 292

Feedback describes the modification of a process by its results or effects. 101

First law of thermodynamics states that during a conversion of internal energy into mechanical work, energy is conserved. 47

Free energy is a thermodynamic quantity equivalent to the capacity of a system to do work. 60

Generation rate is the rate by which free energy of a particular form is generated (same as power). 50

Geostrophic flow is a hypothetical, non-dissipative flow without friction. It is determined by a momentum balance consisting of the pressure gradient and Coriolis forces. 157

Greenhouse effect refers to the additional surface warming resulting from the absorption and re-emission of terrestrial radiation within the atmosphere. 124

Grey atmosphere is an approximation for radiative transfer in which an atmosphere partially absorbs terrestrial radiation independent of wavelength. 138

Habitability refers to planetary conditions that are suitable to sustain life. 332

Heat is a form of energy arising from the random motion of molecules that is reflected by a body's temperature and pressure. Instead of heat, the terms thermal energy and uncompensated heat are used in this book, while "heat flux" refers to energy transfers induced by temperature gradients that do not involve radiation. 30

Heat engine is an abstract device for producing motive power from heat. 72

Human activity is used here to describe the activities of human beings in terms of how they relate to energy consumption by metabolic or externalized activity. 292

Intensive variable is a variable that does not depend on the size of the system. 24

Irreversibility describes a change that is unable to be undone without the performance of work, that is, processes that dissipate free energy. 54

Isentropic describes a condition in which the entropy of the system remains unchanged. 78

Isolated system is a thermodynamic system which does not exchange any of its properties with the surroundings. 23

Isothermal describes a condition in which the temperature of a system remains unchanged. 78

Lapse rate is the rate by which air cools when it is lifted. 201

Le Châtelier's principle states that if a change is applied to a system in equilibrium, the equilibrium will shift so as to counteract the effect of the change. 230

Macroscopic scale is the scale at which thermodynamic variables, such as temperature, density, and pressure, can be defined. 20

Maximum entropy production (MEP) is a proposed thermodynamic principle which states that an irreversible process in steady state evolves to a maximum rate of entropy production. 94

Maximum power limit is a thermodynamic limit of the power that can be derived from a driving gradient that is set, for instance, by the combination of the Carnot limit with energy balances. 86

Microscopic scale is the scale of photons, atoms, electrons, and molecules at which processes are described by quantum physics. 19

Minimum entropy production (MinEP) is a thermodynamic principle which states that the steady state of an irreversible process is represented by a minimum rate of entropy production. 96

Momentum flux is a flux of momentum that is better known as a mechanical force of some type. 66

Open system is a thermodynamic system which exchanges physical properties such as energy, mass, or momentum with its surroundings. 23

Photolysis refers to the decomposition or separation of molecules by the action of radiation. Particular forms include photodissociation and photoionization. 233

Photon is a particle representing a quantum amount of energy of electromagnetic radiation. 125

Potential temperature is a hypothetical temperature that refers to the temperature that a system would attain if it is brought adiabatically to a reference pressure. 80

Power describes the rate of performing work, measured in Watt. It represents a generation rate of free energy of a particular process at the expense (i.e., depletion) of another gradient. 72

Radiative temperature is the temperature of an object that characterizes the spectrum of its emitted radiation. 122

Respiration describes the process in living organisms by which chemical free energy is dissipated by metabolic reactions and is associated with gas exchange of oxygen and carbon dioxide. 247

Saturation is a state of thermodynamic equilibrium between two phases of a substance, such as liquid and gas, in contact with another. 189

Second law of thermodynamics states that processes are directed to result in an overall increase in entropy. 47

Solar radiation describes the electromagnetic radiation emitted by the Sun at a high temperature of about 5760 K that consists mostly of visible light. 121

Solid angle is the three-dimensional analogue of an angle. 127

State variable is a variable that describes the state of a system. 22

Steady state is a condition for a physical process in which the properties, or their temporal means, do not change in time. 4

System is a set of processes working together in a defined volume. 21

Terrestrial radiation describes the electromagnetic radiation emitted by the Earth at temperatures around 255 K that consists mostly of infrared radiation. 122

Thermodynamic limit describes a limit that follows directly from the laws of thermodynamics, in some cases in combination with a conservation law. 2

Transpiration is a plant-mediated form of evaporation taking place inside leaves. 273

Waste-heat flux is a heat flux leaving a heat engine that exports entropy. 72

References

Aiello, L. C., and Wheeler, P. 1995. The expensive-tissue hypothesis: the brain and the digestive system in human and primate evolution. *Curr. Anthropol.*, **36**, 199–221.

Allen, M. R., and Ingram, W. J. 2002. Constraints on future changes in climate and the hydrologic cycle. *Nature*, **419**, 224–232.

Allman, J. M. 1999. *Evolving Brains*. New York: Scientific American Library.

Ambaum, M. H. P. 2010. *Thermal Physics of the Atmosphere*. Chichester: J. Wiley and Sons.

Amend, J. P., and Shock, E. L. 2001. Energetics of overall metabolic reactions of thermophilic and hyperthermophilic Archea and Bacteria. *FEMS Microbiol. Rev.*, **25**, 175–243.

Amthor, J. S. 1995. Terrestrial higher-plant response to increasing atmospheric CO_2 in relation to the global carbon cycle. *Glob. Ch. Biol.*, **1**, 243–274.

Amthor, J. S. 2010. From sunlight to phytomass: on the potential efficiency of converting solar radiation to phyto-energy. *New Phytol.*, **188**, 939–959.

Andreas, E. L. 2011. Fallacies of the enthalpy transfer coefficient over the ocean in high winds. *J. Atmos. Sci.*, **68**, 1435–1445.

Andresen, B., Berry, R. S., Nitzan, A., and Salamon, P. 1977. Thermodynamics in finite time. I: the step-Carnot cycle. *Phys. Rev. A*, **15**, 2086–2093.

Andresen, B., Salamon, P., and Berry, R. S. 1984. Thermodynamics in finite time. *Phys. Today*, **37**, 62–70.

Aoki, I. 1983. Entropy productions on the Earth and other planets of the solar system. *J. Phys. Soc. Japan*, **52**, 1075–1078.

Archer, C. L., and Caldeira, K. 2009. Global assessment of high-altitude wind power. *Energies*, **2**, 307–319.

Arens, S., and Kleidon, A. 2008. Global sensitivity of weathering rates to atmospheric CO_2 under the assumption of saturated river discharge. *Mineral. Mag.*, **72**, 301–304.

Arya, S. P. 1998. *Introduction to Micrometeorology*. San Diego, CA: Academic Press.

Ashton, K. G., Tracy, M. C., and de Queiroz, A. 2000. Is Bergmann's rule valid for mammals? *Am. Nat.*, **156**, 390–415.

Atkins, P, and de Paula, J. 2010. *Physical Chemistry*. 9th edn. Oxford and New York: Oxford University Press.

Ayres, R., and Kneese, A. 1969. Production, consumption, and externalities. *Am. Econ. Rev.*, **59**, 282–297.

Ayres, R. U. 1989. Industrial metabolism. Pages 23–49 of: Ausubel, J.H., and Sladovich, H.E. (eds) *Technology and Environment*. Washington DC: National Academy Press.

Ayres, R. U., and Nair, I. 1984. Thermodynamics and economics. *Phys. Today*, **37**, 62–71.

Ayres, R. U., Ayres, L. W., and Warr, B. 2003. Exergy, power and work in the US economy, 1900–1998. *Energy*, **28**, 219–273.

Backus, G. E. 1975. Gross thermodynamics of heat engines in deep interior of Earth. *Proc. Natl. Acad. Sci. USA*, **72**, 1555–1558.

Bala, G., Duffy, P. B., and Taylor, K. E. 2008. Impact of geoengineering schemes on the global hydrologic cycle. *Proc. Natl. Acad. Sci. USA*, **105**, 7664–7669.

Barber, J. 2009. Photosynthetic energy conversion: natural and artificial. *Chem. Soc. Rev.*, **38**, 185–196.

Beer, C., Reichstein, M., Tomelleri, E., Ciais, P., Jung, M., Carvalhais, N., Roedenbeck, C., Arain, M. A., Baldocchi, D., and Bonan, G. B. 2010. Terrestrial gross carbon dioxide uptake: global distribution and covariation with climate. *Science*, **329**, 834–838.

Beerling, D. J., and Royer, D. L. 2002. Reading a CO_2 signal from fossil stomata. *New Phytol.*, **153**, 387–397.

Bejan, A. 1996. Entropy generation minimization: the new thermodynamics of finite-size devices and finite-time processes. *J. Appl. Phys.*, **79**, 1191–1218.

Bejan, A. 1997. *Advanced Engineering Thermodynamics*. New York: Wiley.

Bejan, A. 2002. Fundamentals of exergy analysis, entropy generation minimization, and the generation of flow architecture. *Int. J. Energy Res.*, **26**, 545–565.

Bejan, A. 2007. Constructal theory of pattern formation. *Hydrol. Earth Syst. Sci.*, **11**, 753–768.

Bering, E. A., Few, A. A., and Benbrook, J. R. 1998. The global electric circuit. *Phys. Today*, 24–30.

Betts, A. K., and Ridgway, W. 1989. Climatic equilibrium of the atmospheric convective boundary layer over a tropical ocean. *J. Atmos. Sci.*, **46**, 2621–2641.

Betts, A. K., Ball, J. H., Beljaars, A. C. M., Miller, M. J., and Viterbo, P. A. 1996. The land surface-atmosphere interaction: a review based on observational and global modeling perspectives. *J. Geophys. Res.*, **101**, 7209–7225.

Betz, A. 1920. Das Maximum der theoretisch möglichen Ausnützung des Windes durch Windmotoren. *Z. Gesamte Turbinenwesen*, **26**, 307–309.

Bister, M., and Emanuel, K. A. 1998. Dissipative heating and hurricane intensity. *Meteorol. Atmos. Phys.*, **65**, 233–240.

Bister, M., Renno, N., Pauluis, O., and Emanuel, K. 2011. Comment on Makarieva et al. 'A critique of some modern applications of the Carnot heat engine concept: the dissipative heat engine cannot exist'. *Proc. R. Soc. London A*, **467**, 1–6.

Bobe, R., and Behrensmeyer, A. K. 2004. The expansion of grassland ecosystems in Africa in relation to mammalian evolution and the origin of the genus Homo. *Paleogeogr. Paleocl.*, **207**, 399–420.

Bohren, C. F., and Albrecht, B. A. 1998. *Atmospheric Thermodynamics*. New York: Oxford University Press.

Bolton, J. R., and Hall, D. O. 1991. The maximum efficiency of photosynthesis. *Photochem. Photobiol.*, **53**, 545–548.

Boltzmann, L. 1886. Der zweite Hauptsatz der mechanischen Wärmetheorie. *Almanach der kaiserlichen Akademie der Wissenschaften*, **36**, 225–259.

Bonan, G B. 2008. Forests and climate change: forcings, feedbacks, and the climate benefits of forests. *Science*, **320**, 1444–1449.

Bony, S., Bellon, G., Klocke, D., Sherwood, S., Fermepin, S., and Denvil, S. 2013. Robust direct effect of carbon dioxide on tropical circulation and regional precipitation. *Nature Geosci.*, **6**, 447–451.

Borucki, W. J., and Chameides, W. L. 1984. Lightning: estimates of rates of energy dissipation and nitrogen fixation. *Rev. Geophys.*, **22**, 363–372.

Bowring, S. P. K., Miller, L. M., Ganzeveld, L., and Kleidon, A. 2014. Applying the concept of "energy return on investment" to desert greening of the Sahara/Sahel using a global climate model. *Earth Syst. Dynam.*, **5**, 43–53.

Boyce, C. K., Brodribb, T. J., Feild, T. S., and Zwieniecki, M. A. 2009. Angiosperm leaf vein evolution was physiologically and environmentally transformative. *Proc. R. Soc. Lond. B*, **276**, 1771–1776.

BP. 2014. *BP Statistical Review of World Energy 2014*. Tech. rept. London, UK: BP P.L.C.

Brodribb, T. J., and Feild, T. S. 2010. Leaf hydraulic evolution led a surge in leaf photosynthetic capacity during early angiosperm diversification. *Ecol. Lett.*, **13**, 175–183.

Brodribb, T. J., Feild, T. S., and Jordan, G. J. 2007. Leaf maximum photosynthetic rate and venation are linked by hydraulics. *Plant Physiol.*, **144**, 1890–1898.

Brovkin, V. 2002. Climate-vegetation interaction. *J. Phys-Paris*, **12**, 57–72.

Brunsell, N. A., Schymanski, S. J., and Kleidon, A. 2011. Quantifying the thermodynamic entropy budget of the land surface: is this useful? *Earth Syst. Dynam.*, **2**, 87–103.

Brunt, D. 1941. *Physical and Dynamical Meteorology*. London: Cambridge University Press.

Budyko, M. I. 1974. *Climate and Life. Translated from the Original Russian Edition*. New York: Academic Press.

Callies, U., and Herbert, F. 1988. Radiative processes and non-equilibrium thermodynamics. *Z. Angew. Math. Phys.*, **39**, 242–266.

Campbell, G. S., and Norman, J. M. 1998. *An Introduction to Environmental Biophysics*. 2nd edn. New York: Springer.

Campbell, I. H., and Taylor, S. R. 1983. No water, no granites - no oceans, no continents. *Geophys. Res. Lett.*, **10**, 1061–1064.

Carnot, S. 1824. *Reflections on the Motive Power of Fire and on Machines Fitted to Develop that Power*. Paris: Bachelier.

Carpenter, S. R., and Kitchell, J. F. 1984. Plankton community structure and limnetic primary production. *Am. Nat.*, **124**, 159–172.

Catling, D. C. 2005. Coupled evolution of Earth's atmosphere and biosphere. Pages 191–206 of: Kleidon, A., and Lorenz, R. D. (eds), *Non-Equilibrium Thermodynamics and the Production of Entropy: Life, Earth, and Beyond*. Heidelberg: Springer.

Cawood, P. A., Hawkesworth, C. J., and Dhuime, B. 2013. The continental record and the generation of continental crust. *GSA Bulletin*, **125**, 14–32.

Chaisson, E. J. 1998. The cosmic environment for the growth of complexity. *Biosystems*, **46**, 13–19.

Charru, F., Andreotti, B., and Claudin, P. 2013. Sand ripples and dunes. *Annu. Rev. Fluid Mech.*, **45**, 469–493.

Christensen, U. R., and Tilgner, A. 2004. Power requirements of the geodynamo from ohmic losses in numerical and laboratory dynamos. *Nature*, **429**, 169–171.

Ciais, P., Sabine, C., Bala, G., Bopp, L., Brovkin, V., Canadell, J., Chhabra, A., DeFries, R., Galloway, J., Heimann, M., Jones, C., Quéré, C. Le, Myneni, R. B., Piao, S., and Thornton, P. 2013. Carbon and other biogeochemical cycles. In: Stocker, T. F., Qin, D., Plattner, G.-K., Tignor, M., Allen, S. K., Boschung, J., Nauels, A., Xia, Y., Bex, V., and Midgley, P. M. (eds), *Climate Change 2013: The Physical Science Basis. Contribution of Working Group I to the Fifth Assessment Report of the Intergovernmental Panel on Climate Change*. Cambridge, UK and New York: Cambridge University Press.

Ciamician, G. 1912. The photochemistry of the future. *Science*, **36**, 385–394.

Cicerone, R. J. 1987. Changes in stratospheric ozone. *Science*, **237**, 35–42.

Clark, T. L., Jenkins, M. A., Coen, J., and Packham, D. 1996. A coupled atmosphere-fire model: convective feedback on fire-line dynamics. *J. Appl. Meteor.*, **35**, 875–901.

Clausius, R. 1850. Ueber die bewegende Kraft der Wärme und die Gesetze, welche sich daraus für die Wärmelehre selbst ableiten lassen. *Ann. Phys. Chem.*, **155**, 368–397.

Cleveland, C. J., and Ruth, M. 1997. When, where, and by how much do biophysical limits constrain the economic process? A survey of Nicholas Georgescu-Roegen's contribution to ecological economics. *Ecol. Econ.*, **22**, 203–223.

Cleveland, C. J., Costanza, R., Hall, C. A. S., and Kaufmann, R. 1984. Energy and the US economy: a biophysical perspective. *Science*, **225**, 890–897.

Cohen, J. E. 1995. Population growth and Earth's human carrying capacity. *Science*, **269**, 341–346.

Corten, G. 2001. Novel views on the extraction of energy from wind-heat generation concentration and terrain. Page 5 of: *Proceedings of the 2001 EWEC Conference*. available at www.ecn.nl/docs/library/report/2001/rx01054.pdf. Accessed May 2, 2014.

Count, E. W. 1947. Brain and body weight in man. *Ann. N. Y. Acad. Sci.*, **46**, 993–1122.

Cross, M, and Hohenberg, P. 1993. Pattern formation outside of equilibrium. *Rev. Mod. Phys.*, **65**, 851–1112.

Crutzen, P. J. 2002. Geology of mankind. *Nature*, **415**, 23.

Curzon, F. L., and Ahlborn, B. 1975. Efficiency of a Carnot engine at maximum power output. *Am. J. Phys.*, **43**, 22–24.

Daly, H. E. 1992. Is the entropy law relevant to economics of natural resource scarcity? Yes, of course it is! *J. Environ. Econ. Manage.*, **23**, 91–95.

Davies, J. H., and Davies, D. R. 2010. Earth's surface heat flux. *Solid Earth*, **1**, 5–24.

de Arellano, J. Vila-Guerau, Ouwersloot, H. G., Baldocchi, D., and Jacobs, C. M. J. 2014. Shallow cumulus rooted in photosynthesis. *Geophys. Res. Lett.*, **41**, 1796–1802.

de Bruin, H. A. R., and Lablans, W. N. 1998. Reference crop evapotranspiration determined with a modified Makkink equation. *Hydrol. Process.*, **12**, 1053–1062.

Denmead, O. T., Raupach, M. R., Dunin, F. X., Cleugh, H. A., and Leuning, R. 1996. Boundary layer budgets for regional estimates of scalar fluxes. *Glob. Ch. Biol.*, **2**, 255–264.

Desjardins, R. L., Brach, E. J., Alvo, P., and Schuepp, P. H. 1982. Aircraft monitoring of surface carbon dioxide exchange. *Science*, **216**, 733–735.

Dewar, R. C. 2003. Information theory explanation of the fluctuation theorem, Maximum Entropy Production, and self-organized criticality in non-equilibrium stationary states. *J. Physics A*, **36**, 631–641.

Dewar, R. C. 2005a. Maximum entropy production and non-equilibrium statistical mechanics. In: Kleidon, A., and Lorenz, R. D. (eds), *Non-Equilibrium Thermodynamics and the Production of Entropy: Life, Earth, and Beyond*. Heidelberg: Springer.

Dewar, R. C. 2005b. Maximum entropy production and the fluctuation theorem. *J. Physics A*, **38**, L371–L381.

Dewar, R C. 2010. Maximum entropy production as an inference algorithm that translates physical assumptions into macroscopic predictions: don't shoot the messenger. *Entropy*, **11**, 931–944.

Dewar, R C, Lineweaver, C H, Niven, R K, and Regenauer-Lieb, K. 2014. Beyond the second law: an overview. Pages 3–27 of: Dewar, R C, Lineweaver, C H, Niven, R K, and Regenauer-Lieb, K (eds), *Beyond the Second Law: Entropy Production and Non-equilibrium Systems*. Heidelberg: Springer.

Dincer, I., and Rosen, M. A. 2005. Thermodynamic aspects of renewables and sustainable development. *Renew. Sust. Energ. Rev.*, **9**, 169–189.

Dukes, J. S. 2003. Burning buried sunshine: human consumption of ancient solar energy. *Clim. Ch.*, **61**, 31–44.

Duysens, L. N. M. 1958. The path of light energy in photosynthesis. Pages 10–25 of: *Brookhaven Symposia in Biology 1: The Photochemical Apparatus, its Structure & Function*. Upton, NY, USA: Brookhaven Natl. Lab.

Dyke, J. G., Gans, F., and Kleidon, A. 2011. Towards understanding how surface life can affect interior geological processes: a non-equilibrium thermodynamics approach. *Earth Syst. Dynam.*, **2**, 139–160.

Dyson, F J. 1960. Search for artificial stellar sources of infrared radiation. *Science*, **131**, 1667–1668.

Eddington, A. S. 1928. *The Nature of the Physical World*. New York: Macmillan.

Edlefsen, N. E., and Anderson, A. B. C. 1943. Thermodynamics of soil moisture. *Hilgardia*, **15**, 31–298.

Ehleringer, J. R., and Cerling, T. E. 1995. Atmospheric CO_2 and the ratio of intercellular to ambinent CO_2 concentrations in plants. *Tree Physiol.*, **15**, 105–111.

Elder, J. 1976. *The Bowels of the Earth*. London: Oxford University Press.

Elimelech, M., and Phillip, W. A. 2011. The future of seawater desalination: energy, technology and the environment. *Science*, **333**, 712–717.

Emanuel, K. 1987. The dependence of hurricane intensity on climate. *Nature*, **326**, 483–485.

Emanuel, K. 2005. Increasing destructiveness of tropical cyclones over the past 30 years. *Nature*, **436**, 686–688.

Emanuel, K. 2007. Environmental factors affecting tropical cyclone power dissipation. *J. Clim.*, **20**, 5497–5509.

Emanuel, K. 2013. Downscaling CMIP5 climate models shows increased tropical cyclone activity over the 21st century. *Proc. Natl. Acad. Sci. USA*, **110**, 12219–12224.

Emanuel, K. A. 1986. An air-sea interaction theory for tropical cyclones. Part I: steady-state maintenance. *J. Atmos. Sci.*, **43**, 585–604.

Emanuel, K. A. 1999. Thermodynamic control of hurricane intensity. *Nature*, **401**, 665–669.

Ertel, H., and Köhler, H. 1948. On the thermodynamic efficiency of steady atmospheric circulations. *Geofisica Pura e Applicata*, **13**, 102–108.

Essex, C. 1984. Radiation and the irreversible thermodynamics of climate. *J. Atmos. Sci.*, **41**, 1985–1991.

Falk, D. 1990. Brain evolution in Homo: the "radiator" theory. *Behav. Brain Sci.*, **13**, 333–381.

Farquhar, G. D., and Sharkey, T. D. 1982. Stomatal conductance and photosynthesis. *Ann. Rev. Plant. Physiol.*, **33**, 317–345.

Feild, T. S., Brodribb, T. J., Iglesias, A., Chatelet, D. S., Baresch, A., Upchurch, G. R., Gomez, B., Mohr, B. A. R., Coiffard, C., Kvacek, J., and Jaramillo, C. 2011. Fossil evidence for Cretaceous escalation in angiosperm leaf vein evolution. *Proc. Natl. Acad. Sci. USA*, **108**, 8363–8366.

Ferrari, R., and Wunsch, C. 2009. Ocean circulation kinetic energy: reservoirs, sources, and sinks. *Annu. Rev. Fluid Mech.*, **41**, 253–282.

Feynman, R. P., Leighton, R. B., and Sands, M. 1966. *The Feynman Lectures on Physics*. Reading, MA, USA: Addison-Wesley.

Field, C. B., Behrenfeld, M. J., Randerson, J. T., and Falkowski, P. 1998. Primary production of the biosphere: integrating terrestrial and oceanic components. *Science*, **281**, 237–240.

Field, C. B., Campbell, J. E., and Lobell, D. B. 2007. Biomass energy: the scale of the potential resource. *Trends Ecol. Evol.*, **23**, 65–72.

Fischer-Kowalski, M., and Haberl, H. 1998. Sustainable development: socio-economic metabolism and the colonization of nature. *Int. Soc. Sci. J.*, **50**, 573–587.

Foley, J. A., DeFries, R., Asner, G. P., Barford, C., Bonan, G., Carpenter, S. R., Chapin, F. S., Coe, M. T., Daily, G. C., Gibbs, H. K., Helkowski, J. H., Holloway, T., Howard, E. A., Kucharik, C. J., Monfreda, C., Patz, J. A., Prentice, I. C., Ramankutty, N., and Snyder, P. K. 2005. Global consequences of land use. *Science*, **309**, 570–574.

Foley, R. A., and Lee, P. C. 1991. Ecology and energetics of encephalization in hominid evolution. *Phil. Trans. R. Soc. B*, **334**, 223–232.

Forster, P., Ramaswamy, V., Artaxo, P., Berntsen, T., Betts, R., Fahey, D.W., Haywood, J., Lean, J., Lowe, D.C., Myhre, G., Nganga, J., Prinn, R., Raga, G., Schulz, M., and Dorland, R. Van. 2007. Changes in atmospheric constituents and in radiative forcing. In: Solomon, S., Qin, D., Manning, M., Chen, Z., Marquis, M., Averyt, K.B., M.Tignor, and Miller, H.L. (eds), *Climate Change 2007: The Physical Science Basis. Contribution of Working Group I to the Fourth Assessment Report of the Intergovernmental Panel on Climate Change*. Cambridge, UK and New York: Cambridge University Press.

Franck, S., von Bloh, W., Müller, C., Bondeau, A., and Sakschweski, B. 2011. Harvesting the Sun: new estimations of the maximum population of planet Earth. *Ecol. Mod.*, **222**, 2019–2026.

Gans, F., Miller, L. M., and Kleidon, A. 2012. The problem of the second wind turbine: a note on a common but flawed wind power estimation method. *Earth Syst. Dynam.*, **3**, 79–86.

Garratt, J. R. 1992. *The Atmospheric Boundary Layer*. Cambridge, UK: Cambridge University Press.

Garrett, C., and Cummins, P. 2007. The efficiency of a turbine in a tidal channel. *J. Fluid Mech.*, **588**, 243–251.

Garrett, T J. 2009. Are there basic physical constraints on future anthropogenic emissions of carbon dioxide? *Clim. Change*, **104**, 437–455.

Georgescu-Roegen, N. 1971. *The Entropy Law and the Economic Process*. Cambridge, MA: Harvard University Press.

Glaser, P. E. 1968. Power from the Sun: its future. *Science*, **162**, 857–861.

Gnanadesikan, A., Slater, R. D., Swathi, P. S., and Vallis, G. K. 2005. The energetics of ocean heat transport. *J. Clim.*, **18**, 2604–2616.

Goldstein, B., Hiriart, G., Bertani, R., Bromley, C., Gutierrez-Negrin, L., Huenges, E., Muraoka, H., Ragnarsson, A., Tester, J., and Zui, V. 2011. Geothermal energy. Pages 401–436 of: Edenhofer, O., Pichs-Madruga, R., Sokona, Y., Seyboth, K., Matschoss, P., Kadner, S., Zwickel, T., Eickemeier, P., Hansen, G., Schlömer, S., and von Stechow, C. (eds), *IPCC Special Report on Renewable Energy Sources and Climate Change Mitigation*. Cambridge, UK and New York: Cambridge University Press.

Goody, R. 2000. Sources and sinks of climate entropy. *Q. J. R. Meteorol. Soc.*, **126**, 1953–1970.

Goody, R. 2003. On the mechanical efficiency of deep, tropical convection. *J. Atmos. Sci.*, **60**, 2827–2832.

Goody, R. 2007. Maximum entropy production in climate theory. *J. Atmos. Sci.*, **64**, 2735–2739.

Goody, R., and Abdou, W. 1996. Reversible and irreversible sources of radiation entropy. *Q. J. Roy. Meteorol. Soc.*, **122**(530), 483–494.

Gould, S. J. 1966. Allometry and size in ontogeny and phylogeny. *Biol. Rev.*, **41**, 587–640.

Gyftopoulos, E. P., and Beretta, G. P. 1991. *Thermodynamics: Foundations and Applications*. New York: Macmillan.

Haaf, W., Friedrich, K., Mayr, G., and Schlaich, J. 1983. Solar chimneys. Part I: principle and construction of the pilot plant in Manzanares. *Int. J. Solar Energy*, **2**, 3–20.

Haberl, H, Erb, K H, Krausmann, F, Gaube, V, Bondeau, A, Pluttzar, C, Gingrich, S, Lucht, W, and Fischer-Kowalski, M. 2007. Quantifying and mapping the human appropriation of net primary productivity in earth's terrestrial ecosystems. *Proc. Natl. Acad. Sci. USA*, **104**, 12942–12947.

Haberl, H., Erb, K.-H., and Krausmann, F. 2014. Human appropriation of net primary production: patterns, trends, and planetary boundaries. *Annu. Rev. Environ. Resour.*, **39**, 363–391.

Haff, P. K. 2010. Hillslopes, rivers, plows, and trucks: mass transport on Earth's surface by natural and technological processes. *Earth Surf. Process. Landforms*, **35**, 1157–1166.

Haff, P. K. 2013. Technology as a geological phenomenon: implications for human well-being. In: Waters, C. N., Zalasiewicz, J. A., Williams, M., Ellis, M. A., and Snelling, A. M. (eds), *A Stratigraphic Basis for the Anthropocene*. Special Publications, vol. 395. London: Geological Society.

Haken, H. 1975. Cooperative phenomena in systems far from thermal equilibrium and in nonphysical systems. *Rev. Mod. Phys.*, **47**, 67–121.

Hall, C., Lindenberger, D., Kümmel, R., Kroeger, T., and Eichhorn, W. 2001. The need to reintegrate the natural sciences with economics. *Bioscience*, **51**, 663–673.

Hammond, K. A., and Diamond, J. 1997. Maximal sustained energy budgets in humans and animals. *Nature*, **386**, 457–462.

Hansen, J., Lacis, A., Rind, D., Russell, G., Stone, P., Fung, I., Ruedy, R., and Lerner, J. 1984. Climate sensitivity: analysis of feedback mechanisms. In: *Climate Processes and Climate Sensitivity*, Geophysical Monograph 29. Washington, DC: American Geophysical Union.

Hartmann, D. L. 1994. *Global Physical Climatology*. San Diego: Academic Press.

Held, I. M., and Soden, B. J. 2006. Robust responses of the hydrological cycle to global warming. *J. Clim.*, **19**, 5686–5699.

Herrmann, W. A. 2006. Quantifying global exergy resources. *Energy*, **31**, 1685–1702.

Hewitt, J. M., McKenzie, D. P., and Weiss, N. O. 1975. Dissipative heating in convective flows. *J. Fluid Mech.*, **68**, 721–738.

Hill, R., and Rich, P. R. 1983. A physical interpretation for the natural photosynthetic process. *Proc. Natl. Acad. Sci. USA*, **80**, 978–982.

Hitchcock, D. R., and Lovelock, J. E. 1967. Life detection by atmospheric analysis. *Icarus*, **7**, 149–159.

Hobbs, P. V. 2000. *Introduction to Atmospheric Chemistry*. Cambridge, UK: Cambridge University Press.

Hoening, D., Hansen-Goos, H., Airo, A., and Spohn, T. 2014. Biotic vs. abiotic Earth: a model for mantle hydration and continental coverage. *Planet. Space Sci.*, **98**, 5–13.

Hoffert, M. I., Caldeira, K., Benford, G., Criswell, D. R., Green, C., Herzog, H., Jain, A. K., Kheshgi, H. S., Lackner, K. S., Lewis, J. S., Lightfoot, H. D., Manheimer, W., Mankins, J. C., Mauel, M. E., Perkins, L. J., Schlesinger, M. E., Volk, T., and Wigley, T. M. L. 2002. Advanced technology paths to global climate stability: energy for a greenhouse planet. *Science*, **298**, 981–987.

Holdaway, R. J., Sparrow, A. D., and Coomes, D. 2010. Trends in entropy production during ecosystem development in the Amazon basin. *Phil. Trans. R. Soc. B*, **365**, 1437–1447.

Holland, H. H. 2006. The oxygenation of the atmosphere and ocean. *Phil. Trans. Roy. Soc. London B*, **361**, 903–915.

Holton, J. R. 1992. *An Introduction to Dynamic Meteorology*. 3rd edn. San Diego: Academic Press.

Howard, A. D. 1990. Theoretical model of optimal drainage networks. *Water Resour. Res.*, **26**, 2107–2117.

Huang, S-S. 1959. Occurrence of life in the universe. *Am. Sci.*, **47**, 397–402.

Hubbert, M. King. 1981. The world's evolving energy system. *Am. J. Phys.*, **49**, 1007–1029.

Hyde, W. T., Crowley, T. J., Baum, S. K., and Peltier, R. 2000. Neoproterozoic 'snowball Earth' simulations with a coupled climate/ice-sheet model. *Nature*, **405**, 425–429.

Imhoff, M. L., Bounoua, L., Ricketts, T., Loucks, C., Harriss, R., and Lawrence, W. T. 2004. Global patterns in human consumption of net primary production. *Nature*, **429**, 870–873.

IPCC. 2011. IPCC Special Report on Renewable Energy Sources and Climate Change Mitigation. Prepared by Working Group III of the Intergovernmental Panel on Climate Change [O. Edenhofer, R. Pichs-Madruga, Y. Sokona, K. Seyboth, P. Matschoss, S. Kadner, T. Zwickel, P. Eickemeier, G. Hansen, S. Schlömer, C. von Stechow (eds)]. Cambridge, UK and New York: Cambridge University Press.

Isaacs, J D, and Schmitt, W R. 1980. Ocean energy: forms and prospects. *Science*, **207**, 265–273.

Jacobsen, M. Z., and Archer, C. L. 2012. Saturation wind power potential and its implications for wind energy. *Proc. Natl. Acad. Sci. USA*, **109**, 15679–15684.

Jerison, H. J. 1955. Brain to body ratios and the evolution of intelligence. *Science*, **121**, 447–449.

Jerison, H. J. 1977. The theory of encephalization. *Ann. N. Y. Acad. Sci.*, **299**, 146–160.

Jerison, H. J. 1985. Animal intelligence as encephalization. *Phil. Trans. R. Soc. B*, **308**, 21–35.

Jupp, T., and Schultz, A. 2000. A thermodynamic explanation for black smoker temperatures. *Nature*, **403**, 880–883.

Jupp, T. E., and Cox, P. M. 2010. MEP and planetary climates: insights from a two-box climate model containing atmospheric dynamics. *Phil. Trans. R. Soc. B*, **365**, 1355–1365.

Jupp, T. E., and Schultz, A. 2004. Physical balances in subseafloor hydrothermal convection cells. *J. Geophys. Res.*, **109**.

Kabelac, S. 1994. *Thermodynamik der Strahlung*. Braunschweig and Wiesbaden: Vieweg.

Kagan, B. A., and Sündermann, Jürgen. 1996. Dissipation of tidal energy, paleotides, and evolution of the Earth-Moon system. *Adv. Geophys.*, **38**, 179–266.

Kasting, J. F. 1993. Earth's early atmosphere. *Science*, **259**, 920–926.

Kasting, J. F., and Catling, D. 2003. Evolution of a habitable planet. *Annu. Rev. Astron. Astrophys.*, **41**, 429–463.

Kasting, J. F., Whitmire, D. P., and Reynolds, R. T. 1993. Habitable zones around main sequence stars. *Icarus*, **101**, 108–128.

Keith, D., DeCarolis, J., Denkenberger, D., Lenschow, D., Malyshev, S. L., Pacala, S., and Rasch, P. 2004. The influence of large-scale wind power on global climate. *Proc. Natl. Acad. Sci. USA*, **101**, 16116–16120.

Klausmeier, C A. 1999. Regular and irregular patterns in semiarid vegetation. *Science*, **284**, 1826–1828.

Kleiber, M. 1932. Body size and metabolism. *Hilgardia*, **6**, 315–353.

Kleidon, A. 2004a. Beyond Gaia: thermodynamics of life and Earth system functioning. *Clim. Ch.*, **66**, 271–319.

Kleidon, A. 2004b. Optimized stomatal conductance of vegetated land surfaces and its effects on simulated productivity and climate. *Geophys. Res. Lett.*, **31**, L21203.

Kleidon, A. 2006. The climate sensitivity to human appropriation of vegetation productivity and its thermodynamic characterization. *Glob. Planet. Ch.*, **54**, 109–127.

Kleidon, A. 2007. Optimized stomatal conductance and the climate sensitivity to carbon dioxide. *Geophys. Res. Lett.*, 34, L14709.

Kleidon, A. 2008. Entropy production by evapotranspiration and its geographic variation. *Soil Water Res.*, **3**, S89–S94.

Kleidon, A. 2009a. Climatic constraints on maximum possible levels of human activity and their relation to human evolution and global change. *Clim. Ch.*, **95**, 405–431.

Kleidon, A. 2009b. Maximum entropy production and general trends in biospheric evolution. *Paleontol. J.*, **43**, 130–135.

Kleidon, A. 2009c. Non-equilibrium thermodynamics and maximum entropy production in the Earth system: applications and implications. *Naturwissenschaften*, **96**, 653–677.

Kleidon, A. 2010. Life, hierarchy, and the thermodynamic machinery of planet Earth. *Phys. Life Rev.*, **7**, 424–460.

Kleidon, A. 2012. How does the Earth system generate and maintain thermodynamic disequilibrium and what does it imply for the future of the planet? *Phil. Trans. R. Soc. A*, **370**, 1012–1040.

Kleidon, A., and Fraedrich, K. 2005. Biotic entropy production and global atmosphere-biosphere interactions. Pages 173–190 of: Kleidon, A., and Lorenz, R. D. (eds), *Non-Equilibrium Thermodynamics and the Production of Entropy: Life, Earth, and Beyond*. Heidelberg: Springer.

Kleidon, A., and Heimann, M. 1998. A method of determining rooting depth from a terrestrial biosphere model and its impacts on the global water- and carbon cycle. *Global Change Biol.*, **4**, 275–286.

Kleidon, A., and Lorenz, R. D. (eds). 2005. *Non-Equilibrium Thermodynamics and the Production of Entropy: Life, Earth, and Beyond*. Heidelberg: Springer.

Kleidon, A., and Renner, M. 2013a. A simple explanation for the sensitivity of the hydrologic cycle to climate change. *Earth Syst. Dynam.*, **4**, 455–465.

Kleidon, A, and Renner, M. 2013b. Thermodynamic limits of hydrologic cycling within the Earth system: concepts, estimates and implications. *Hydrol. Earth Syst. Sci.*, **17**, 2873–2892.

Kleidon, A., Fraedrich, K., and Heimann, M. 2000. A green planet versus a desert world: estimating the maximum effect of vegetation on land surface climate. *Clim. Ch.*, **44**, 471–493.

Kleidon, A., Fraedrich, K., Kunz, T., and Lunkeit, F. 2003. The atmospheric circulation and states of maximum entropy production. *Geophys. Res. Lett.*, **30**, 2223.

Kleidon, A., Fraedrich, K., Kirk, E., and Lunkeit, F. 2006. Maximum entropy production and the strength of boundary layer exchange in an atmospheric general circulation model. *Geophys. Res. Lett.*, **33**, L06706.

Kleidon, A., Schymanski, S., and Stieglitz, M. 2009. Thermodynamics, irreversibility and optimality in land surface hydrology. Pages 107–118 of: Strelcova, K., Matyas, C., Kleidon, A., Lapin, M., Matejka, F., Skvarenina, J., and Holecy, J. (eds), *Bioclimatology and Natural Hazards*. Heidelberg: Springer.

Kleidon, A., Malhi, Y., and Cox, P. M. 2010. Maximum entropy production in environmental and ecological systems. *Phil. Trans. R. Soc. B*, **365**, 1297–1302.

Kleidon, A, Zehe, E, Ehret, U, and Scherer, U. 2013. Thermodynamics, maximum power, and the dynamics of preferential river flow structures on continents. *Hydrol. Earth Syst. Sci.*, **17**, 225–251.

Kleidon, A., Zehe, E., Ehret, U., and Scherer, U. 2014a. Earth system dynamics beyond the second law: maximum power limits, dissipative structures, and planetary interactions. Pages 163–182 of: Dewar, R. C., Lineweaver, C. H., Niven, R. K., and Regenauer-Lieb, K. (eds), *Beyond the Second Law: Entropy Production and Non-equilibrium Systems*. Berlin and Heidelberg: Springer.

Kleidon, A., Renner, M., and Porada, P. 2014b. Estimates of the climatological land surface energy and water balance derived from maximum convective power. *Hydrol. Earth Syst. Sci.*, **18**, 2201–2218.

Kleidon, A., Miller, L., and Gans, F. 2016. Physical limits of solar energy conversion in the Earth system. *Top. Curr. Chem.*, **371**, 1–22.

Klein, M. J. 1967. Thermodynamics in Einstein's thought. *Science*, **157**, 509–516.

Klein Goldewijk, K., Beusen, A., and Janssen, P. 2010. Long term dynamic modeling of global population and built-up area in a spatially explicit way, HYDE 3 .1. *Holocene*, **20**, 565–573.

Klein Goldewijk, K., Beusen, A., de Vos, M., and van Drecht, G. 2011. The HYDE 3.1 spatially explicit database of human induced land use change over the past 12,000 years. *Glob. Ecol. Biogeog.*, **20**, 73–86.

Knox, R. S. 1969. Thermodynamics and the primary processes of photosynthesis. *Biophys J*, **9**, 1351–1362.

Kok, J. F., Parteli, E. J. R., Michaels, T. I., and Karam, D. B. 2012. The physics of wind-blown sand and dust. *Rep. Prog. Phys.*, **75**, 106901.

Kondepudi, D., and Prigogine, I. 1998. *Modern Thermodynamics: From Heat Engines to Dissipative Structures*. Chichester: Wiley.

Köppen, W. 1923. *Die Klimate der Erde*. Berlin: de Gruyter.

Kump, L. R., Brantley, S. L., and Arthur, M. A. 2000. Chemical weathering, atmospheric CO_2, and climate. *Annu. Rev. Earth Planet. Sci.*, **28**, 611–667.

Lambert, F. L. 2002. Entropy is simple, qualitatively. *J. Chem. Ed.*, **79**, 1241–1246.

Landsberg, P T, and Tonge, G. 1979. Thermodynamics of the conversion of diluted radiation. *J. Phys. A*, **12**, 551–562.

Landsberg, P. T., and Tonge, G. 1980. Thermodynamic energy conversion efficiencies. *J. Appl. Phys.*, **51**, R1.

Law, B. E., Falge, E., Gu, L., Baldocchi, D. D., Bakwin, P., Berbigier, P., Davis, K., Dolman, A. J., Falk, M., Fuentes, J. D., Goldstein, A., Granier, A., Grelle, A., Hollinger, D., Janssens, I. A., Jarvis, P., Jensen, N. O., Katul, G., Malhi, Y., Matteucci, G., Meyers, T., Monson, R., Munger, W., Oechel, W., Olson, R., Pilegaard, K., U, K. T. Paw, Thorgeirsson, H., Valentini, R., Verman, S., Vesala, T., Wilson, K., and Wofsy, S. 2002. Environmental controls over carbon dioxide and water vapor exchange of terrestrial vegetation. *Agric. For. Meteor.*, **113**, 97–120.

Leff, H. S. 1996. Thermodynamic entropy: the spreading and sharing of energy. *Am. J. Phys.*, **64**, 1261–1271.

Leff, H. S. 2007. Entropy, its language, and interpretation. *Found. Phys.*, **37**, 1744–1766.

Lejeune, O, and Tlidi, M. 1999. A model for the explanation of vegetation stripes (Tiger Bush). *J. Veg. Sci.*, **10**, 201–208.

Lejeune, O., Tlidi, M., and Couteron, P. 2002. Localized vegetation patches: a self-organized response to resource scarcity. *Phys. Rev. E*, **66**, 010901.

Lenardic, A, Moresi, L.-N., Jellinek, A. M., and Manga, M. 2005. Continental insulation, mantle cooling, and the surface area of oceans and continents. *Earth Planet. Sci. Lett.*, **234**, 317–333.

Lenton, T. M. 1998. Gaia and natural selection. *Nature*, **394**, 439–447.

Lenton, T. M., Schellnhuber, H. J., and Szathmary, E. 2004. Climbing the co-evolutionary ladder. *Nature*, 913.

Lenton, T. M., Held, H., Kriegler, E., Hall, J. W., Lucht, W., Rahmstorf, S., and Schellnhuber, H. J. 2008. Tipping elements in the Earth's climate system. *Proc. Natl. Acad. Sci. USA*, **105**, 1786–1793.

Li, L., Ingersoll, A. P., Jiang, X., Feldman, D., and Yung, Y. L. 2007. Lorenz energy cycle of the global atmosphere based on reanalysis datasets. *Geophys. Res. Lett.*, **34**, L16813.

Liao, W., Heijungs, R., and Huppes, G. 2012. Thermodynamic analysis of human–environment systems: a review focused on industrial ecology. *Ecol. Mod.*, **228**, 76–88.

Lineweaver, C. H., and Chopra, A. 2012. Earth and other Earths: astrophysical, geochemical, geophysical, and biological limits on planet habitability. *Annu. Rev. Earth Planet. Sci.*, **40**, 597–623.

Lineweaver, C. H., and Egan, C. A. 2008. Life, gravity and the second law of thermodynamics. *Phys. Life Rev.*, **5**, 225–242.

Lobell, D. B., Cassman, K. G., and Field, C. B. 2009. Crop yield gaps: their importance, magnitudes, and causes. *Ann. Rev. Environ. Resour.*, **34**, 179–204.

Lorenz, E. N. 1955. Available potential energy and the maintenance of the general circulation. *Tellus*, **7**, 157–167.

Lorenz, E. N. 1960. Generation of available potential energy and the intensity of the general circulation. Pages 86–92 of: Pfeffer, R. C. (ed), *Dynamics of Climate*. Oxford: Pergamon Press.

Lorenz, R. D. 2002. Planets, life and the production of entropy. *Intl. J. Astrobiol.*, **1**, 3–13.

Lorenz, R. D. 2010. The two-box model of climate: limitations and applications to planetary habitability and maximum entropy production studies. *Phil. Trans. R. Soc. B*, **365**, 1349–1354.

Lorenz, R. D., and Renno, N. O. 2002. Work output of planetary atmospheric engines: dissipation in clouds and rain. *Geophys. Res. Lett.*, **29**, 1023.

Lorenz, R. D., Lunine, J. I., Withers, P. G., and McKay, C. P. 2001. Titan, Mars and Earth: entropy production by latitudinal heat transport. *Geophys. Res. Lett.*, **28**, 415–418.

Lotka, A. J. 1921. Note on the economic conversion factors of energy. *Proc. Natl. Acad. Sci. USA*, **7**, 192–197.

Lotka, A. J. 1922a. Contribution to the energetics of evolution. *Proc. Natl. Acad. Sci. USA*, **8**, 147–151.

Lotka, A. J. 1922b. Natural selection as a physical principle. *Proc. Natl. Acad. Sci. USA*, **8**, 151–154.

Lotka, A. J. 1925. *Elements of Physical Biology*. Baltimore: Williams and Wilkins.

Lovelock, J. E. 1965. A physical basis for life detection experiments. *Nature*, **207**, 568–570.

Lovelock, J. E. 1972a. *Gaia: A New Look at Life on Earth*. Oxford: Oxford University Press.

Lovelock, J. E. 1972b. Gaia as seen through the atmosphere. *Atmos. Environ.*, **6**, 579–580.

Lovelock, J. E. 1975. Thermodynamics and the recognition of alien biospheres. *Proc. Roy. Soc. Lond. B*, **189**, 167–181.

Lovelock, J. E., and Margulis, L. 1974. Atmospheric homeostasis by and for the biosphere: the Gaia hypothesis. *Tellus*, **26**, 2–10.

Lunine, J. I., and Lorenz, R. D. 2002. A simple prescription for calculating day–night temperature contrasts on synchronously rotating planets. In: *33rd Annual Lunar and Planetary Science Conf., Houston, TX, 11–15 March, 2002*.

Lyons, T. J. 2002. Clouds prefer native vegetation. *Met. Atmos. Phys.*, **80**, 131–140.

Magnus, G. 1844. Versuche über die Spannkräfte des Wasserdampfs. *Ann. Phys. Chem.*, **61**, 225–248.

Makkink, G. F. 1957. Testing the Penman formula by means of lysimeters. *J. Inst. Wat. Engrs.*, **11**, 277–288.

Manabe, S., and Moeller, F. 1961. On the radiative equilibrium and heat balance of the atmosphere. *Mon. Wea. Rev.*, **89**, 503–532.

Marais, D. J. Des. 2000. When did photosynthesis emerge on Earth? *Science*, **289**, 1703–1705.

Marchetti, C. 1979. 10^{12}: a check on the Earth-carrying capacity for man. *Energy*, **4**, 1107–1117.

Margules, M. 1905. Über die Energie der Stürme. *Jahrb. Zentralanst. Meteorol.*, **40**, 1–26.

Martin, R. D. 1981. Relative brain size and basal metabolic rate in terrestrial vertebrates. *Nature*, **293**, 57–60.

Martin, W. F., Sousa, F. L., and Lane, N. 2014. Energy at life's origin. *Science*, **344**, 1092–1093.

Martyushev, L. M., and Seleznev, V. D. 2006. Maximum entropy production principle in physics, chemistry, and biology. *Phys. Rep.*, **426**, 1–45.

Masuda, K. 1988. Meridional heat transport by the atmosphere and the ocean: analysis of FGGE data. *Tellus A*, **40**, 285–302.

McDonald, J. E. 1960. Direct absorption of solar radiation by atmospheric water vapor. *J. Meteor.*, **17**, 319–328.

McNaughton, K. G., and Spriggs, T. W. 1986. A mixed-layer model for regional evaporation. *Bound. Lay. Meteorol.*, **34**, 243–262.

McNaughton, S. J. 1979. Grazing as an optimization process: grass-ungulate relationships in the Serengeti. *Am. Nat.*, **113**, 691–703.

Meadows, D. H., Meadows, D. L, Randers, J., and III, W. W. Behrens. 1972. *The Limits to Growth*. New York: Universe Books.

Metchnik, V. I., Gladwin, M. T., and Stacey, F. D. 1974. Core convection as a power source for the geomagnetic dynamo: a thermodynamic argument. *J. Geomag. Geoelectr.*, **26**, 405–415.

Meyer, H.J. 1886. *Meyers Konversations-Lexikon: Eine Encyklopädie des allgemeinen Wissens*. No. Bd. 4. Leipzig: Bibliographisches Institut.

Meysman, F. J. R., and Bruers, S. 2007. A thermodynamic perspective on food webs: quantifying entropy production within detrital-based ecosystems. *J. Theor. Biol.*, **249**, 124–139.

Miller, L M, Gans, F, and Kleidon, A. 2011a. Estimating maximum global land surface wind power extractability and associated climatic consequences. *Earth Syst. Dynam.*, **2**, 1–12.

Miller, L. M., Gans, F., and Kleidon, A. 2011b. Jet stream wind power as a renewable energy resource: little power, big impacts. *Earth Syst. Dynam.*, **2**, 201–212.

Milly, P. C. D. 1994. Climate, soil water storage, and the average annual water balance. *Water Resour. Res.*, **30**, 2143–2156.

Monsi, M., and Saeki, T. 1953. Über den Lichtfaktor in den Pflanzengesellschaften und seine Bedeutung für die Stoffproduktion. *Jap. J. Bot.*, **14**, 22–52.

Monteith, J. L. 1972. Solar radiation and productivity in tropical ecosystems. *J. Appl. Ecol.*, **9**, 747–766.

Monteith, J. L. 1977. Climate and the efficiency of crop production in Britain. *Phil. Trans. R. Soc. B*, **281**, 277–294.

Monteith, J. L. 1978. Reassessment of maximum growth rates for C3 and C4 crops. *Expl. Agric.*, **14**, 1–5.

Mortimer, R. G., and Mazo, R. M. 1961. Irreversible thermodynamics of systems containing radiation. Application to photochemical reactions. *J. Chem. Phys.*, **35**, 1013–1018.

Mueller, N. D., Gerber, J. S., Johnston, M., Ray, D. K., Ramankutty, N., and Foley, J. A. 2012. Closing yield gaps through nutrient and water management. *Nature*, **490**, 254–257.

Mulligan, J. F., and Hertz, H. G. 1997. An unpublished lecture by Heinrich Hertz: "On the energy balance of the Earth." *Am. J. Phys.*, **65**, 36–45.

Munk, W., and Wunsch, C. 1998. Abyssal recipes II: energetics of tidal and wind mixing. *Deep-Sea Res.*, **45**, 1977–2010.

Murphy, D. J., and Hall, C. A. S. 2011a. Energy return on investment, peak oil, and the end of economic growth. *Ann. N. Y. Acad. Sci.*, **1219**, 52–72.

Murphy, D. J., and Hall, C. A. S. 2011b. Year in review: EROI or energy return on (energy) invested. *Ann. N. Y. Acad. Sci.*, **1185**, 102–118.

Nair, U. S., Wu, Y., Kala, J., Lyons, T. J., Sr., R. A. Pielke, and Hacker, J. M. 2011. The role of land use change on the development and evolution of the west coast trough, convective clouds, and precipitation in southwest Australia. *J. Geophys. Res.*, **116**, D07103.

Nepstad, D. C., de Carvalho, C. R., Davidson, E. A., Jipp, P. H., Lefebvre, P. A., Negreiros, H. G., da Silva, E. D., Stone, T. A, Trumbore, S. E., and Vieira, S. 1994. The role of deep roots in the hydrological and carbon cycles of Amazon forests and pastures. *Nature*, **372**, 666–669.

Nepstad, D. C., Stickler, C. M., Soares-Filho, B., and Merry, F. 2008. Interactions among Amazon land use, forests and climate: prospects for a near-term forest tipping point. *Phil. Trans. R. Soc. B*, **363**, 1737–1746.

Nihous, G. C. 2006. A preliminary assessment of ocean thermal energy conversion resources. *J. Energy Resour. Technol.*, **129**, 10–17.

Nisbet, E. G., and Sleep, N. H. 2001. The habitat and nature of early life. *Nature*, **409**, 1083–1091.

Novikov, I. I. 1958. The efficiency of atomic power stations (A review). *J. Nuclear Energy II*, **7**, 125–128.

Odum, E. P. 1969. The strategy of ecosystem development. *Science*, **164**, 262–270.

Odum, H. T. 1973. Energy, ecology and economics. *Ambio*, **2**, 220–227.

Odum, H. T. 1988. Self-organization, transformity, and information. *Science*, **242**, 1132–1139.

Odum, H. T., and Pinkerton, R. C. 1955. Time's speed regulator: the optimum efficiency for maximum power output in physical and biological systems. *Am. Sci.*, **43**, 331–343.

Oke, T R. 1987. *Boundary Layer Climates*. London: Methuen.

Oki, T, and Kanae, S. 2006. Global hydrological cycle and world water resources. *Science*, **313**, 1068–1072.

Oort, A. H. 1989. Angular momentum cycle in the atmosphere-ocean-solid Earth system. *Bull. Amer. Meteorol. Soc.*, **70**, 1231–1242.

Oort, A. H., Ascher, S. C., Levitus, S., and Peixoto, J. P. 1989. New estimates of the available potential energy in the world ocean. *J. Geophys. Res.*, **94**, 3187–3200.

Oort, A. H., Anderon, L. A., and Peixoto, J. P. 1994. Estimates of the energy cycle of the oceans. *J. Geophys. Res.*, **99**, 7665–7688.

Ornstein, L, Aleinov, I, and Rind, D. 2009. Irrigated afforestation of the Sahara and Australian Outback to end global warming. *Clim. Change*, **97**, 409–437.

Ostwald, W. 1909. *Energetische Grundlagen der Kulturwissenschaften*. Leipzig: Klinkhardt.

Ou, H. W. 2001. Possible bounds on the Earth's surface temperature: from the perspective of a conceptual global-mean model. *J. Clim.*, **14**, 2976–2988.

Ou, H. W. 2006. Thermal properties of a coupled ocean-atmosphere: a conceptual model. *Tellus*, **58**(3), 404–415.

Ou, H. W. 2007. Hydrological cycle and ocean stratification in a coupled climate system: a theoretical study. *Tellus*, **59A**, 683–694.

Ozawa, H., and Ohmura, A. 1997. Thermodynamics of a global-mean state of the atmosphere: a state of maximum entropy increase. *J. Clim.*, **10**, 441–445.

Ozawa, H., Shimokawa, S., and Sakuma, H. 2001. Thermodynamics of fluid turbulence: A unified approach to the maximum transport properties. *Phys. Rev. E*, **64**, 026303.

Ozawa, H., Ohmura, A., Lorenz, R. D., and Pujol, T. 2003. The second law of thermodynamics and the global climate system: a review of the maximum entropy production principle. *Rev. Geophys.*, **41**, 1018.

Paltridge, G. W. 1975. Global dynamics and climate: a system of minimum entropy exchange. *Q. J. Roy. Meteorol. Soc.*, **101**, 475–484.

Paltridge, G. W. 1978. The steady-state format of global climate. *Q. J. Roy. Meteorol. Soc.*, **104**, 927–945.

Paltridge, G. W. 1979. Climate and thermodynamic systems of maximum dissipation. *Nature*, **279**, 630–631.

Pascale, S., Gregory, J. M., Ambaum, M. H. P., and Tailleux, R. 2012. A parametric sensitivity study of entropy production and kinetic energy dissipation using the FAMOUS AOGCM. *Clim. Dyn.*, **38**, 1211–1227.

Pascale, S., Ragone, F., Lucarini, V., Wang, Y., and Boschi, R. 2013. Nonequilibrium thermodynamics of circulation regimes in optically thin, dry atmospheres. *Planet. Space Sci.*, **84**, 48–65.

Pauluis, O., and Dias, J. 2012. Satellite estimates of precipitation-induced dissipation in the atmosphere. *Science*, **335**, 953–956.

Pauluis, O., and Held, I. M. 2002a. Entropy budget of an atmosphere in radiative convective equilibrium. Part I: maximum work and frictional dissipation. *J. Atmos. Sci.*, **59**, 126–139.

Pauluis, O., and Held, I. M. 2002b. Entropy budget of an atmosphere in radiative convective equilibrium. Part II: latent heat transport and moist processes. *J. Atmos. Sci.*, **59**, 140–149.

Pauluis, O., Balaji, V., and Held, I. M. 2000. Frictional dissipation in a precipitating atmosphere. *J. Atmos. Sci.*, **57**, 987–994.

Peixoto, J. P., and Oort, A. H. 1992. *Physics of Climate*. New York: American Institute of Physics.

Peixoto, J. P., Oort, A. H., de Almeida, M., and Tome, A. 1991. Entropy budget of the atmosphere. *J. Geophys. Res.*, **96**, 10,981–10,988.

Penman, H. L. 1948. Natural evaporation from open water, bare soil, and grass. *Proc. R. Soc. London A*, **193**, 120–146.

Petela, R. 1964. Exergy of heat radiation. *J. Heat Transfer*, **86**, 187–192.

Petela, R. 2003. Exergy of undiluted thermal radiation. *Solar Energy*, **74**, 469–488.

Pierrehumbert, R. T. 2002. The hydrologic cycle in deep-time climate problems. *Nature*, **419**, 191–198.

Pimentel, D., Hurd, L. E., Bellotti, A. C., Forster, M. J., Oka, I. N., Sholes, O. D., and Whitman, R. J. 1973. Food production and the energy crisis. *Science*, **182**, 443–449.

Planck, M. 1906. *Theorie der Wärmestrahlung*. Leipzig: Barth.

Porada, P., Kleidon, A., and Schymanski, S. J. 2011. Entropy production of soil hydrological processes and its maximization. *Earth Syst. Dynam.*, **2**, 179–190.

Press, W. H. 1976. Theoretical maximum for energy from direct and diffuse sunlight. *Nature*, **264**, 734–735.

Price, C., Penner, J., and Prather, M. 1997. NO_x from lightning. 1: global distribution based on lightning physics. *J. Geophys. Res.*, **102**, 5929–5941.

Priestley, C. H. B., and Taylor, R. J. 1972. On the assessment of surface heat flux and evaporation using large-scale parameters. *Mon. Wea. Rev.*, **100**, 81–92.

Prigogine, I. 1947. *Etude Thermodynamique des Phenomenes Irreversibles*. Liege, Belgium: Desoer.

Prigogine, I. 1962. *Introduction to Non-equilibrium Thermodynamics*. New York: Wiley Interscience.

Prigogine, I. 1978. Time, structure, and fluctuations. *Science*, **201**, 777–785.

Prigogine, I., Nicolis, G., and Babloyantz, A. 1972. Thermodynamics of evolution. *Phys. Today*, **25**, 23–28.

Rabinowitch, E., and Govindjee. 1969. *Photosynthesis*. New York: Wiley.

Radmer, R., and Kok, B. 1977. Photosynthesis: limited yields, unlimited dreams. *Bioscience*, **27**, 599–605.

Ramanathan, V., Cess, R. D., Harrison, E. F., Minnis, P., Barkstrom, B. R., Ahmad, E., and Hartmann, D. 1989. Cloud-radiative forcing and climate: results from the Earth radiation budget experiment. *Science*, **243**, 57–63.

Ramankutty, N., Evan, A. T., Monfreda, C., and Foley, J. A. 2008. Farming the planet: 1. Geographic distribution of global agricultural lands in the year 2000. *Global Biogeochem. Cy.*, **22**, GB1003.

Raupach, M. R. 1998. Influences of local feedbacks on land-air exchanges of energy and carbon. *Glob. Ch. Biol.*, **4**, 477–494.

Renno, N O, and Ingersoll, A P. 1996. Natural convection as a heat engine: a theory for CAPE. *J. Atmos. Sci.*, **53**, 572–585.

Retallack, G. J. 2001. Cenozoic expansion of grasslands and climatic cooling. *J. Geology*, **109**, 407–426.

Retallack, G. J. 2007. Coevolution of life and Earth. Pages 295–230 of: Schubert, G. (ed), *Treatise of Geophysics*. Amsterdam: Elsevier.

Rietkerk, M., Dekker, S., Wassen, M., Verkroost, A., and Bjerkens, M. 2004. A putative mechanism for bog patterning. *Am. Nat.*, **163**, 699–708.

Rinaldo, A., Rodriguez-Iturbe, I., Rigon, R., Bras, R. L., Ijjasz-Vasquez, E., and Marani, A. 1992. Minimum energy and fractal structures of drainage networks. *Water Resour. Res.*, **28**, 2183–2195.

Rinaldo, A., Rodriguez-Iturbe, I., and Rigon, R. 1998. Channel networks. *Ann. Rev. Earth Planet. Sci.*, **26**, 289–327.

Rochetin, N., Lintner, B. R., Findell, K. L., Sobel, A. H., and Gentine, P. 2014. Radiative-convective equilibrium over a land surface. *J. Clim.*, **27**, 8611–8629.

Rockström, J., Steffen, W., Noone, K., Persson, A., Chapin, F. S., Lambin, E. F., Lenton, T. M., Scheffer, M., Folke, C., Schellnhuber, H. J., Nykvist, B., de Wit, C. A., Hughes, T., van der Leeuw, S., Rodhe, H., Sörlin, S., Snyder, P. K., Constanza, R., Svedin, U., Falkenmark, M., Karlberg, L., Corell, R. W., Fabry, V. J., Hansen, J., Walker, B., Liverman, D., Richardson, K., Crutzen, P., and Foley, J. A. 2009. A safe operating space for humanity. *Nature*, **461**(472–475).

Rodgers, C. D. 1976. Minimum entropy exchange principle: reply. *Q. J. Roy. Meteorol. Soc.*, **102**, 455–457.

Rodriguez-Iturbe, I., and Rinaldo, A. 1997. *Fractal River Basins: Chance and Self-Organization*. Cambridge, UK: Cambridge University Press.

Rogner, H. H., Barthel, F., Cabrera, M., Faaij, A., Girouc, M., Hall, D., Kagramanian, V., Kononov, S., Lefevre, T., Moreira, R., Nötstaller, R., Odell, P., and Taylor, M. 2000. Energy resources. In: *World Energy Assessment: Energy and the Challenge of*

Sustainability. New York, USA: United Nations Development Programme, United Nations Department of Economic and Social Affairs, and World Energy Council.

Rojstaczer, S., Sterling, S. M., and Moore, N. J. 2001. Human appropriation of photosynthesis products. *Science*, **294**, 2549–2552.

Romps, D. M., Seeley, J. T., Vallaro, D., and Molinari, J. 2014. Projected increase in lightning strikes in the United States due to global warming. *Science*, **346**, 851–854.

Romps, David M., and Charn, Alexander B. 2015. Sticky thermals: evidence for a dominant balance between buoyancy and drag in cloud updrafts. *J. Atmos. Sci.*, **72**, 2890–2901.

Rosen, M. A., and Scott, D. S. 2003. Entropy production and exergy destruction: Part I - hierarchy of Earth's major constituencies. *Int. J. Hydrogen Energ.*, **28**, 1307–1313.

Rosen, P. 1954. Entropy of radiation. *Phys. Rev.*, **96**, 555.

Rosenzweig, M. L. 1968. Net primary productivity of terrestrial communities: prediction from climatological data. *Am. Nat.*, **102**, 67–74.

Rosing, M. T., Bird, D. K., Sleep, N. H., Glassley, W., and Albarede, F. 2006. The rise of continents: an essay on the geologic consequences of photosynthesis. *Paleogeogr. Paleocl.*, **232**, 99–113.

Ross, R. T. 1966. Thermodynamic limitations on the conversion of radiant energy into work. *J. Chem. Phys.*, **45**, 1–7.

Rotenberg, E., and Yakir, D. 2010. Contribution of semi-arid forests to the climate system. *Science*, **327**, 451–454.

Rotenberg, E., and Yakir, D. 2011. Distinct patterns of changes in surface energy budget associated with forestation in the semiarid region. *Glob. Ch. Biol.*, **17**, 1536–1548.

Rubin, D. M., and Hunter, R. E. 1987. Bedform alignment in directionally varying flows. *Science*, **237**, 276–278.

Ruimy, A., Jarvis, P. G., Baldocchi, D. D., and Saugier, B. 1995. CO_2 fluxes over plant canopies and solar radiation: a review. *Adv. Ecol. Res.*, **26**, 1–68.

Russell, M. J., and Hall, A. J. 1997. The emergence of life from iron monosulphide bubbles at a submarine hydrothermal redox and pH front. *J. Geol. Soc. London*, **154**, 377–402.

Sagan, C., and Mullen, G. 1972. Earth and Mars: evolution of atmospheres and surface temperatures. *Science*, **177**, 52–56.

Schlaich, J., Bergermann, R., Schiel, W., and Weinrebe, G. 2005. Design of commercial solar updraft tower systems: utilization of solar induced convective flows for power generation. *J. Solar Energy Eng.*, **127**, 117–124.

Schlesinger, W H. 1997. *Biogeochemistry: An Analysis of Global Change*. San Diego: Academic Press.

Schmidt, W. 1915. Strahlung und Verdunstung an freien Wasserflächen; ein Beitrag zum Wärmehaushalt des Weltmeeres und zum Wasserhaushalt der Erde. *Ann. d. Hydrogr. u. maritimen Meteorol.*, **43**, 111–178.

Schneider, E. D., and Kay, J. J. 1994a. Complexity and thermodynamics: towards a new ecology. *Futures*, **26**, 626–647.

Schneider, E. D., and Kay, J. J. 1994b. Life as a manifestation of the second law of thermodynamics. *Math. Comput. Modeling*, **19**, 25–48.

Schrenk, M. O., Brazelton, W. J., and Lang, S. Q. 2013. Serpentinization, carbon, and deep life. *Rev. Mineral. Geochem.*, **75**, 575–606.

Schrödinger, E. 1944. *What is Life? The Physical Aspect of the Living Cell*. Cambridge, UK: Cambridge University Press.

Schumann, U., and Huntrieser, H. 2007. The global lightning-induced nitrogen oxides source. *Atmos. Chem. Phys.*, **7**, 3823–3907.

Schwartzman, D. W., and Middendorf, G. 2000. Biospheric cooling and the emergence of intelligence. Pages 425–429 of: Lemarchand, G., and Meech, K. (eds), *A New Era in Bioastronomy*. ASP Conference Series, Vol. 213.

Schwartzman, D. W., and Volk, T. 1989. Biotic enhancement of weathering and the habitability of Earth. *Nature*, **340**, 457–460.

Schymanski, S. J., Kleidon, A., Stieglitz, M., and Narula, J. 2010. Maximum entropy production allows a simple representation of heterogeneity in semiarid ecosystems. *Phil. Trans. R. Soc. B*, **365**(1449–1455).

Seager, S. 2013. Exoplanet habitability. *Science*, **340**, 577–581.

Shock, E. L., and Holland, M. E. 2007. Quantitative habitability. *Astrobiology*, **7**, 839–851.

Shukla, J., and Mintz, Y. 1982. The influence of land-surface-evapotranspiration on the Earth's climate. *Science*, **247**, 1322–1325.

Shutts, G. J. 1981. Maximum entropy production states in quasi-geostrophic dynamical models. *Q. J. Roy. Meteorol. Soc.*, **107**(453), 503–520.

Simoncini, E., Russell, M. J., and Kleidon, A. 2011. Modeling free energy availability from Hadean hydrothermal systems to first metabolism. *Orig. Life Evol. Biosph.*, **41**, 529–532.

Simoncini, E., Virgo, N., and Kleidon, A. 2013. Quantifying drivers of chemical disequilibrium: theory and application to methane in the Earth's atmosphere. *Earth Syst. Dynam.*, **4**, 317–331.

Smil, V. 1999. *Energies: An Illustrated Guide to the Biosphere and Civilization*. Cambridge, MA, USA: MIT Press.

Smil, V. 2000. Energy in the twentieth century: resources, conversions, costs, uses, and consequences. *Annu. Rev. Energy Environ.*, **25**, 21–51.

Smith, W. K., Cleveland, C. C., Reed, S. C., and Running, S. W. 2014. Agricultural conversion without external water and nutrient inputs reduces terrestrial vegetation productivity. *Geophys. Res. Lett.*, **41**, 449–455.

Sperry, J. S. 2000. Hydraulic constraints on plant gas exchange. *Agric. Forest Meteorol.*, **104**, 13–23.

Spohn, T. 2012. Planets and life. *Spatium*, **30**, 3–15.

Stacey, F. D. 1967. Convecting mantle as a thermodynamic engine. *Nature*, **214**, 476–477.

Stephens, G. L., and O'Brien, D. M. 1993. Entropy and climate. 1. ERBE observations of the entropy production of the Earth. *Q. J. Roy. Meteorol. Soc.*, **119**(509), 121–152.

Stephens, G. L., Li, J., Wild, M., Clayson, C. A., Loeb, N., Kato, S., L'Ecuyer, T., Stackhouse, P. W., Lebsock, M., and Andrews, T. 2012. An update on Earth's energy balance in light of the latest global observations. *Nature Geosci.*, 691–696.

Stevens, B. 2005. Atmospheric moist convection. *Annu. Rev. Earth Planet. Sci.*, **33**, 605–643.

Stevens, B., and Bony, S. 2013. What are climate models missing? *Science*, **340**, 1053–1054.

Stevenson, D. J. 1983. Planetary magnetic fields. *Rep. Prog. Phys.*, **46**, 555–620.

Stevenson, D. J., Spohn, T., and Schubert, G. 1983. Magnetism and thermal evolution of the terrestrial planets. *Icarus*, **54**, 466–489.

Stone, P. H. 1978. Constraints on dynamical transports of energy on a spherical planet. *Dynamics of Atmospheres and Oceans*, **2**, 123–139.

Stull, R B. 1989. *An Introduction to Boundary Layer Meteorology*. Boston: Kluwer Academic Press.

Tailleux, R. 2010. Entropy versus APE production: on the buoyancy input in the oceans energy cycle. *Geophys. Res. Lett.*, **37**, L22603.

Tailleux, R. 2013. Available potential energy and exergy in stratified fluids. *Annu. Rev. Fluid Mech.*, **45**, 35–58.

Tamura, S. T. 1905. Doctor Margules on the energy of storms. *Mon. Wea. Rev.*, **33**, 519–521.

Taylor, S. R., and McLennan, S. M. 1996. The evolution of continental crust. *Sci. Am.*, **274**, 76–81.

Tennekes, H. 1973. A model for the dynamics of the inversion above a convective boundary layer. *J. Atmos. Sci.*, **30**, 558–566.

Thompson, A. M. 1992. The oxidizing capacity of the Earth's atmosphere: probable past and future changes. *Science*, **256**, 1157–1165.

Thomson, W. 1881. On the sources of energy in nature available to man for the production of mechanical effect. *Science*, **2**, 475–478.

Trenberth, K. E., and Caron, J. M. 2001. Estimates of meridional atmosphere and ocean heat transports. *J. Clim.*, **14**, 3433–3443.

Ulanowicz, R. E., and Hannon, B. M. 1987. Life and the production of entropy. *Proc. R. Soc. Lond. B*, **232**, 181–192.

Unrean, P., and Srienc, F. 2011. Metabolic networks evolve towards states of maximum entropy production. *Metabolic Engineering*, **13**, 666–673.

Vallino, J. J. 2010. Ecosystem biogeochemistry considered as a distributed metabolic network ordered by maximum entropy production. *Phil. Trans. R. Soc. B*, **365**, 1417–1427.

Vance, E. 2009. High hopes. *Nature*, **460**, 564–566.

Vanyo, J. P., and Paltridge, G. W. 1981. A model for energy dissipation at the mantle-core boundary. *Geophys. J.*, **66**(3), 677–690.

Verhoogen, J. 1980. *Energetics of the Earth*. Washington DC: National Academy of Sciences.

Vitousek, P. M., Ehrlich, P. R., Ehrlich, A. H., and Matson, P. A. 1986. Human appropriation of the products of photosynthesis. *Bioscience*, **36**, 368–373.

Vitousek, P. M., Mooney, H. A., Lubchenco, J., and Melillo, J. M. 1997. Human domination of Earth's ecosystems. *Science*, **277**, 494–499.

Volk, T. 1998. *Gaia's Body: Toward a Physiology of Earth*. New York: Springer.

Volk, T, and Pauluis, O. 2010. It is not the entropy you produce, rather, how you produce it. *Phil. Trans. R. Soc. B*, **365**, 1317–1322.

von Storch, J.-S., Eden, C., Fast, I., Haak, H., Hernandez-Deckers, D., Maier-Reimer, E., Marotzke, J., and Stammer, D. 2012. An estimate of the Lorenz energy cycle for the world ocean based on the 1/10" STORM/NCEP simulation. *J. Phys. Oceanogr.*, **42**, 2185–2205.

Walker, D. 1992. *Energy, Plants and Man*. Sausalito, CA, USA: University Science Books.

Walker, J. C., Hays, P. B., and Kasting, J. F. 1981. A negative feedback mechanism for the long-term stabilization of Earth's surface temperature. *J. Geophys. Res.*, **86**, 9776–9782.

Wallace, J. M., and Hobbs, P. V. 1977. *Atmospheric Science: An Introductory Survey*. New York: Academic Press.

Wang, J., Salvucci, G. D., and Bras, R. L. 2004. An extremum principle of evaporation. *Water Resour. Res.*, **40**, W09303.

Wang, J., Bras, R. L., Lerdau, M., and Salvucci, G. D. 2007. A maximum hypothesis of transpiration. *J. Geophys. Res.*, **112**, G03010.

Werner, B. T. 1999. Complexity in natural landform patterns. *Science*, **284**, 102–104.

West, G. B., Brown, J. H., and Enquist, B. J. 1997. A general model for the origin of allometric scaling laws in biology. *Science*, **276**, 122–126.

West, G. B., Brown, J. H., and Enquist, B. J. 1999. A general model for the structure and allometry of plant vascular systems. *Nature*, **400**, 664–667.

Wheeler, P. E. 1991a. The influence of bipedalism on the energy and water budgets of early hominids. *J. Human Evol.*, **21**, 117–136.

Wheeler, P. E. 1991b. The thermoregulatory advantages of hominid bipedalism in open equatorial environments: the contribution of increased convective heat loss and cutaneous evaporative cooling. *J. Human Evol.*, **21**, 107–115.

Wheeler, R. M., Mackowiak, C. L., Stutte, G. W., Yorio, N. C., Ruffe, L. M., Sager, J. C., Prince, R. P., and Knott, W. M. 2008. Crop productivities and radiation use efficiencies for bioregenerative life support. *Adv. Space Res.*, **41**, 706–713.

Wildt, R. 1956. Radiative transfer and thermodynamics. *Astrophys. J.*, **123**, 107–116.

Wildt, R. 1972. Thermodynamics of the gray atmosphere. IV. Entropy transfer and production. *Astrophys. J.*, **174**, 69–77.

Williamson, S. C., Detling, J. K., Dodd, J. L., and Dyer, M. I. 1989. Experimental evaluation of the grazing optimization hypothesis. *J. Range Manage.*, **42**, 149–152.

Wofsy, S. C., Harriss, R. C., and Kaplan, W. A. 1988. Carbon dioxide in the atmosphere over the Amazon basin. *J. Geophys. Res.*, **93**, 1377–1387.

Wong, S. C., Cowan, I. R., and Farquhar, G. D. 1979. Stomatal conductance correlates with photosynthetic capacity. *Nature*, **282**, 424–426.

Woodward, F. I. 1987. Stomatal numbers are sensitive to increases in CO_2 from pre-industrial levels. *Nature*, **327**, 617–618.

World Commission on Environment and Development. 1987. *Our Common Future*. Oxford: Oxford University Press.

Wu, W, and Liu, Y. 2010. Radiation entropy flux and entropy production of the Earth system. *Rev. Geophys.*, **48**, RG2003.

Xu, C.-Y., and Singh, V. P. 2000. Evaluation and generalization of radiation-based methods for calculating evaporation. *Hydrol. Process.*, **14**, 339–349.

Xu, K.-M., Arakawa, A., and Krueger, S. K. 1992. The macroscopic behavior of cumulus ensembles simulated by a cumulus ensemble model. *J. Atmos. Sci.*, **49**, 2402–2420.

Yen, J. D. L., Paganin, D. M., Thomson, J. R., and Macnally, R. 2014. Thermodynamic extremization principles and their relevance to ecology. *Austral Ecology*, **39**, 619–632.

Zehe, E., Ehret, U., Blume, T., Kleidon, A., Scherer, U., and Westhoff, M. 2013. A thermodynamic approach to link self-organization, preferential flow and rainfall–runoff behaviour. *Hydrol. Earth Syst. Sci.*, **17**, 4297–4322.

Zener, C. 1973. Solar sea power. *Phys. Today*, **26**, 48–53.

Zhou, X., Wang, F., and Ochieng, R. M. 2010. A review of solar chimney power technology. *Renew. Sust. Energ. Rev.*, **14**, 2315–2338.

Zhu, X.-G., Long, S. P., and Ort, D. R. 2008. What is the maximum efficiency with which photosynthesis can convert solar energy into biomass? *Curr. Opin. Biotechnol.*, **19**, 153–159.

Zotin, A. I. 1984. Bioenergetic trends of evolutionary progress of organisms. Pages 451–458 of: Lamprecht, I., and Zotin, A. I. (eds), *Thermodynamics and Regulation of Biological Processes*. Berlin, New York: de Gruyter.

Index

For EU product safety concerns, contact us at Calle de José Abascal, 56–1°,
28003 Madrid, Spain or eugpsr@cambridge.org.

www.ingramcontent.com/pod-product-compliance
Lightning Source LLC
LaVergne TN
LVHW062111200926
848484LV00007B/1640

* 9 7 8 1 1 0 7 0 2 9 9 4 1 *